Eco-Capitalism

Robert Guttmann

Eco-Capitalism

Carbon Money, Climate Finance,
and Sustainable Development

Robert Guttmann
Economics Department
Hofstra University
Hempstead, NY, USA

ISBN 978-3-319-92356-7   ISBN 978-3-319-92357-4   (eBook)
https://doi.org/10.1007/978-3-319-92357-4

Library of Congress Control Number: 2018942888

Cover image: © duncan1890/Getty Images
Cover design by Ran Shauli

Printed on acid-free paper

This Palgrave Macmillan imprint is published by the registered company Springer International Publishing AG part of Springer Nature
The registered company address is: Gewerbestrasse 11, 6330 Cham, Switzerland

CONTENTS

CHAPTER 1

The Challenge of Climate Change

In December 2015, hundred and ninety-five countries signed the historic Climate Change Agreement in Paris, committing their governments to steady reductions in the emission of greenhouse gases which are heating up our planet. This crucial first step in the right direction has been a long time coming. For nearly a quarter of a century, ever since the 1992 Earth Summit in Rio de Janeiro, the world community had discussed to no avail how to proceed with a common approach to the global problem of a warming planet. Lacking sufficient consensus, we let several initiatives just peter out ineffectively—from the 1997 Kyoto Protocol to the failed 2009 Copenhagen Climate Change Conference. Now that we have secured the Paris Accord, we will have to see how governments will manage to put into effect their promised emission-reduction targets. These will require fairly ambitious policy initiatives some of which will be politically difficult to implement as they hurt vested interests—for example, reducing the role of coal or oil as energy sources for power plants. Nowhere is this question of political will more urgent and problematic than in the USA, the world's leading emitter of greenhouse gases. Notwithstanding the crucial leadership role of their country in the world, Americans have by and large been quite hesitant to face this challenge. On the contrary, there is a deeply rooted skepticism about climate change which has so far prevented the US government from addressing the issue with measures matching the problem.

© The Author(s) 2018
R. Guttmann, *Eco-Capitalism*,
https://doi.org/10.1007/978-3-319-92357-4_1

The Undue Influence of Climate Deniers

Of the seventeen Republican presidential candidates vying for the party's nomination in 2016, only one—Ohio governor John Kasich—believed that climate change was a serious problem caused by human activity. All the others denied the existence of the problem as such. Donald Trump, the eventual GOP nominee and surprise victor of the election, had referred to the climate threat alternately as "nonexistent," "bullshit," or a "con job" before promising to cancel the Paris Climate Change Agreement of December 2015 if elected president. On several occasions, Trump denounced climate-change mitigation measures, such as the Paris Agreement, as a "tax," or as an issue solely designed for China to gain a competitive advantage, or as a way to give "foreign bureaucrats control over how much energy we use right here in America."[1] The Republican party platform of July 2016, after calling the Democrats "environmental extremists" who are committed to "sustain the illusion of an environmental crisis," went on to "forbid any carbon tax," promised to "do away with" Obama's Clean Power Plan, and proposed "to forbid the EPA to regulate carbon dioxide." The platform also committed the party to boosting domestic oil and coal production by easing the issue of permits, a position strongly endorsed by Trump.

This deeply grounded resistance to take climate change seriously extends to Republican members of the Congress. Jim Inhofe (R-OK), who chairs the Senate's Environment and Public Works Committee, has regularly described the climate-change issue as a "hoax" and characterized the work of the scientists grouped together in the Intergovernmental Panel on Climate Change as a "Soviet style trial."[2] A large majority of Republican senators and representatives in the House are adamantly opposed to any meaningful measure of climate-change mitigation. According to research by the Center of American Progress Action Fund (reported in Ellingboe and Koronowski 2016), fifty-nine percent of the Republican caucus in the House and seventy percent of all Republican senators reject the scientists' overwhelming consensus that climate change is occurring and human activity is its major cause. This strong opposition to climate change among Republicans in Congress made it impossible for President Obama and his allies in the Democratic Party to pass wide-ranging legislation on that issue when he first got elected. Most notably, Obama's push to pass a nationwide cap-and-trade system under which the federal government would limit the emission of greenhouse gases with the support of a market-friendly incentive approach, the American Clean Energy and Security Act of 2009, failed in the Senate after barely passing in the House. When the Republicans regained majority control of the House in 2010, any chance for meaningful legislation died. At that point, Obama opted to advance climate-change mitigation measures through the regulatory apparatus under his direct control, notably the Environmental Protection Agency's (EPA) regulation of carbon dioxide aimed specifically at coal-fired plants or setting ambitious fuel-efficiency standards for cars, while at the same time also taking executive action in promotion of international agreements not subject to Senate ratification, as was the case with the aforementioned Paris Agreement of December 2015.

Obama's unilateral measures swiftly became subject to lawsuits by Republican governors who got sympathetic judges to block some of his key measures. This was especially true for Obama's ambitious Clean Power Plan of 2015, at the heart of America's carbon-emission-reduction program he brought to the table in the run-up to Paris. This initiative obliged states to accelerate the use of cleaner power plants using renewables (or gas if swapped for coal) and improve power-generation efficiency. Four days before the death of leading conservative Justice Antonin Scalia in February 2016, the Supreme Court blocked enforcement of the plan in a 5-4 decision until a lower court rules in a lawsuit brought against it by eighteen Republican governors. This was the first time ever the Supreme Court had stayed a regulation *before* a judgment by the lower Court of Appeals, clear indication how politicized the question of climate-change mitigation had become to engulf the country's judiciary in such openly partisan fashion. Trump's election victory in November 2016 prompted in short order his unilateral canceling of Obama's power-plant initiative, executive orders to boost domestic fossil-fuel production (including coal), plans for other rollbacks of environmental regulation (such as relaxing fuel-efficiency standards for cars), and—in a stunning move defying domestic majority opinion and pleas from other world leaders—the unilateral decision on June 1, 2017, to take the USA out of the Paris Agreement. All these initiatives of Trump and his Republican backers in US Congress jeopardize that treaty's effective implementation. If the USA as the world's largest emitter of carbon dioxide per capita backslides, this gives license for other countries to do so as well. We have already seen this happening with the Kyoto Protocol of 1997 which the US Senate never ratified and which consequently failed to meet its initial (decidedly modest) objectives.[3] While the time frame of the Paris Agreement extends beyond Trump's first term, his reversal of Obama's initiatives may well endanger that treaty's long-term viability.

Why is it that Republican leaders are so hostile to the issue of climate change? One might be tempted to place their opposition into the context of the current *Zeitgeist*. We are living these days through a period of more polemical politics where large pockets of post-crisis anger among the electorate feed a discourse of denigrating elites (and that includes scientists), where emotion often crowds out facts, and where belief in conspiracies often appeals more than any other explanation. Still, one has to wonder why hundreds of the world's greatest scientists would conspire to invent this "hoax" of the planet's steady warming if they must know that they are bound to be found out eventually. That does not make much sense. Republican resistance to climate change may also be intimately tied to political influence-seeking by some of America's most powerful lobbies, notably the gas and oil industry recently rendered even stronger by the oil shale boom of the late 2000s and early 2010s across large parts of the country. Big Oil, by sector the fifth-largest lobby in the USA, typically gives 80% of its political contributions to Republicans. Just take a look at the massive funding of political campaigns and conservative think tanks by the Koch Brothers, who control energy firm Koch Industries and for whom climate-change denial has long been a crucial objective! There is also a widespread feeling among Americans, shared by its political leaders especially on the political Right, that any global governance structure, such as the Paris Agreement of 2015, is automatically a matter of other countries exploiting US generosity and/or international bureaucrats restricting American sovereignty—a paranoid predisposition of wrong-headed "nationalism" that flies in the face of the truth to the extent that most of these global governance structures are profoundly shaped by American policy-makers pursuing the national interest in the global context. Finally, Republicans may also be hostile to the notion of man-made climate change for profoundly ideological reasons. It must not be easy for apostles of the "free" market to recognize such a huge market failure and accept a large role for government policy in combating this problem.[4] But as long as a dedicated minority of climate deniers exercises such a stranglehold on US policy, it will be impossible for the world community to address the issue of climate change effectively. Americans need to understand with a greater sense of urgency what is at stake here.

(In)Action Bias

It is in the very nature of the problem to make it difficult, if not impossible, to address. Climate change is largely invisible, very abstract as a notion, and extremely slow moving. It is hard to imagine, easy to ignore, and lacks immediate urgency for action, hence tempting to set aside for later. Addressing it also has uncertain pay-offs which even in the best of circumstances will only bear fruit much later while initially causing quite a bit of pain. The problem thus requires a long-term, intergenerational vision where the current generation is willing to bear sacrifice for its children and grandchildren. And the problem also depends on collective action, necessitating coordination among many players with divergent interests and requiring enough sanctions in place to discourage "free riders" not willing to do their fair share while benefitting unfairly from the efforts of the others. So when looking at it from all those angles, it becomes clear that doing something meaningful about climate change is a tremendous challenge. Perhaps the climate deniers reflect just a grudging admission that the problem is too difficult to address and hence more easily wished away, especially when there is no convincing reason why this problem should be tackled right now rather than ignored a bit longer.

The climate-change challenge reminds me of the dilemma facing a heavy smoker who is still quite young and thus not yet really worried about his/her health. You know smoking is not good for you and eventually will cause you health problems. But that is later, and right now you are more inclined to enjoy the calming effects of a cigarette. So you keep smoking as long as the habits of today outweigh worries about the future in the back of your mind. This, after all, is addictive behavior and as such takes a lot of effort to break, effort not worth undertaking unless obliged to. Pushing this metaphor one step further, add to this the wrinkle that my partner smokes too which makes it that much harder for me to stop the bad habit. We would both have to stop at the same time to succeed, making it twice as unlikely that this will happen any time soon. Thus, we are more inclined to tell ourselves every day that we will get back to the challenge later, one day for sure, but not now. And who knows anyway whether, when, and how the long-term consequences of smoking will kick in. With all this uncertainty, why bother worrying so much? Better to enjoy the pleasure while it lasts! In this calculation, we subconsciously remove ourselves as actors and instead allow ourselves to be shaped by circumstance, in the hope that nothing too grave will happen in the wake of our inaction. Unfortunately, we no longer have the luxury of such wishful thinking when it comes to climate change!

THE INTERGOVERNMENTAL PANEL ON CLIMATE CHANGE

We are now gradually coming to the point where the threat must be taken seriously. Looking back, we have long suspected that the climate can be subject to large variations playing out over centuries. In the nineteenth-century climatologists became obsessed with timing and explaining the "Ice Age," and in the process they grew increasingly aware that human activity can also contribute to climate change. In 1896, the great Swedish scientist Svante Arrhenius predicted that the emission of carbon dioxide from the burning of fossil fuel or other sources could, if large enough, raise the air temperature significantly.[5] Arrhenius' identification of this "greenhouse effect" triggered much debate, and it ultimately required more accurate measurement instruments as well as computers becoming available in the 1950s to validate his prediction. But during the 1960s and 1970s, climatologists focused on other priorities crowding out concern for global warming, notably smog and ozone-layer depletion both of which they thought would cool the atmosphere. As these environmental concerns became addressed and worries about them abated subsequently, attention shifted in the late 1970s to the potentially powerful warming effect of chlorofluorocarbons (CFC) to revive concern over global warming. At that point, it had become undeniably clear that there were several greenhouse gases—carbon dioxide, CFC, and methane—all of which were emitted into the atmosphere at an accelerating pace. During the same period, evidence accumulated that rapid and dramatic climate-change variation was possible in response to relatively small atmospheric alterations. These two concerns fused during the 1980s into a new framework for global temperature analysis, sponsored by NASA's Goddard Institute for Space Studies. In June 1988, the project's chief scientist, James Hansen, gave testimony to the US Congress that greenhouse gas emissions had already begun to alter climate patterns in potentially dangerous ways. His widely discussed presentation put global warming on the radar screen of public attention like no other event had before!

That same year, the United Nations set up the Intergovernmental Panel on Climate Change (IPCC) to report ongoing research and findings pertaining to climate change. As the internationally accepted authority on climate change, it produces reports agreed to by the world's leading climate scientists and approved by the hundred and twenty or so participating governments, thus providing the scientific view on climate change, its evolutionary dynamic, its economic impact, and political repercussions. The IPCC tries to engage the world's leading scientists from a variety of disciplines in an ongoing discussion about the evidence of climate change and its consequences while at the same time obliging government officials with policy-making authority over the issue to form a consensus as to the evolving nature of the problem. That dual objective gets crystallized around the highly formalized and tightly structured process of the panel's successive reports on climate change whose every word gets discussed and approved collectively. Given their laborious nature, it is actually quite amazing that the IPCC has been able to write five reports between 1988 and 2015 and so provide us with a detailed account of the climate-change challenge which all the major thinkers and policy-makers on the subject have agreed on.[6]

The five assessment reports produced by the IPCC so far all follow pretty much the same structure. Each report begins with a thorough review of the scientific evidence, looking at thousands of peer-reviewed articles across disciplines and a selection of non-peer-reviewed writings. A second section discusses the impact of climate change, primarily framed in terms of economic and social consequences. Finally, a concluding section proposes policy response options, in terms of both mitigation (focused on reducing greenhouse gas emissions) and adaptation (focused on measures limiting the damage already done). Each of these three sections contains at the end a so-called Summary for Policymakers which, when taken together, form an agreed-upon framework for understanding and addressing the problem. We can look at this elaborate report structure, coupled with the consensus-based process of its elaboration, as an innovative example of global governance aimed at interdisciplinary consideration of a complex worldwide problem and cooperation between experts and policy-makers in the search for adequate solutions. To the extent that the IPCC has succeeded in those tasks, it has prepared the ground for global policy initiatives such as the aforementioned Paris Agreement of 2015.

Other than the occasional controversy being immediately blown out of proportion by climate deniers intent on disqualifying the IPCC (see, for instance, "Climategate" in 2009 comprising hacked emails written by four climatologists), the panel has been amazingly disciplined and largely transparent in its work and findings. This has given it much needed credibility. True, its consensus-based approach comes with some inherent drawbacks. The requirement for unanimity imposes a bias toward caution which may thwart adequate consideration of more controversial findings over which there is not yet sufficient agreement. Alternative (minority) viewpoints may be suppressed. And perhaps worst of all, it is very difficult, if not impossible, to deal properly with the uncertainty of future scenarios when in the end everybody will have to have agreed on the same interpretation of what is likely to happen. In other words, uncertainties tend to get downplayed in such a search for consensus opinion. But other than these inherent drawbacks, the consensual approach of the IPCC has brought many advantages, notably in support of its reputation as a serious body of credible experts capable of defining and analyzing a long-term problem we shall all have to face eventually in collective fashion.

Greenhouse Gases (GHG)

What the five IPCC reports have established beyond reasonable doubt is that anthropogenic (i.e., caused by human activity) concentrations of certain gases are heating our planet above what normal atmospheric activity would suggest the average temperature to be, and that this "greenhouse effect" runs the risk of potentially dramatic changes in our climate pattern in the not-so-distant future. Earth, like a few other planets, is naturally warmed. The interaction between the rays from the sun, our atmosphere, and the earth's surface gives rise to thermal radiation which is absorbed in the atmosphere to warm our planet to an average temperature of about 15 °C (from what otherwise would be a much colder planet at about −18 °C). Key to this process are various atmospheric gases, such as water vapor or ozone, absorbing and emitting radiation within the thermal infrared range that dominates on the surface of the earth and in the atmospheric layers just above it. Those gases tend to radiate energy in all directions, including toward the earth's surface, thereby warming it. This natural greenhouse effect is absolutely critical to supporting life on our planet. Human activity, however, has caused progressively larger concentrations of certain gases, which have given rise to an enhanced greenhouse effect warming the atmosphere more than would or should naturally be the case.

There are several greenhouse gases prone to becoming increasingly concentrated in the atmosphere thanks to human activity. The most important in terms of volume emitted, absorbing about two-thirds of all anthropogenic GHG emissions, is carbon dioxide (CO_2), which gets emitted in the wake of fossil-fuel combustion (e.g., producing electricity, driving cars, heating buildings), industrial processes (e.g., producing cement, other chemicals), deforestation, and other land use. This gas can stay in the atmosphere anywhere from fifty years to thousands of years. The second most important GHG, representing 17% of the total, is methane (CH_4) which arises in the mining of coal, the production and transportation of natural gas, from landfills (as trash breaks down over time), and in raising livestock (when cows or sheep digest food, when manure decays). Methane typically stays in the atmosphere only about a dozen years, but traps about twenty times more heat than the same quantity of carbon dioxide would. A third atmospheric gas, nitrous oxide (N_2O) which amounts to about 6% of all anthropogenic GHG emissions, also arises from burning of fossil fuels and some industrial processes. Its most important sources, however, are farming practices relying heavily on fertilizers. Remaining in the atmosphere for 114 years, N_2O traps about three-hundred times more heat than the same quantity of CO_2. Chlorofluorocarbons (CFC) are another greenhouse gas, widely used at some point as solvents, propellants in aerosol applications, and refrigerants. The discovery in 1973 of CFC's role in the depletion of the ozone layer, which caused a strong public reaction across the planet, prompted in 1987 a global accord ratified by all members of the United Nations, the Montreal Protocol, to phase out all CFC by 2000—the most spectacularly successful example of global environmental cooperation to date![7]

Once emitted into the atmosphere by human activity, these various gases remain there for a very long time. That is why global warming would continue as a trend, even if we somehow managed miraculously to halt all anthropogenic GHG emissions from one day the next. As they get pumped into the atmosphere, those gases move around with the air circulation and so get mixed up as well as evened out across the planet. Even though some places emit more GHG than others, there will in the end be pretty much the same evenly distributed level of GHG concentrations across the globe, making this a truly planetary phenomenon necessitating a globally coordinated response. Add to this the challenge of positive feedback loops which intensify the global warming trend! For instance, as the global temperature rises, the tundra—a vast flat treeless zone covering North America and Eurasia south of the Arctic whose subsoil is permanently frozen—may well thaw and in the process release huge quantities of methane currently trapped by the permafrost. This alone would make the global warming trend much worse in a hurry! Furthermore, global warming also causes more water to evaporate from the Earth's surface and end up as water vapor in the atmosphere. But water vapor is nothing but yet another greenhouse gas further warming the planet. These multiple interactions threaten to accelerate global warming over time.

To get a sense of how much these anthropogenic greenhouse emissions have already reached a critical level where they have begun to impact adversely on our environment, it suffices to take a look at the latest assessment report of the IPCC (known as AR5) from 2014.[8] Ancient air bubbles trapped in ice allow us to see what Earth's climate was like in the distant past and trace its evolution over time. We know that before the start of the Industrial Revolution in the mid-eighteenth-century CO_2 concentrations in the atmosphere had ranged between 180 and 280 parts per million for millennia. Since then, this number has risen steadily, with half of the increase coming just in the last thirty years and rising above 400 ppm for the first time in 2013. Driving these growing atmospheric concentrations is the accelerating pace of CO_2 emissions, whose annual increases averaged 6.8% during the 2000s compared to 2.9% in the 1980s and just 2.4% in the 1990s. Much of that increased pace came about because of the rapid growth and income gains in emerging-market economies, such as China or Brazil. We went from annual increases of 10 gigatons (a gigaton equals one billion metric tons) in the 1960s and 20 gt/year in the mid-1980s to about 40 gt/year presently.[9] If we continue at that pace, we run the distinct risk of making our planet uninhabitable by the time this century comes to a close.

These increased concentrations of GHG in our atmosphere have already significantly warmed the planet. The IPCC (2014) report noted that each of the last three decades has been successively warmer at the Earth's surface than any decade since 1850. The intensification of this trend is manifest in now routinely having pretty much each consecutive month being the hottest on average ever recorded. The combined global average air and sea temperature is at this point already almost 1 °C higher than it was in 1850, with half of that increase occurring just in the last three decades. You may ask yourself what can a one-degree hike in temperature do to the climate, environment, or weather patterns. The correct answer is a lot. Just that amount of warming of the air and sea, still quite modest compared to what we can expect to occur over the rest of the century if we do not reverse the trend dramatically (with predicted current-baseline average temperature increases of up to 4.8 °C by 2100), has already had an array of rather dramatic effects in terms of destabilizing weather patterns, damaging our eco-system, and endangering our living conditions.

ALREADY NOTICEABLE CONSEQUENCES OF CLIMATE CHANGE

The steady warming of the planet has above all encouraged more extreme-weather events, especially in the Northern hemisphere where there is proportionately more landmass than in the ocean-dominated Southern hemisphere. We are talking here about principally drier weather in already dry regions as well as increased precipitation in typically wet areas especially around the equator and near the Arctic circle. Both droughts and more intense storms have become more likely. Extreme-weather events put pressure on existing populations having to cope with chronic water shortages, shrinking food supplies, and more devastating wildfires in drought-stricken areas or floods, storm damage, and mudslides in rain-soaked areas. The five-year drought in California, Hurricane Sandy hitting New York in October 2012, or the once-in-a-millennium rainfall of >20 inches in central Louisiana during August 2016 come to mind if you are an American. But such unprecedented storms or droughts have also hit many other areas of the world in recent years. Climatologists understand well how changes in the atmospheric dynamic in the wake of greater GHG concentrations produce longer periods of extremely hot weather, self-reinforcing drought conditions, more intense storms, and more extensive heavy-rainfall periods. While none of those events can be individually attributed in direct fashion to climate change, those scientists have framed the "proof" of this connection in terms of steadily rising probabilities of extreme-weather events depending on the region and season, which can be (and have been) statistically verified.[10]

Those same climatologists are also providing growing evidence that observed changes in weather patterns are unevenly distributed across the globe in such a way as to reinforce regionally distinct conditions already amply prevailing there, meaning in essence that they are making dry regions drier and wet regions wetter. Such climactic polarization can have devastating effects on already vulnerable areas and local populations living in them, as for instance evidenced by accelerating desertification rendering certain regions increasingly uninhabitable. As a matter of fact, the devastating civil war in Syria, triggering millions to flee their war-torn country and causing a huge migration crisis in Europe, was preceded by a four-year drought causing local tensions over the shrinking supply of arable land to come to a boil when the Arab Spring swept Syria and endangered the Alawite minority's control of the country. We can expect many more such failing states unable to deal with the confluence of environmental degradation and demographic tensions, followed by mass migrations. Richer nations, themselves in the grip of post-crisis surges of nationalism and isolationism, are not ready to absorb drought-induced and violence-driven mass migrations meaningfully as the European Union's failure to cope with the massive influx of Syrian refugees has so starkly demonstrated. It also goes without saying that droughts have the most devastating impact in poorer, more vulnerable areas and so hurt the already dispossessed the most.

The trend toward growing numbers of more intense storms poses an entirely different set of problems. Storm damage to property and loss of life are immediate concerns, especially in regions exposed to tornadoes, hurricanes, typhoons, or monsoon rain all of which are bound to grow in frequency and intensity. Of course, their intensification also interacts dramatically with rising sea levels, another manifestation of climate change, to produce far more dangerous storm surges that can do a huge amount of damage as they hit land. Both Hurricane Katrina destroying New Orleans' levies in 2005 and Hurricane Sandy's damage to low-lying areas of New York City (e.g., Staten Island, Hoboken) as well as its power supply and subway system in 2012 have demonstrated in shocking fashion how vulnerable coastal cities are to such storm surges. While there are different types of floods (flash floods, overbank flooding, mudflows, etc.)—and we are rapidly becoming more experienced with regard to these distinctions as our climate changes—they all typically involve very slow and difficult recoveries. Flooding is rapidly becoming more common in certain areas of the world with a pattern of more intense storms carrying greater wind speeds and far more rainfall amounts. This also holds true for winter storms bringing record amounts of snow all at once, even though average snowfall amounts are steadily decreasing over time. Urban planners and meteorologists are also becoming aware that sustained downpours put infrastructure, such as bridges, roads, irrigation systems, or power plants, under special stress which may well require those structures to be reinforced.

One of the most profound consequences of global warming already observable is the melting of the polar ice caps in the Arctic and, less obviously so, Antarctica. This phenomenon also extends to Greenland, another large area covered by a huge ice sheet. As warmer temperatures melt icebergs or, more dramatically, accelerate the breaking off of ice masses from glaciers ("calving"), the sea level rises in proportion to the receding of land ice. Add to this the melting of sea ice in the wake of rising temperatures. Suffice it to say that, as ocean temperatures rise as well in the wake of global warming, the warmer water expands to make the sea level rise even further. The sea level has already risen on average by 8 inches since 1880, with that trend nearly doubling its speed since 1993 in the wake of much more rapid melting of land ice.[11] Irrespective of the large variations between different regions due to local factors such as currents or ocean topography, the oceans are rising all over the world and can be expected to do so in accelerating fashion into the unforeseeable future. Scientists, such as those grouped together in the Union of Concerned Scientists or the Intergovernmental Panel on Climate Change, project possibly devastating further sea-level rises of anywhere between 16 and 24 inches by 2050 and from 48 to 78 inches by 2100, if current GHG emission trends continue. Of course, it is quite likely that we will succeed in slowing the pace of these emissions considerably in which case global average sea-level rises should be kept to a range of 6–16 inches by 2050 and 12–48 inches by 2100 above current levels. Even that projection is quite disturbing. Imagine an ocean that is four feet, or more than a meter, higher than what it is today!

Rising sea levels threaten low-lying coastal areas in a variety of ways. Shorelines erode as waves penetrate farther inland. Storm surges get amplified and so end up doing a lot more damage. Saltwater reaches further into coastal groundwater sources. Coastal ecosystems, such as coral reefs, mangroves, salt marshes, or sea grass meadows, upon which the livelihood of many coastal communities depends, get undermined. And finally, after a period of growing incidences of flooding, exposed coastal areas face permanent inundation as they literally sink into the sea. That ultimate threat is already acutely felt today by several small-island nations, such as the archipelago of coral atolls known as Kiribati in the Pacific Ocean, the Maldives in the Indian Ocean, Palau south of the Philippines, or the Solomon Islands east of Indonesia, all of whom have seen part of their territory submerged. Not surprisingly, these small-island nations have become leading voices in the fight against climate change. River deltas are also very exposed, as can be seen even today with the Nile and Rhone deltas in the Mediterranean, the Danube delta at the Black Sea, or the Mississippi delta in the Gulf of Mexico. Some densely populated areas along the coasts of Eastern England, Belgium, the Netherlands, and Germany lie already below the mean high-water mark and are thus in need of coastal protection, as are parts of Italy's Po Valley all the way to the now perennially flooded Venice. Similar fates await large areas of Asia (e.g., Bangladesh, Vietnam), Africa (e.g., Egypt, Senegal, or Nigeria), or Latin America (e.g., Caribbean islands, Costa Rica, Columbia, Brazil) all of whom contain large cities within the danger zones.[12] When it comes to the USA, much of Florida is acutely threatened by rising sea levels as is a portion of the Atlantic coast (New York City and Boston included), low-lying coastal areas all along the Gulf of Mexico, and significant coastal pockets in California (e.g., San Diego, San Francisco).

Much of the human species lives on or near the sea, attracted by the economic benefits arising from ocean navigation (see the importance of ports for global trade, for instance), tourism, fisheries, or recreation. Over a billion people worldwide live today within twenty meters of mean sea levels, and they all have to worry about varying degrees of exposure to rising sea levels. Forty percent of the world's population, according to estimates by the United Nations, lives within 100 kilometers from the coastline or in low elevation coastal zones, including river deltas, of less than 10 meters elevation whichever is closer to the sea.[13] Thanks to rapid urbanization over the coming decades, coastal megacities are expected to grow disproportionately faster than other urban centers and so ironically push up the percentage of the total population living on or near the coast just when rising sea levels pose a growing threat to their livelihood. Of course, there are reasonably effective adaptation measures available to protect coastal populations, such as the construction of levees, of mechanical barriers that can be opened and closed (as already in place in the Netherlands or Southeast England), or of sea walls, coupled with restoring wetlands and/or beaches as natural protection systems, while also emphasizing improved flood evacuation maps and stronger coastal permitting. But none of these counteractive protections will suffice if the most catastrophic sea-level rise predictions, as recently projected in disturbingly dire updates (see, for instance, DeConto and Pollard 2016), come to bear. It all depends to what extent we can contain the melting of the gigantic ice sheets of Greenland or Western Antarctica, making it once again abundantly clear that mitigation outweighs adaptation. If sea levels are allowed to rise by two meters at the end of the century, then we can expect the twenty-second century to be consumed by huge population movements away from the coast.

Melting ice amidst global warming creates an additional environmental stressor of the highest order inasmuch as it threatens to wipe out glaciers in the Himalayas, Alps, Andes, and other major mountain ranges. Everywhere, across all regions of the world, scientists have noted that glaciers are receding at an accelerating pace.[14] Initially, the runoff from melting glaciers increases stream flow and even creates new mountain lakes, both of which increase the danger of floods in lower-lying areas. We have already had instances when glacial lakes suddenly overflowed to flood valleys downstream, as happened in 1996 in Bhutan or in 2005 in Peru. But as the glaciers continue to shrink, they start providing less runoff until they disappear altogether. Together with steadily shrinking snowpacks which global warming also causes to melt earlier, there is less and less freshwater supply coming down from the mountains on which millions of people depend in any given catchment area for drinking, irrigation, and hydropower production of electricity. Droughts will further intensify as water runs increasingly scarce. And what about the long-term health of rivers once they become deprived of natural replenishment from glaciers and snowpack? Reduced river flow and increased sediment render rivers more polluted and erode their ecological contribution to the well-being of humans and animals who depend on them. Many of the world's great cities lie along rivers whose water flow and quality are threatened to deteriorate steadily in the wake of climate change. With that in mind, it is hard to escape the conclusion that densely populated areas in, say, California, the Pacific Coast of Latin America (e.g., Lima in Peru), the heartland of Europe (along, say, the Danube), or along the Ganges in Northern India will find life more difficult and their livelihood threatened as the rivers they depend on see gradual erosion.

Going back to the oceans, we need to consider further how global warming affects their modus operandi. We know that accelerating GHG emissions end up raising the temperature of the ocean surface, a fact that adds to both rising sea levels and the frequency of more extreme-weather events in the form of heat waves, droughts, and storms. The same CO_2 absorption process in oceans also renders them more acidic.

And this progressive acidification of our oceans contributes to coral bleaching, thereby greatly adding to the stress the world's coral reefs face already from invasive tourism practices, agricultural runoff adding to water pollution, and aggressive fishing practices. Our coral reefs are undergoing a slow-motion death by many cuts! Often referred to as the "rain forests of the sea," these incredibly diverse ecosystems provide a home to about a quarter of all marine species and at the same time offer crucial services to tourism, fisheries, and shoreline protection. Their progressive destruction is bound to have a devastating impact on ocean life. In this sense, climate change is aggravating a situation of rapidly declining fish stocks which have been decimated by human practices of overfishing, ranging from fish piracy to various destructive fishing techniques (e.g., wasteful bycatch, bottom trawling). At the current rate of marine species destruction, we risk wiping out the fish population of our planet in forty years! Over one and a half billion people depend on fish as the main staple in their daily diet.

Fisheries are not the only natural resource under pressure. Climate change threatens our ecosystems at large. For instance, agriculture is bound to be negatively affected by global warming and more extreme-weather events. We can already observe lower crop yields for wheat and maize across many regions, with similar declines to be expected for other crops, such as rice and soybeans, if and when the temperature continues to rise across the globe. One aspect of this problem has to do with changes in seasonal growth patterns, as longer summers encourage faster maturing of food and earlier harvests. The other dimension of climate-induced declines in agricultural productivity stems from the greater likelihood of more and longer extreme-weather events whether droughts or storms. The IPCC's AR5 reports several instances of rapid, sudden price hikes affecting cereals and other foodstuffs following extreme climate events in key producing regions, and this indicates how sensitive food markets are to weather-related disruptions. Food security is thus bound to emerge as a major worldwide concern in the wake of climate change.[15] The most vulnerable rural populations engaged in food production as well as the urban poor depending for their survival on affordable food are both bound to be highly vulnerable to this problem worldwide, adding to the already insupportably large income inequality undermining societal cohesion. In certain areas, especially in Africa and Asia, the climate-induced collapse of food production will threaten to set off mass migrations of the kind we have already witnessed recently with the millions fleeing war-torn Syria. In more gradual fashion, we will also witness geographic shifts in food production patterns, as we are already beginning to see with wine production. Some classic and famed wine-producing regions, such as Burgundy in France, are beginning to feel the heat while other regions not known for outstanding wines, such as Quebec in Canada, will find themselves with steadily improving vineyards.

Such climate-induced geographic shifts may apply more broadly to animal species and vegetation, threatening existing ecosystems. Some plants, trees, or flowers typically found in one region will at some point cease to grow there and will have moved to somewhere else. The same should be true for animals whose natural habitats might become untenable in certain regions undergoing profound climate change. They too will have to move! This is not a linear process, because of the ways ecosystems depend on the interactions between different species and their surroundings. When there is a lot of movement, those once-established interaction patterns get disrupted and transformed. Will bees in the future find their flowers to pollinate? Will grizzly bears still find their salmon to catch in live streams? At some point, we can imagine such dislocations to go to the point of specie extinction. We all have become aware in recent years of the incredibly sad images depicting the majestic polar bears desperately hanging on to ever-shrinking floats of sea ice while slowly wasting away for lack of food. The beautiful mountain flowers of Europe's Alpine region are moving up the mountains right now, but may ultimately very well disappear for lack of moisture. Our changing climate thus represents an existential threat to our planet's biodiversity.

Finally, it is no exaggeration to consider climate change as life-threatening to humans. Extreme weather can and will kill! France's heat wave of 2003 killed 14,800 people. Hurricane Katrina and its destruction of New Orleans cost 1836 lives in August 2005. More than 3000 people were killed in November 2007 by Cyclone Sidr in Bangladesh. In the great six-week heat wave hitting Russia in the summer of 2010, over 55,000 people perished while wildfires destroyed over 2000 buildings. That year alone Munich Re, the world's largest reinsurance company, reported 874 weather and climate-related disasters resulting in 68,000 deaths and $99 billion in damages worldwide.[16] And the list goes on, bound to get worse in coming years. According to a report by Britain's financial regulator Prudential Regulation Authority (2015), inflation-adjusted losses from natural events, most of which triggered by extreme weather, rose from an annual average of $10 billion, during the 1980s, to some $50 billion, over the past decade. But climate change can also kill in less obviously violent ways. Higher temperatures, particularly heat waves, threaten to worsen smog and pollution in urban areas with potentially negative repercussions for vulnerable populations (children, elderly, people with asthma). We can also expect more intense combinations of heat and humidity to facilitate the spread of contagious diseases, especially those transmitted by insect bites such as malaria or dengue fever, and so also create conditions favoring new pandemics as we have recently witnessed with Ebola fever or the Zika virus.

Climate change is real, and so are its many destabilizing consequences. We are only at the beginning of this process, and already its many-faceted effects have raised very troubling questions. If a one-degree temperature hike can be so destructive, what will happen when the globe is an average of 5 degrees warmer? Hard to imagine, but we have to assume this to be a catastrophic prospect. We are becoming more aware as well that the global warming process begins to feed on itself in accelerating fashion beyond a certain level, as land ice begins to melt more rapidly when the polar caps have less ice left to absorb the rays of the sun or the thawing of permafrost across Canada or Siberia starts releasing huge quantities of methane gas. Scientists have therefore agreed that the world community must do everything it can, starting now, to keep the global average temperature from rising beyond two degrees. Any rise beyond that could be so destructive as to threaten our very existence!

THE LONG AND WINDING ROAD TO PARIS

The IPCC's maiden report in 1990 laid out in meticulous fashion for the first time what climate change was all about and how it might disrupt living conditions on our planet. For those who paid attention, this made shocking reading. Something had to be done. Two years later—in June 1992—174 countries, with 116 of them sending their heads of state, gathered in Rio de Janeiro (Brazil) under the auspices of the United Nations for an Earth Summit.[17] Amidst several important agreements dealing with biodiversity, desertification, and sustainable development, that historic meeting also concluded a treaty known as the United Nations Framework Convention on Climate Change (UNFCCC) which entered into force in March 1994. Initially signed by 154 countries, the convention has today 197 members. This nearly universal membership gives it the credibility it needs for promotion of worldwide collective action in the face of a truly planetary challenge, a necessary but not sufficient condition to deal with the problem at hand.

The UNFCCC defined its principal objective very succinctly in its Article 2, namely to "stabilize greenhouse gas concentrations in the atmosphere at a level that would prevent dangerous anthropogenic interference with the climate system." In Article 3(3), the UNFCCC also drew explicitly on the Precautionary Principle according to which the uncertainty associated with potential future harm from GHG should not be used as an excuse for inaction, a "better safe than sorry" prescription clearly aimed at climate deniers. But its precise wording implied a weak version of this principle rather than a strong interpretation requiring action in the face of potential damage. This more cautious commitment reflected the political reality of lacking sufficient consensus to take decisive action immediately.[18] Instead of specifying at that point binding limits on GHG emissions or imposing an enforcement mechanism, the agreement instead provided a framework how countries would negotiate such limits in the future. To begin with, signatory parties agreed to meet annually, in so-called Conferences of the Parties (COP), to assess progress in their collective battle against climate change and conclude emission-limitation agreements on a rolling basis going forward. As a first broad goal, the UNFCCC proposed to stabilize GHG emissions at 1990 levels by the year 2000 and establish toward that objective national greenhouse gas inventories among member states on the basis of which the 1990 benchmark levels would be set. Those inventories would get updated regularly to help define future emission targets. All the signatories to this framework committed themselves to both climate-change mitigation (i.e., lowering emissions) and adaptation (i.e., coping responses).

One of the more contested issues surrounding the Rio Declaration as well as future climate-change agreements arising in its wake concerned the distribution of burden sharing. The GHG emissions driving up air and sea temperatures have for the most part been caused by the rich, highly industrialized nations whose excessive contribution to global warming has been going on for decades. This imbalance concerns not only production patterns in terms of energy, industry, transportation, or agriculture. It also applies to consumption patterns of rich populations in industrial nations, as expressed, for example, in their disproportionately high demand for fossil fuels from oil-exporting nations or the percentage of their cars carrying only one person. It stood therefore to reason right from the very beginning that efforts at climate-change mitigation should be placed predominantly on the shoulders of the advanced capitalist countries. Lesser developed countries had played much less of a role in global warming and in addition had other development goals of greater urgency, such as fighting hunger and poverty, which should not be crowded out or rendered more difficult to pursue by their efforts at fighting climate change. Yet at the same time, many of the poorer nations were likely to be affected earlier and more heavily by the nefarious consequences of a warming planet, as most dramatically evidenced by the small-island nations in the Pacific or the Indian Ocean. Richer nations should therefore also commit themselves to helping the poorer, developing countries cope with this challenge by providing both financial assistance and technology transfers.

The UNFCCC crafted a carefully balanced compromise in this regard by distinguishing between three different groups of countries—the advanced capitalist nations, the least-developed nations, and transition economies of the former Soviet Bloc which were mostly middle-income countries relying too much on heavy industry and other sources of much pollution. The framework's Article 3(1) made it clear that the burden of the fight against climate change would not be shared equally, as it called for "common but differentiated responsibilities" in which the rich advanced capitalist nations would "take the lead." The latter, of which there are 24 members (including the European Union) all grouped together as so-called Annex II Parties, would commit themselves first to emission-reduction targets and also provide assistance to both the 14 transition economies of the former Soviet Bloc (grouped together in Annex 1) as well as 49 least-developed countries in their efforts to combat climate change. It was made clear that these poorer countries, which had contributed proportionately much less to the global warming problem than the more industrialized countries, could not be expected to sacrifice their social and economic development priorities by lowering their GHG emissions from current levels. To the extent that they lacked the means to do their fair share toward climate-change adaptation, they could expect financial as well as technological assistance from the rich Annex II Parties. Implied here was the ultimately very controversial notion that developing countries, including such huge ones as China or India, would not be expected to cut their GHG emissions at all for the time being.

Subsequent annual COP meetings primarily dealt with putting teeth into this agreement. It became clear relatively soon that the Framework Convention's goal of "stabilizing" GHG emissions at levels low enough to avoid "dangerous" human-caused interference with the planet's climate would require explicit commitments by the largest possible number of countries as soon as possible. Already in 1995, the initial target of keeping emissions in 2000 at 1990 levels was deemed inadequate, triggering intense discussions over a more ambitious plan which ended up known as the 1997 Kyoto Protocol. Taking effect in 2005, that international treaty saw the rich Annex II countries undertake legally binding commitments to reduce their GHG emissions by an average of 5.3% below 1990 levels over an initial commitment period lasting from 2008 to 2012 to which the Doha Amendment (of 2012) added a second commitment period from 2013 to 2020. The emission cuts would also include net GHG absorption by carbon "sinks," such as forests or land-use change, whose importance this treaty elevated strategically.

The Kyoto Protocol introduced three so-called flexibility mechanisms each of which aimed at making the promised GHG emission cuts more easily achieved. One such mechanism were national emissions trading schemes limiting producers to certain emission levels and allowing those doing better than their respective cap to sell off their surplus allowances to others wanting to emit beyond their assigned limit. Aggressive emission cutters were thus to be directly rewarded for their successful efforts while others had to pay for the privilege of excessive pollution. Another advantage of those trading schemes would be to establish a market-driven price for carbon. The protocol also foresaw trading of national Kyoto obligations so that countries in deficit could acquire surpluses of allowances from other countries. This turned out to be especially important for the transition economies of the former Soviet Bloc who had been assigned a huge number of emission allowances which the European Union countries and Japan were eager to acquire. Another important flexibility mechanism introduced by the Kyoto Protocol was the Clean Development Mechanism (CDM) through which rich countries helped finance emission-reduction projects (e.g., adoption of cleaner energy sources) in developing economies as part of their national emission-reduction goals. And finally, there was Joint Implementation which allowed any Annex I country to invest in an emission-reduction project of another Annex I country as an alternative to lowering emissions at home on its way to meeting Kyoto targets for emission reductions.

The story of the Kyoto Protocol is a sad one, highlighting the enormous difficulties in putting an asymmetric framework into an effect that explicitly endorses unequal burden sharing. The US Senate refused to ratify the treaty, complaining that more than half of the planet was left off the hook, in particular India and China, two hugely populous nations that were spared any efforts to reduce their GHG emissions. The "poor country" claim by the leaders of these two giants belied the fact that both had started to grow much more rapidly after 1994, and with it also their GHG emissions. Congressional leaders had begun to consider especially China as a competitor and did not want to burden the US economy unfairly in its battle over market shares. While many industrial nations and especially most transition economies made significant progress in reducing their GHG emissions from the 1990 baseline, this improvement was more than counterbalanced by huge emission growth among the rapidly expanding "emerging market" economies such as Mexico, Brazil, and above all China.[19] The absence of US participation led other countries to renege on their commitments as well, most notably Canada's decision in 2011 to drop out of the Kyoto Protocol in protest against inadequate contributions by other major economies. This withdrawal by a most strategic country, a major energy producer who also possessed very large carbon sinks while containing half of the world's tundra within its borders, mortally wounded Kyoto's extension via the Doha Amendment when Japan, Russia, and New Zealand opted out of the latter in short order during 2012. The Doha Amendment failed to get ratified by enough countries and so ended up never put into effect.

Dissension among the world's leaders over climate-change action and burden sharing reached its peak during the disastrous Copenhagen Climate Summit in 2009. That meeting, engaging many heads of state in fruitless last-minute efforts to save some sort of consensual agreement, was riddled by many divisions and ended in stalemate. Despite intense preparatory negotiations, there were no outlines of a workable deal agreed to before the summit started. Developing countries were suspicious of Western leaders, like President Obama, arriving last minute to impose an agreement without adequate consultation, a fear reinforced right at the beginning of the meeting amidst a leak that the Danish hosts had already prepared the final text. The mistrust and recriminations dominating the conference hid much deeper disagreements concerning the size and distribution of emission cuts, over the amounts to be transferred by rich countries to the poor nations, and as to when poorer countries should start cutting their emissions. The sharpest arguments arose around the insistence of Americans and Europeans that fast-growing emerging-market economies like China, India, or Brazil agree to peak their GHG emissions by 2020. Those countries resisted this request strenuously, arguing that such a commitment would lock them into poverty. A second layer of tension arose among the Group of Seventy-Seven representing the (actually one hundred and thirty) poorer nations as smaller island nations led by Tuvalu pushed hard for massive and legally binding cuts while others hesitated to commit to any obligatory emission cuts. In the end, an anti-Western alliance comprising Venezuela, Bolivia, Sudan, and Tuvalu called for a boycott of any deal concluded on the last day when the heads of state would arrive. As chances for a last-minute deal fell apart, President Obama salvaged a declaration agreed to by him and the leaders of Brazil, China, India, and South Africa which reconfirmed their intention of limiting temperature rises to below 2 degrees and setting up a fund in support of the poorer countries. But this agreement contained no details pertaining to baselines, no specific cuts in emissions, nor any concrete timetable while lacking at the same time any legal force.[20]

The failure of the Copenhagen Summit, at the time considered a debacle for international diplomacy, ultimately served the purpose of focusing minds on the task of coming up with a global accord on climate change to replace the expiring Kyoto Protocol. In addition, the fiasco helped leaders learn from their mistakes so that diplomatic initiatives soon thereafter improved in building gradual consensus where previously discord had reigned. Subsequent COP meetings addressed the most contentious issues rocking the Copenhagen Summit and so set the stage for successful conclusion of the Paris Conference in December 2015. The first post-Copenhagen meeting, known as COP16 and held in Cancún (Mexico), set up a Green Climate Fund and a Climate Technology Centre to help developing countries launch effective mitigation and adaptation projects. In turn, those developing countries were now expected to start planning their efforts against global warming. The idea was for all countries to submit annual greenhouse gas inventory reports and so-called Nationally Appropriate Mitigation Actions (NAMAs) for which the developing countries could request international financing, technology, and capacity-building support. Once support was assured, the NAMAs would be entered into a global action registry as

"Internationally Supported Mitigation Actions" (ISMAs) whereupon they would become subject to international measurement, reporting, and verification procedures. One year later, in December 2011, COP17 adopted the Durban Platform for Enhanced Action which committed members to seek a new global accord for the post-Kyoto period following 2020.

COP21: The Paris Climate Agreement of 2015

That accord emerged at the conclusion of COP21 in December 2015 and became known as the Paris Climate Agreement. This meeting, masterfully orchestrated by French Foreign Minister Laurent Fabius and his ambassador for international climate negotiations Laurence Tubiana, managed to get every single country on board (except for Syria and Nicaragua, the only two countries not signing the agreement; both have joined since). This is a tremendous achievement which can be rightfully considered a major breakthrough twenty-some years in the making.[21] Crucial in the run-up to Paris was a bilateral agreement between US President Obama and China's premier Xi Jinping spelling out their respective emission-reduction targets together which removed a major point of contention involving the world's two largest GHG emitters and signaled to others that a global deal was thus finally within reach. It also helped that the European Union, long a leading voice in the fight against climate change, announced already early on (in March 2015) an ambitious target of cutting emissions 40% below 1990 levels by 2030. This spurred many other countries to follow suit with their own climate-change mitigation and adaptation plans.

The Paris Agreement, which covers the period after 2020, went into effect in November 2016 after ratification by the needed quorum of countries comprising a minimum of fifty-five countries representing at least fifty-five percent of global emissions. It set a long-term goal of keeping the increase in the average global temperature to "well below" 2 °C above pre-industrial levels, with an explicit aspiration of trying to limit the temperature rise to just 1.5 °C. Toward that objective, all 196 signatory countries submitted comprehensive national climate action plans known as "Intended Nationally Determined Contributions" (INDCs) comprising both mitigation projects and adaptation planning. Those initial INDCs, covering the period 2020–2025 but possibly stretching all the way to 2030, were converted into official "Nationally

Determined Contributions" (NDCs) with ratification of the Paris Agreement. Those NDCs will have to be replaced every five years with new action plans. Laying out in great detail baselines, targets, specific measures, methodologies, and justifications in light of nationally specific circumstances, the NDCs of all parties would be subject to ongoing collective assessment within a robust transparency, communication, and accountability system so as to assess each party's progress toward meeting its declared goals.

Taken together, it is clear that NDCs announced so far do not suffice to limit the cumulative temperature increase to below 2 °C. Best estimates by the IPCC see these contributions, if all put into effect as intended, limit the temperature increase to 2.7 °C from pre-industrial 1850 levels. That is surely better than the "business as usual" baseline of at least a 3.5° rise if we continue along current trends, but still a far cry from the "well below 2 °C" goal announced in Paris.[22] Hence, that agreement also contained an explicit commitment to progressively tougher goals and more ambitious plans to be spelled out from 2023 onward, a so-called ambition cycle. Global emissions would have to peak "as soon as possible," with developing countries given a bit more time to reverse course. There would have to be rapid reductions thereafter so as to achieve carbon neutrality by mid-century or shortly thereafter at which point a necessary minimum of carbon emissions would be matched with an equivalent amount of carbon being sequestered or offset (including credit purchases from emissions trading schemes).

Recognizing the damage already done to the planet by global warming and the growing likelihood of worse to come irrespective of subsequent containment efforts, the Paris Agreement subtly put greater emphasis on adaptation than earlier treaties and declarations had done. Beyond emphasizing a needed balance between mitigation and adaptation, it also included for the first time significant provisions concerning finance. An explicit commitment was made in Paris by the developed countries to scale up the Green Climate Fund to $100 billion per year by 2025 in support of the efforts by the developing countries and to involve a much larger group of donors (including China and other fast-growing emerging-market economies) so as to insure sustained funding of the GFC at an adequate level. Cognizant of the fact that the goal of a carbon-neutral economy by, say, 2060 would involve trillions in new funding and huge shifts in existing investment flows, the Paris Agreement also aims to "making finance flows consistent with a pathway towards low greenhouse gas emissions and climate-resilient development" (Article 2.c). This explicit "financial-flow consistency" goal is new and represents a major step forward in aligning global finance to the new realities of a major overhaul of the world's economic structure as we face the existential threat of global warming—the birth of what the Paris Agreement itself has termed *climate finance* (in Article 9).[23] While this goal would undoubtedly involve a large chunk of public funds and government-sponsored development banks, the new framework for climate finance made explicit the mobilization of private financial institutions as well, including new risk management facilities and a crucial role for insurance. This was part of the Paris Agreement's remarkable effort to go beyond governments ("Parties") and invite all kinds of "non-Party stakeholders, including civil society, the private sector, financial institutions, cities and other subnational authorities, local communities and indigenous peoples" (paragraphs 134 and 136) to do their share in the effort.

Finally, the Paris Agreement also included a reform of the emission trading schemes that had been launched nearly two decades earlier in the Kyoto Protocol. The international trading of emission credits, as crystallized in Kyoto's CDM and Joint Implementation schemes, had until now presumed homogenous caps allowing for easy transfers of credits between countries as well as two groups of countries with and without caps. The Paris Agreement's "nationally determined contributions" (NDCs) did away with both of those conditions, applying some sort of cap to all countries and making these caps much more heterogeneous from country to country, hence less transferable. Article 3 of the 2015 Agreement revamped the international swapping of emission-reduction credits to avoid double counting and provide a better balance between countries representing demand for such credits and other countries supplying them—a complex subject which we will have to revisit when discussing how carbon markets work (see Chapters 5 and 7).

The Paris Agreement has been criticized by environmentalists as "too little, too late."[24] Yes, it is true that the NDCs of all the parties combined are far too modest to move us onto a GHG-emission-reduction path that would prevent temperatures from rising beyond 2 degrees. There is no enforcement mechanism to assure that countries keep even to these modest promises, let alone make the politically difficult decisions awaiting them later if and when they want to give the Paris goals a real chance. No single country can be stopped from opting out of the agreement, and there have already been bad precedents with compliance under the Kyoto Protocol and signing up to the Doha Amendment.

Trump's July 2017 decision to take the USA out of the Paris Agreement, while possibly reversed by the next US administration, bodes ill in this context. We continue to invoke alternative-energy options and untested technologies that are not really ready-made solutions to address our climate-change problems, such as huge dams for hydropower, nuclear power, fracking, "clean" coal, or carbon storage. The emphasis in terms of mitigation is still on the market-based emission trading schemes which in the end allow emitters to continue their pollution and crowd out more effective initiatives to cut emissions. And notwithstanding the $100 billion pledge, there is insufficient recognition how incredibly vulnerable poorer regions and dispossessed populations will be in their exposure to more extreme-weather events and rapidly rising sea levels. We still can see a frightening lack of urgency in terms of the timetable adopted in Paris for more ambitious goals and measures, crystallizing perhaps a decade from now. When looking at the carbon budget we can safely burn before moving toward a 2° rise in average global temperatures, it becomes pretty obvious that we actually do not have much time left before starting to do irreversible damage to our environment of the kind our children and grandchildren will blame us forever.

Yet at the same time, it is also true that the Paris Agreement was a tremendous achievement. For the first time, the entire world, without exception, acknowledged that there was a huge problem brewing which needs to be addressed. This is like the first step in a twelve-step program toward rehabilitation, and a huge first step it was. The Paris Agreement laid out a sensible road map of how to proceed, especially in terms of assuring a pragmatic framework for collective action that has a chance to prompt everyone into intensifying effort over time. Even though the scientists tell us how urgent a problem climate change is already, we also have to account for the political reality of divergent national interests, powerful lobbies in favor of the status quo, and a bias toward short-term thinking. Amidst these tangible obstacles, it is actually very impressive that the world's nations have all committed themselves to a consensual framework for collective action that so far each nation-state has engaged in (with the possible exception of the USA under Trump). This is nothing short of a breakthrough, clearly a game changer. I actually believe that the Paris Agreement of 2015 is a turning point in human history and will in retrospect be seen for what it is—a paradigm shift capable of pushing us to transform our social, economic, and political system. We are nowhere ready for such, but we have at least agreed to start engaging in preparations for such a transformation!

Notes

1. For more on Trump's musings about climate change during his run for the presidency and their contrast to candidate Clinton's more measured and grounded approach, see League of Conservation Voters (2016).
2. Senator Inhofe's (R—OK) many questionable statements about climate change have been well documented by the blog *Skeptical Science* (2016).
3. The Kyoto Protocol, the first international emission-reduction accord and as such a pace-setter for future agreements, ran afoul in the US Congress not least due to its much more lenient treatment of developing countries, such as China or India, that have contributed much less to the problem for lack of industrialization. For more on the Kyoto Protocol, go to the United Nations Framework Convention on Climate Change (2014).
4. Like their leaders, conservative Republican voters largely reject (at a rate of 85% of the survey sample) the experts' consensus of climate change being caused by human activity as reported in a recent poll by the prestigious Pew Research Center whose results have been well summarized by D. Nuccitelli (2016). For a nuanced discussion of Americans' ambiguous attitude toward climate change, see also B. Jopson (2015). M. Wolf (2014) provides a useful summary of these reasons explaining the Republicans' denial of the climate threat.
5. See S. Arrhenius (1896). See also the fascinating account of how scientists came to grasp the implications of greenhouse gases for global warming by S. Weart (2008).
6. These reports are called First Assessment Report (1990), Second Assessment Report (1995), Third Assessment Report (2001), and so forth. Each assessment report can be accessed at the panel's Web site ipcc.ch. It should be noted that the IPCC was awarded the Nobel Peace Prize in 2007, shared with American environmentalist and politician Al Gore whose documentary on climate change entitled *An Inconvenient Truth* won the Oscar that same year.
7. Details about the various greenhouse gases discussed here are from the informative students' guide on climate change provided by the US Environmental Protection Agency (2016). The Trump Administration has systematically removed useful information pertaining to global warming from the EPA's Web site.
8. We used the "Summary for Policymakers" projections and recommendations in IPCC (2014) to inform this section's discussion of GHG emissions and their climate-change effects.
9. Another factor in the acceleration of CO_2 emissions since 2000 has been a reversal in the de-carbonization trend brought about by worldwide increases in the use of coal, as fast-growing developing countries like

India found this to be a cheap source with which to meet their surging energy-supply needs. The result has been a slight increase in the carbon intensity of energy across the globe, whereas in previous decades that trend had moved in the opposite direction.

10. Good examples of such probability-based proof that the world is trending toward an accelerating pace of extreme-weather events are National Climate Assessment (2014) or E. M. Fischer and R. Knutti (2015).

11. The NASA Earth Observatory (e.g., Scott and Hansen 2016) provides good summaries of its polar-ice observations. For more details on rising see levels, see the Union of Concerned Scientists (2013).

12. The threat of rising sea levels in Africa is well summarized in J. Hinkel et al. (2012). Relevant information for Latin America and Caribbean can be found in UN-HABITAT (2008). The four reports of World Ocean Review (worldoceanreview.com) contain a more general discussion of sea-level rise and its possibly devastating consequences for coastal populations.

13. The UN's Division for Sustainable Development has since 2010 mapped the evolution of coastal population trends using the indicator "Percentage of total population living in coastal areas." See sedac.ciesin.columbia.edu/es/papers/Coastal_Zone_Pop_Method.pdf for an elaboration of this new demographic indicator.

14. The accelerating pace of mountain glacier melting has been noted, for instance, by V. Radic and R. Hock (2011), E. Struzik (2014), or J. Qiu (2016).

15. For more details on how climate change affects agriculture and food security, see E. Van Ommen Kloeke (2014), including her extensive literature survey, as well as Food and Agriculture Organization (2008) and World Food Programme (2016).

16. See D. Huber and J. Gulledge (2011).

17. The official title of the Rio Summit was the United Nations Conference on Environment and Development, connecting in explicit fashion the two objectives underpinning sustainability.

18. The Precautionary Principle and criticisms it has provoked are discussed by D. Kriebel et al. (2001), COMEST (2005), as well as J. P. Van Der Sluijs and W. Turkenberg (2007).

19. For an assessment of the Kyoto Protocol's effects per country, see D. Clark (2012).

20. J. Cao (2010) provides a comprehensive assessment of the Copenhagen Agreement, whereas D. Bodansky (2010) sheds light on the underlying divisions causing the COP15 meeting to fail.

21. The provisions of the Paris Agreement can be found in United Nations Framework Convention on Climate Change (2015). Most commentators, such as M. Wolf (2015) or F. Harvey (2015), celebrated the Paris

Agreement as a historic breakthrough of great significance, but reminded their readers that this was only "a first step" to be followed by many more.

22. As reported by P. Clark (2015), the collective impact of the INDCs announced so far by 146 countries in connection with the Paris Agreement would basically stabilize annual CO_2 emissions slightly below the current level of 40 gt/year, to around 36–38 gt/year, by 2030. This compares with a rise to over 50 gt/year under a "business as usual" scenario whereas the Paris commitment of "well below 2 degrees" requires emission reduction cuts to about 25 gt/year.

23. The provisions of the Paris Agreement pertaining to finance, especially paragraphs 53–65, are nicely summarized in H. Chen (2015) or in J. Thwaites et al. (2015).

24. Typical criticisms of the deal can be found in M. Le Page (2015) or T. Odendahl (2016).

References

Arrhenius, S. (1896). On the Influence of Carbonic Acid in the Air Upon the Temperature on the Ground. *Philosophical Magazine and Journal of Science, 41*(5), 237–276.

Bodansky, D. (2010). *The International Climate Change Regime: The Road from Copenhagen*. Policy Brief. Harvard Project on International Climate Agreements, Belfer Center for Science and International Affairs, Harvard Kennedy School.

Cao, J. (2010). *Beyond Copenhagen: Reconciling International Fairness, Economic Development, and Climate Protection*. Discussion Paper 2010-44, Harvard Project on International Climate Agreements, Belfer Center for Science and International Affairs, Harvard Kennedy School.

Chen, H. (2015, December 12). Paris Climate Agreement Explained: Climate Finance. *National Resources Defence Council Blog*. https://www.nrdc.org/experts/han-chen/paris-climate-agreement-explained-climate-finance. Accessed 8 Oct 2016.

Clark, D. (2012, November 26). Has the Kyoto Protocol Made Any Difference to Carbon Emissions? *The Guardian*. www.theguardian.com/environment/blog/2012/nov/26/kyoto-protocol-carbon-emissions.

Clark, P. (2015, December 15). Climate Deal: Carbon Dated? *Financial Times*.

COMEST (World Commission on the Ethics of Scientific Knowledge and Technology). (2005). *The Precautionary Principle*. Paris: UNESCO. http://unesdoc.unesco.org/images/0013/001395/139578e.pdf.

DeConto, R., & Pollard, D. (2016, March 31). Contribution of Antarctica to Past and Future Sea-Level Rise. *Nature, 531*, 591–611.

Ellingboe, K., & Koronowski, R. (2016, March 8). Most Americans Disagree with Their Congressional Representative On Climate Change. *ThinkProgress*. https://thinkprogress.org/most-americans-disagree-with-their-congressional-representative-on-climate-change-95dc0eee7b8f#.7swbc9wd3.

Environmental Protection Agency. (2016). *A Student's Guide to Global Warming*. https://www3.epa.gov/climatechange/kids/index.html. Accessed 24 Sept 2016.

Fischer, E. M., & Knutti, R. (2015). Anthropogenic Contribution to Global Occurrence of Heavy-Precipitation or High-Temperature Extremes. *Nature Climate Change, 5*(6), 560–565. http://iacweb.ethz.ch/staff/fischer/download/etc/fischer_knutti_15.pdf.

Food and Agriculture Organization. (2008). *Climate Change and Food Security: A Framework Document*. Rome: United Nations. http://www.fao.org/forestry/15538-079b31d45081fe9c3dbc6ff34de4807e4.pdf.

Harvey, F. (2015, December 14). Paris Climate Change Agreement: The World's Greatest Diplomatic Success. *The Guardian*. https://www.theguardian.com/environment/2015/dec/13/paris-climate-deal-cop-diplomacy-developing-united-nations. Accessed 16 Dec 2015.

Hinkel, J., Brown, S., Exner, L., et al. (2012). Sea-Level Rise Impacts on Africa and the Effects of Mitigation and Adaptation: An Application of DIVA. *Regional Environmental Change, 12*(1), 207–224.

Huber, D., & Gulledge, J. (2011). *Extreme Weather and Climate Change: Understanding the Link and Managing the Risk*. Arlington, VA: Center of Climate and Energy Solution. http://www.c2es.org/publications/extreme-weather-and-climate-change.

IPCC. (2014). *Fifth Assessment Report: Climate Change 2014—Synthesis Report*. Cambridge, UK: Cambridge University Press. http://www.ipcc.ch/report/ar5/.

Jopson, B. (2015, December 30). US Views on Climate Change Pose Test for 2016 Candidates. *Financial Times*.

Kriebel, D., Tickner, J., Epstein, P., et al. (2001). The Precautionary Principle in Environmental Science. *Environmental Health Perspectives, 109*(9), 871–876. https://www.ncbi.nlm.nih.gov/pmc/articles/PMC1240435/.

League of Conservation Voters. (2016). *In Their Own Words: 2016 Presidential Candidates on Climate Change*. http://www.lcv.org/assets/docs/presidential-candidates-on.pdf. Accessed 15 Sept 2016.

Le Page, M. (2015, December 19). Paris Climate Deal Is Agreed—But Is It Really Good Enough? *New Scientist*, Issue 3052. https://www.newscientist.com/issue/3052%20/. Accessed 9 Oct 2016.

National Climate Assessment. (2014). *Climate Change Impacts in the United States*. Washington, DC: U.S. Global Change Research Program. http://nca2014.globalchange.gov.

Nuccitelli, D. (2016, October 6). Pew Survey: Republicans Are Rejecting Reality on Climate Change. *The Guardian.* https://www.theguardian.com/environment/climate-consensus-97-per-cent/2016/oct/06/pew-survey-republicans-are-rejecting-reality-on-climate-change. Accessed 10 Dec 2016.

Odendahl, T. (2016, January 22). The Failures of the Paris Climate Change Agreement and How Philantropy Can Fix Them. *Stanford Social Innovation Review.* https://ssir.org/articles/entry/the_failures_of_the_paris_climate_change_agreement_and_how_philanthropy_can. Accessed 10 Oct 2016.

Prudential Regulation Authority. (2015). *The Impact of Climate Change on the UK Insurance Sector.* London: Bank of England. http://www.bankofengland.co.uk/pra/Documents/supervision/activities/pradefra0915.pdf. Accessed 4 June 2017.

Qiu, J. (2016, June 2). Investigating Climate Change the Hard Way at Earth's Icy "Third Pole." *Scientific American.* www.scientificamerican.com.

Radic, V., & Hock, R. (2011). Regionally Differentiated Contribution of Mountain Glaciers and Ice Caps to Future Sea-Level Rise. *Nature Geoscience.* www.nature.com/naturegeoscience.

Scott, M., & Hansen, K. (2016, September 16). Sea Ice Overview. *NASA Earth Observatory.* http://earthobservatory.nasa.gov/Features/SeaIce/?src=features-hp&eocn=home&eoci=feature.

Skeptical Science. (2016). *Quotes by James Inhofe vs. What the Science Says.* https://www.skepticalscience.com/skepticquotes.php?s=30. Accessed 15 Sept 2016.

Struzik, E. (2014, July 10). Loss of Snowpack and Glaciers in Rockies Poses Water Threat. *Yale Environment 360.* http://e360.yale.edu/feature/loss_of_snowpack_and_glaciers_in_rockies_poses_water_threat/2785/. Accessed 29 Sept 2016.

Thwaites, J., Amerasinghe, N. M., & Ballesteros, A. (2015, December 18). What Does The Paris Agreement Do For Finance? *World Resources Institute Blog.* http://www.wri.org/blog/2015/12/what-does-paris-agreement-do-finance. Accessed 8 Oct 2016.

Union of Concerned Scientists. (2013). *Causes of Sea Level Rise: What the Science Tells Us.* http://www.ucsusa.org/global_warming/science_and_impacts/impacts/causes-of-sea-level-rise.html#.V-rp9TKPDBI. Accessed 27 Sept 2016.

United Nations Framework Convention on Climate Change. (2014). *Kyoto Protocol.* http://unfccc.int/kyoto_protocol/items/2830.php.

United Nations Framework Convention on Climate Change. (2015). *Adoption of the Paris Agreement.* http://unfccc.int/resource/docs/2015/cop21/eng/l09r01.pdf.

UN-HABITAT. (2008). *Latin American and Caribbean Cities at Risk Due to Sea-Level Rise.* http://www.preventionweb.net/english/professional/maps/v.php?id=5649.

Van Der Sluijs, J., & Turkenburg, W. (2007). Climate Change and the Precautionary Principle. In E. Fisher, J. Jones, & R. von Schomberg (Eds.), *Implementing the Precautionary Principle: Perspectives and Prospects* (pp. 245–269). Cheltenham, UK: Edgar Elgar.

Van Ommen Kloeke, E. (2014). How Will Climate Change Affect Food Security? *Elsevier Connect.*

Weart, S. (2008). *The Discovery of Global Warming.* Cambridge, MA: Harvard University Press.

Wolf, M. (2014, November 11). An Unethical Bet in the Climate Casino. *Financial Times.*

Wolf, M. (2015, December 15). The Paris Climate Change Summit Is But One Small Step for Humankind. *Financial Times.*

World Food Programme. (2016). *Climate Impacts on Food Security.* Rome: United Nations. https://www.wfp.org/climate-change/climate-impacts.

CHAPTER 2

Moving Toward an Ecologically Oriented Capitalism ("Eco-Capitalism")

The Paris Agreement sets in motion an accelerating worldwide effort to reduce GHG emissions, one that must be aimed at intensifying and speeding up over time. We need to lower those emissions substantially over the next couple of decades if we are to achieve its goal of keeping global warming below 2 °C (or 3.6 °F). Eventually, we would hopefully arrive at more or less zero net emissions by mid-century, which implies matching an unavoidably minimal level of necessary man-made emissions with carbon-absorbing "sinks" like forests. These ambitious objectives, designed to keep our planet livable, will require fairly rapid and massive changes across large parts of our economy, notably energy mix, transportation, industrial processing, construction, urban planning, and agriculture. Are we going to be able to do this? Our answer to this question touches on a deeper quandary. Can our socioeconomic system save us from the prospect of environmental disaster whose conditions it has created in the first place? Certainly, not unless its modus operandi and structure both change substantially, and throughout the rest of this book, we shall elaborate some of these needed changes. Can we envisage an ecologically oriented capitalism for whom the sustainability of our ecosystem remains of paramount importance, or is this a contradiction in terms? Only time will tell! In the meantime, we can already begin outlining the challenges such an *eco-capitalism* (i.e., ecologically oriented capitalism) will face imminently as the world community begins to mobilize its fight against climate change in compliance with the goals and road map set forth in Paris at the end of 2015.

© The Author(s) 2018
R. Guttmann, *Eco-Capitalism*,
https://doi.org/10.1007/978-3-319-92357-4_2

A DIFFICULT PATH TO CHOOSE

If we want to boil down the multifaceted implications of the Paris Agreement to a single phrase, it would be "carbon neutrality by some-time shortly after 2050." If indeed achieved, we would at that point have a world economy capable of zero net emissions. Of course, there would (and could) be still some GHG emissions as is inevitable at that minimal level. But those emissions would be low enough in the aggregate to be matched by available carbon sinks, such as forests or improved land use. While those natural sinks may be very helpful in absorbing some of the CO_2 emitted, they cannot in all likelihood take care of more than a fairly limited amount of greenhouse gases emitted into the air, perhaps up to a quarter of current levels. There may be some further GHG absorption capacity added through effective carbon capture and storage technolo-gies, but their prospects are still highly uncertain as of now.

As a matter of fact, radical uncertainty rules here in more general fash-ion! We cannot predict very well how much the level of greenhouse gases in the atmosphere will grow if we do not do anything about its steady rise. And we have no way of knowing in advance what the consequences of such a "business as usual" (BAU) baseline will be for our global tem-perature level or its distribution among regions. Nor can we foresee with any reasonable degree of certainty how that temperature rise will exactly impact on our climate and, by extension, harm our ecosystems. Least of all do we know for sure how much we will be able to reduce GHG emis-sions at different possible effort levels over what period of time.

Yet all that uncertainty must not stop us from acting! This is with-out a doubt a case in which the so-called *Precautionary Principle* applies a crucial point yet to be driven home to all the climate deniers. As the principle states, the issue at hand is too important to forego action just because of the uncertainty involved. We have a moral obligation to leave behind a livable planet for future generations, just as much as we have benefitted from the work done by previous generations. And if there is any chance that our current growth pattern undermines the well-being of our children and grandchildren, we cannot take the chance of inac-tion. There is no excuse for doing nothing, just because the science is too uncertain and/or preventive action too costly. As a matter of fact, the risk posed by inaction is so immense, the possible consequences of a "business as usual" approach so devastating, that we cannot even accept the weak version of the Precautionary Principle which justifies the possibility of action despite the science not being settled yet. Instead, we should hold to the strong version of the Precautionary Principle which requires us to act despite the uncertainty involved. We have to act, and decisively so! Our ethos obliges us to prevent disaster to the best of our abilities, if there is even just a minor possibility that we could harm the well-being of future generations because of our unwillingness to address the danger when there was still time to do so effectively.[1]

This is why the Paris Agreement's specific objective of carbon neutrality by, say, around 2060, consensually agreed to by all nation-states on this planet, is of such significance. It gives us a concrete goal to strive for. Of course, we can take issue with the goal itself, especially considering that it is quite ambitious and hence difficult to achieve. One could argue that the carbon-neutrality goal is based on computer simulations whose results about the pace of global warming depend crucially on negative feedback effects not yet ascertained sufficiently for reliable forward projection over long periods of time. But it is also equally likely that those feedback effects are currently underestimated in our models rather than overestimated. Given how long greenhouse gases stay in our atmosphere and the likely destabilizing effects of continuous ecosystem erosion, we again need to apply the strong version of the Precautionary Principle—better be safe than sorry! The goal of becoming carbon neutral at the onset of the second half of this century takes into account the cumulative and self-reinforcing nature of climate change. It is a goal designed to keep our global average temperature from rising beyond 2 °C from its pre-industrial level (of 1850), based on current trends of GHG emission quantities, accumulated GHG levels present in our atmosphere, and already observed increases in the average global temperature.[2]

The goal of eventual carbon neutrality within the next half century or so to limit temperature hikes to "well below" 2 °C implies capping the concentration of greenhouse gases in the atmosphere to 450 parts per million of CO_2 (up from the current level of 400 ppm). And this in turn imposes, according to estimates by the International Energy Agency, a very tight *carbon budget* limiting the total remaining cumulative energy-related CO_2 emissions to 1000 gigatons of CO_2 equivalents between 2015 and 2100.[3] To get there we would have to cut those emissions by at least 60% between 2015 and 2050 and continue lowering them at an accelerating pace until reaching carbon neutrality as soon as possible during the second half of the twenty-first century. According to the same IEA estimates, if we assume current trends to continue, we can expect the atmospheric GHG stock to rise to a possible 700 ppm as we burn up an energy-related carbon budget of 1700 $GtCO_2$ by 2050. In this case, we might see our temperature go up by anywhere between 4° and 5.5° as we approach 2100—a recipe for likely disaster!

Which of these two scenarios we are more likely to approach depends ultimately on how rapidly we can cut GHG emissions by either changing our energy mix radically and/or improving our energy efficiency greatly. Not making much progress on either energy front within a reasonably short period of time threatens to put us on a ruinous path of no return. As noted in an important report by Carbon Tracker (2011), total known reserves of coal, oil, and gas across the globe are nearly five times the size of the carbon budget associated with a temperature hike of 2 degrees. Even if only 20% of those reserves are burnt over the next four decades, this by itself blows up our entire carbon budget for the two-degree goal.[4] What we therefore urgently need to consider is a fairly swift and radical energy revolution, a very difficult political decision to make.

The Age of Fossil Fuels

The biggest focus in our upcoming battle with climate change will have to be on transforming our energy mix. It is imperative that we move away from fossil fuels, such as coal, oil, or gas, and into renewables, such as solar, wind, hydro, biomass, and other biofuels. This is a huge undertaking! But it is also a task that is quintessential if we are to keep our planet livable for future generations. One-quarter of our current GHG emissions comes from burning fossil fuels as we turn those into energy. If we want to make a serious dent in current levels of GHG emissions, we have to start with our reliance on fossil fuels as our favorite source of energy.

Just as we have characterized earlier phases in human evolution as the "Stone Age" or the "Bronze Age" or the "Iron Age," we can say that we are currently living in the "Fossil-Fuel Age." Ever since the first commercial discoveries of oil in the USA (Pennsylvania's Drake Well in 1859), we have built our economy around petroleum and its many byproducts—gasoline, diesel, jet fuel, heating oil, kerosene, and so forth. It is also a key raw material in the production of many chemicals, including asphalt, plastic, solvents, fertilizers, pharmaceuticals, and a range of other petrochemicals. Petroleum has high energy density, is easily transported, allows itself to be refined and further processed at low additional cost, and can be found all over the planet in sufficiently abundant supplies. These characteristics make it a literally irresistible source of cheap energy. Today, the world is consuming about 95 million barrels of petroleum a day. When it comes to power plants for electricity generation, only about 5% of the world's total comprises oil, compared to 39% using coal and 22% using natural gas. These two other fossil fuels are a better fit than oil to generate the heat needed for the steam turbines driving the generators, because they are more combustible. Coal in particular has the added advantage of burning easily and being locally supplied in ample quantities all over the world. Natural gas, a fossil fuel often found next to petroleum and composed partially of the greenhouse gases methane and carbon dioxide, is also used for cooking, heating, and in the production of various chemicals.

Nuclear power is another major energy source in electricity generation, with an overall global market share of about 11%. In some advanced economies that share is quite a bit higher, notably in Japan (30%) and France (40%). The latter's heavy reliance on nuclear power coincides with comparatively low electricity prices and less pollution, thus allowing France to consider its experiment with nuclear power entirely worthwhile. Notwithstanding this rare example of a success story, nuclear power has failed its original promise of providing cheap, reliable, and long-lived generation of electricity which had made it such a hot prospect during its birth years in the late 1950s and early 1960s. It has turned out that nuclear power plants are difficult to manage and have a tendency to deteriorate more rapidly than initially foreseen. Moreover, they generate radioactive waste much of which has to be stored over long periods of time or reprocessed. To the extent that such waste includes also uranium or plutonium, its presence has raised legitimate nuclear-weapons proliferation concerns among a fearful public. Those fears have been accentuated by a history of spectacular and frightening accidents at nuclear power stations which stoked public resistance to that technology—Three Mile Island 1979, Chernobyl 1986, and the Fukushima I meltdown following an earthquake and tsunami in 2011. Ever since that last incidence, governments all over the world have had to scale back their nuclear power ambitions, even take some of the more vulnerable plants out of commission. This retreat has further slowed the already difficult transition from nuclear fission to nuclear fusion, with delays only adding to the costs of the already comparatively expensive fusion technology. While the exigencies of the fight against climate change may once again revive the fortunes of nuclear power as a viable alternative to fossil fuels, we are definitely not there yet.

This leaves us with about 23% of total worldwide electricity generation using renewable energy sources derived from natural processes capable of constant replenishment which may involve sunlight, wind, water, plant growth, or geothermal resources. Of all these renewable energy sources, in which our future lies, hydroelectric power comprises currently three-quarters of the total, for a global market share of 17%.[5] That energy source depends on large rivers that have significant drops in elevation for the construction of dams within which water flows from the reservoirs to move the turbines turning the generator producing electricity. While this technology is clean and responds very well to fluctuations in electricity demand thanks to pumped storage of recycled water, it is geographically restricted. And the construction of large dams may provoke the resistance of local populations opposed to displacement, flooding of hitherto useable areas of land, and having an eyesore imposed on them.

Be that as it may, hydropower has been up to now by far the most widely used renewable energy source, leaving solar and wind power far behind. Those two energy sources of the future have only very recently had the chance to achieve the scale needed to become economically viable on their own, especially in comparison with cheap coal. They have yet to obtain more reliable storage technologies to provide regular electricity generation even when there is not sufficient amount of sunlight or wind available at the moment. These constraints notwithstanding, both solar power and wind power offer major advantages. Their geographic omnipresence and continuous renewability make either easily deployable. They are also more energy-efficient, not least because their use in the production of electricity wastes less heat.[6] Besides providing energy without emitting greenhouse gases, solar and wind power are also "clean" in the sense that they do not pollute the air like coal-fired or oil-fired power plants would. Renewable energy sources, and here I would also include biomass and biofuels, can be used on a small scale in decentralized fashion which makes them very useful in connection with sustainable development, such as promoting the electrification of rural areas. And they have the flexibility of widespread application, used not just in power generation, but also heating/cooling or transportation. Given all these advantages, it is not surprising to see the public increasingly willing to endorse these new energy sources as the way forward toward an environmentally friendly future.

Yet we are nowhere near the needed energy revolution whereupon we replace fossil fuels with renewables to meet the Paris goals. On the contrary, the coal, oil, and natural gas industries continue to be well entrenched and remain politically powerful in shaping the climate-change policies of key nation-states, such as the USA or China, let alone countries in which these industries predominate, as in Russia. This is one reason why it will be difficult for the NDCs of various countries to be made sufficiently more ambitious and tougher in coming years, in compliance with the spirit and letter of the Paris Agreement. The much-discussed notion of peak oil is not yet in sight, with oil companies still adding new reserves under their control at a rapid pace. As we have already mentioned earlier, we cannot afford to burn but a small fraction of these reserves if we want to keep within the carbon budget needed for modest temperature rises. Hence, a large amount of already declared reserves may never be burned if we wish to realize the climate goals set in Paris, turning them into "stranded assets" which will render the oil companies much less attractive than their current market valuations imply. The London-based non-governmental organization Carbon Tracker, known for its aforementioned reports on unburnable carbon, refers to this prospect as a "carbon bubble." Because they may suffer potentially large losses in the wake of stranded assets they can no longer use for income creation, oil companies have come recently under a lot of pressure from activist shareholder groups and government regulators to take account of climate-change regulations on their balance sheets and act accordingly. Here, we see very mixed reactions among the oil giants, from recalcitrant Exxon becoming subject to probes by the US Securities and Exchange Commission and New York State Attorney General for not going far enough in revaluing its assets to British Petroleum or Royal Dutch Shell announcing that from now on they would factor a putative carbon price (of $40 per ton) into their investment decisions. Chevron has written down the value of its reserves by $5 billion over the last couple of years, and Shell took a $8.3 billion write-down of its assets in October 2015.[7]

While the oil companies will be eventually obliged to tackle the issue of climate-change regulation as a strategic challenge of the highest order, they are not yet ready to give sufficient thought to the profound long-term implications of the Paris Agreement. Take, for instance, British Petroleum's *Energy Outlook*, as summarized by Martin Wolf (2015b). The oil giant, ever since its huge 2010 oil spill in the Gulf of

Mexico forced to take environmental considerations more seriously than competitors with lower reputational risks, predicts a 75% increase in global average real output per capita between 2015 and 2035 primarily in the wake of China's and India's continuous catching up with the rich industrial nations. But thanks to major strides in energy-efficiency BP projects the consumption of energy to rise by only 37% during that same 20-year period, while CO_2 emissions should grow by only 25%. Even though renewables other than hydropower will grow rapidly, their total share in primary energy production will only rise from 2.6 to 6.7%.

According to these somewhat self-serving forecasts by a leading oil giant, the world will remain principally bound to fossil fuels for a long time to come even though its geographic distribution will change quite dramatically between now and then. The USA, still importing 12 million barrels per day a decade ago, has already benefitted from its oil shale boom to become independent from the rest of the world and will over the next two decades turn into a major oil exporter. This may very well keep Americans more wedded to fossil fuels for longer than would be desirable for our collective action to counter climate change with cleaner energy. China, currently self-sufficient in energy with its large coal- and hydropower infrastructure, and India, tempted to exploit its huge coal reserves for cheap electrification of its seriously electricity-short economy, will become large importers of oil. Such huge shifts in trade patterns are bound to have significant geopolitical consequences. As China and India seek to reduce their exposure to oil imports, they will have to decide to what extent they wish to commit more aggressively to renewables instead of continuing their traditionally heavy reliance on cheap local coal which has the highest pollution and CO_2 emissions of all energy sources—a truly fateful decision for the entire planet!

The stubborn persistence of fossil fuels is perhaps best crystallized by the huge subsidies available to both their production as well as consumption (through governments keeping domestic oil prices artificially cheap). Those may take the form of direct cash transfers, but are more often afforded indirectly by means of tax breaks, protectionist trade barriers, price controls, or limitations on market access. While such energy subsidies have recently declined somewhat from $610 billion in 2009 to $493 billion in 2014 thanks to much lower oil prices and subsidy reforms, these figures do not take account of all the pollution caused by fossil fuels which raises their effective subsidy level by a factor of ten according to IMF estimates.[8] So governments still spend very large sums in support of industries that we would be better off shrinking back to a minimum as soon as possible. The net effect of these gigantic fossil-fuel subsidies is to reduce the carbon price worldwide to minus fifteen dollars per ton instead of the forty or fifty dollars per ton where it should be! There is now a general agreement among the G-20 countries that such subsidies should be phased out by 2030. Yet it is not clear that there will be sufficient political will to end such subsidies, especially among oil- and gas-producing nations themselves where domestic oil prices are often kept ridiculously low and producers enjoy lavish protection by captured governments. Still, these subsidies back the wrong sectors and technologies!

The large subsidies for fossil fuels make it more difficult to move our energy mix toward a "greener" composition within a reasonable time horizon. Such changes are by definition very slow, like turning an ocean liner. This has much to do with the long-lived nature of our energy infrastructure, notably the power plants themselves. They easily last forty years! So the power plants we build today will shape our energy mix for decades to come. In that context, it is very troubling to have witnessed the renaissance of coal over the last two decades, by far our dirtiest and most carbon-intense energy source. While the USA has seen a long-term decline of coal's share not least in response to regulatory restrictions such as Obama's Clean Power Plan, fast-growing emerging-market economies especially in Southern and Eastern Asia have found coal irresistibly cheap to meet their surging energy needs. Even though China has lately scaled back its plans for coal-fired power plants in response to rapidly worsening air pollution triggering popular unrest and also in compliance with its burgeoning climate-change ambitions, no slowdown is planned in such coal-reliant fast-growing economies as India, Indonesia, or Vietnam while Australia continues to push its coal exports to the region. This trend alone runs entirely counter to our plans for climate-change mitigation. But the coal problem is just one aspect of a bigger challenge, the fact that the US, the EU, and other industrial nations have many rapidly aging power plants which will need to be replaced soon. If we still use mostly coal-, oil-, or gas-fired plants for the next generation, we will have foregone any chance to keep within the carbon budget for a two-degree rise. This means that we shall face a moment of truth about our future energy path much sooner, within the next five years, than implied by the comparatively relaxed timetable of the Paris Agreement.[9]

Of course, we may able to reduce future emissions of relatively new fossil fuel power plants by endowing them subsequently with carbon capture and storage (CCS) capacity that could be installed after their construction to capture and sequester waste CO_2. This is what is behind the otherwise mystifying notion of "clean" coal. There are a variety of CCS technologies in the works, but as of yet unproven. Irrespective of which approaches will ultimately prove the most attractive, we will have to figure out how to transport large quantities of the captured CO_2 in pipelines or ships and then store that waste gas for longtime periods most likely in deep geological formations—all of this a rather complex undertaking which is likely to cost quite a bit and so add to the electricity generation costs. Even more radical approaches relying on removing CO_2 from the atmosphere by means of such techniques as carbon scrubbing, amounting in effect to climate engineering, are today even more still in their infancy and hence even less foreseeable at this point. All we can say about these technologies is that they hold promise to become widely accessible by mid-century, but are as of now still too far away to make a difference. Better then to maximize the substitution of coal-fired power plants in the works with cleaner gas-fired power plants as a transition strategy and accelerate to the greatest extent possible the adoption of zero-emission renewables for electricity generation from now on. If we fail with either strategy and continue to rely so heavily on coal and the other fossil fuels, then we run a distinct risk in coming decades of having to shut down fossil-fuel power plants before those will have completed their normal life cycle. Such premature scrapping of expensive power-generation capacity represents yet another large category of stranded assets, this one hitting public utilities.[10]

A Push into Renewables

In light of these strategic considerations, it is pretty obvious that the world community should commit to a major push into solar power and wind power, and do so now! Not only are these renewables clean and zero-emission energy sources, but they are also far more efficient than fossil-fuel power plants whose steam cycles waste a lot of heat. Renewables thus have the potential of requiring themselves much less primary energy inputs for a given level of electricity output, offering more competitive transformation relations between electricity, heat, and mechanical energy. These energy-efficiency advantages make renewables also very useful as sources of heating/cooling or transportation, assuring widespread application once installed. They can also be applied in decentralized fashion on a smaller scale, as is the case with solar photovoltaic cells, micro-hydro installations using the natural flow of water and connected to mini-grids, biogas for lighting, or biomass cookstoves. These deployment advantages make renewables especially usable in rural areas and helpful in a variety of development strategies, such as accelerating electrification efforts or providing off-grid power supplies, which help poor countries overcome their energy poverty. Another crucial advantage of renewables is that their production promises to generate a lot of new employment opportunities, as we can already see today with the over eight million jobs tied to renewables worldwide.[11]

The cost of renewables has come down sharply in recent years as their production levels have grown rapidly, a classic example of scale economies taking root and the learning curve bearing fruit. At first entirely dependent on government subsidies to get started, solar power has doubled its size seven times since 2000 while wind power has experienced four doublings over the same fifteen-year period. Whereas solar photovoltaic electricity cost $1200/MWh at its inception in 1990 and $323/MWh in 2009, this cost has fallen to about $180/MWh for solar rooftops and a remarkably low minimum of $72/MWh for large PV solar plants in 2014. Wind power has also become considerably less costly over the last decade with rapidly growing capacity, to about $85/MWh on average (and a minimum of $37/MWh) which is roughly equivalent to the $60–$85/MWh unit cost of either coal or natural gas in electricity generation today. Offshore wind farms, which are becoming more popular thanks to their smaller aesthetic impact and greater wind strength, are still far more expensive to construct and hence cost twice as much, about $175/MWh. But there is a lot of innovation in wind turbine technology and power-generation efficiency to push these costs down significantly in the near future.

By 2020, both solar and wind power will have become more cost-attractive than fossil fuels and therefore should out-compete those traditional energy sources quite easily (even without subsidies, of which they are currently receiving globally $101 billion, a fifth of what fossil fuels get). Renewables are not a fuel but a technology, and as such can be constantly improved for further cost reductions. While capital intense in terms of getting started, wind and solar have negligible marginal costs once the turbines or PV cells are set up. Not dependent on inherently volatile fuel prices, renewables have a more stable cost structure as well. There are still some issues with their intermittent nature disrupting the steady flow of energy supplied, and this challenge requires some improvements in storage capacity as well as backup power reserves for smoother grid integration. Luckily, the intermittencies of solar and wind tend to be complementary, while both are also well combined with hydropower which responds in very elastic fashion to variability in demand for its energy. This implies taking an integrated approach to simultaneous and interactive use of renewables in close proximity, which further adds to their attractiveness relative to fossil fuels. For all these reasons, the high-level Commission on the Economy and Climate (2014) has come to the optimistic conclusion that renewables can and will replace fossil fuels at very little net cost. This sentiment is also shared by other specialists, not least McKinsey (2013) based on its fascinating calculation of a global greenhouse gas abatement cost curve. On top of these possibly modest transition costs, we should also keep in mind that renewable energy sources create a lot of good-paying jobs, further reinforcing their advantages over fossil fuels.[12]

The impending takeoff of renewables has already had an impact worth noting. In thirty countries, renewables contribute today in excess of 20% of total electricity generation, and over 120 countries have set specific target shares for renewables as part of their climate-change mitigation strategies laid out in their NDCs. Some countries, including Denmark, Sweden, New Zealand, Austria, and Brazil, have renewable shares in excess of 50%, all the way to Norway's 98%-share and Iceland's 100%-share. In 2013, we saw for the first time more new electricity generation capacity added worldwide using renewables than fossil fuels whose price collapse over the last few years has seriously squeezed coal, oil, and gas producers into major cutbacks of supplies. And in 2014, the IEA reported for the first time no net increase in CO_2 emissions despite a world economy growing by an average of 3%, thanks not least to important changes in China's energy mix and tougher energy-efficiency standards for power plants, cars, and home appliances kicking in across the globe. Since then trends have only accelerated. 2016 was a banner year for renewables, marking a speedup which may well be characterized as equivalent to a takeoff. Solar power's global capacity grew by over 30% that year, and for the second year in a row over half of the new power-generation capacity added worldwide came from renewables. The trend acceleration toward renewables was deepening beyond energy supplies. Sales of plug-in electric cars grew by 42% worldwide in 2016, eight times faster than the overall car market. That same year, we also witnessed a doubling of the storage capacity of big lithium-ion battery systems.[13] These trends are promising, but still have a long way to go. It is manifestly not enough to stabilize CO_2 emissions at current levels (of, say, 35 gt/year). Because of the long-lived presence of carbon dioxide in the atmosphere during which it contributes persistently to climate change, what ultimately counts much more in how far global average temperatures will rise is the already accumulated stock of CO_2. It is that stock which we have to stabilize, and this goal requires us to slash new emissions toward zero.

Whereas trends in the composition of our global energy mix are promising, their continuation and speed will depend on public policy. Even though the G-20 members have announced every year their intention to phase out "inefficient" fossil-fuel subsidies, so far there has been deplorably little movement in that direction. True, some governments have cut back their subsidies for fossil-fuel consumption by lifting artificially low ceilings on domestic petroleum prices (e.g., Egypt and Indonesia) or lowered their support for coal-fired plants (e.g., China and USA). But others have actually raised subsidies in the face of collapsing petroleum prices. For instance, the UK has recently extended new tax breaks in support of dwindling gas and oil production and exploration in the North Sea while at the same time lowering its fiscal support for renewables as well as carbon capture and storage technology—a policy shift causing massive consternation among its partners and even in the private sector. On the contrary, the right thing to do would be to use this period of low energy prices as a window of opportunity to eliminate fossil-fuel consumption subsidies and even tax that activity as the attendant price hikes would at this point be quite tolerable. Revenues collected from such gasoline and other energy consumption tax hikes, which may usefully be combined with steeper cuts in government support for fossil-fuel producers and utilities, could be used to reduce more damaging taxes (such as payroll taxes) or even expand government subsidies for renewables.

Subsidies are controversial, inasmuch as they distort market mechanisms and price signals, presume that the government is better at picking winners and losers than the private sector, and turn industries into lobbies intent on locking in such public support forever. But these arguments do not apply as clearly to renewables. It should be kept in mind here that throughout the history of capitalism energy sources have always been subsidized by governments during their takeoff phase (and, more troubling, all of these continue to draw subsidies even today, in the declining or twilight phases of their life cycle—coal, oil, natural gas, and nuclear). Such subsidies are aimed at getting a new industry off the ground, with start-ups needing support to get setup and new technologies deserving a chance to prove themselves before their eventual commercial success attracts large, established firms. Normally, it is at this point that we can usefully contemplate phasing out subsidies before they get captured more or less permanently by large, established firms, as we may consider for either wind or solar power now ready for takeoff. But other than well-targeted subsidy cuts for already profitable and competitive solar- or wind-powered electricity generation capacity, we should still contemplate increasing government support for renewables. Such support can be justified both in terms of start-up aid for many still less developed alternative energy applications deserving a chance to scale up to commercial viability and also in terms of the social benefits ("positive externalities") we all may derive from their successful launch, as was the case with wind and solar power.

Most useful in that direction would be raising support for research and development. At this point, government-funded R & D for renewable energy amounts globally to only $5.1 billion in 2014 (equal to 2% of total public support for R & D overall), of which China spends $1.7 billion, Europe $1.4 billion, and the USA a miniscule $0.8 billion. These are ridiculously low figures in need of major boosting. In June 2015, the American billionaire entrepreneur and philanthropist Bill Gates called for a massive public–private research program in support of breakthrough clean energy innovation and set aside $2 billion of his own funds over the next five years for that effort, in particular to explore high-altitude wind (i.e., jet stream) as energy source. That same month an influential group of British scientists, economists, and business leaders called for a "Global Apollo Programme To Combat Climate Change" (see King et al. 2015) which would triple publicly funded R & D in clean energy to $15 billion per year for a decade, still a puny sum if we consider other subsidies vested in energy, the traditionally low R & D spending by the energy firms, and the potentially huge payoffs of energy innovation in the fight against climate change.

The Apollo group wants to spend the $150 billion through a global coordination mechanism for targeted research efforts in all areas of renewable energy currently not yet developed enough. The focus is in particular on solar power as the most promising clean energy, aiming to drive down the cost of solar PV cells still further and making mirror-based concentrated solar power fully viable so that this type of renewable energy can be deployed in the world's deserts for maximum direct sunshine exposure and then transmitted in highly efficient grids to large population concentrations living not too far from those deserts (e.g., India's Rajasthan Desert, China's Gobi Desert). Solar and wind have the problem of intermittency thanks to irregular supply of the natural energy source and thus depend very much on viable storage capacity. Much R & D work should be directed toward various storage options, in particular the development of much better batteries and obtaining hydrogen by electrolysis to use it subsequently in a fuel cell to produce electricity. Hydrogen can play a crucial role in powering electric vehicles using fuel cells, as fuel for internal combustion engines, and in conjunction with other chemicals for production of synthetic gases. Finally, Apollo's research program would focus on smart grids to improve high-grade software and interconnectors for better transmission performance.

Clean energy innovation does not have to stop there. Potential extensions into other types of renewables loom on the horizon. Low-temperature geothermal energy may become very useful one day for heating and cooling. Brazil has demonstrated the great potential of bioenergy in powering its cars with ethanol made out of sugar cane. We can broaden the range of plants and tree species used for biomass and then find better ways to turn that biomass into biofuels, such as ethanol or biodiesel, while hopefully reducing the harm done to the environment during the biomass combustion and transformation processes. Ammonia (NH_3), easily produced and transported while also admirably clean, may well one day become the world's go-to fuel replacing oil. It could also provide an elegant solution to the energy storage problem that has constrained the spread of renewables, assuming that we can stop stripping the needed hydrogen from methane or coal and instead find a "green" way to separate it from oxygen in the water and combine it with nitrogen in the air. At that point, we would be able to absorb excess power from rivers or wind through ammonia-producing plants for storage or transportation to other locales. Another development track worth pursuing is turning waste, which would otherwise end up in landfills, into energy (WtE). So far WtE has relied on incinerators which are not that environmentally friendly. But now better technologies are emerging to make energy from waste without combustion, in conjunction with progress made with regard to biofuels. This could revolutionize how we cope with trash in the future.

DE-CARBONIZING THE ECONOMY

While changing the energy mix from fossil fuels to renewables is arguably our greatest challenge in terms of climate-change mitigation, it only addresses a part of the problem. If we look for instance at the World Bank's sectoral breakdown of GHG emissions reported in 2010 (as measured in gigatons per CO_2 equivalents), we can see electricity and heat, comprising the energy sector, being responsible for about 37% of global GHG emissions. Nearly 20% comes from manufacturing, construction, and industrial processes, 15% from transportation, 14% from agriculture, 9% from other fuel combustion, 7% from residential, and 6% from other sectors whereas forestry and land-use improvements acted as carbon sinks absorbing greenhouse gases to the equivalent of a minus 8% share. While these numbers can fluctuate somewhat from year to year, they are similar enough to sectorial distributions reported elsewhere and constitute therefore roughly accurate shares. If we want to cut overall GHG emissions sharply in coming decades, then we will have to expand our de-carbonization efforts beyond energy to all these other sectors of our economy.[14]

Take, for instance, transportation which emits GHG with the burning of fuel in cars, trucks, buses, and planes. Here, we can improve matters by making the engines of these vehicles more fuel efficient, as we have begun to do recently thanks to more ambitious regulatory standards imposed by governments after decades of neglect. We can also use more biofuels instead of gasoline, even though producing this alternative fuel on a massive scale causes its own environmental stressors in terms of forest loss and energy-intense land use. Ultimately the best solution is switching to electric vehicles, as we are beginning to see. The key here is to make batteries more easily charged and more efficient so that cars can accelerate faster and drive longer without the need for recharge. Tesla, a pioneer in this area, is now developing affordable electric cars whose performance rivals that of gas-fueled cars and which can drive over 300 miles on a single charge thanks to its superior battery technology. Major other car producers are also accelerating their electric-car development efforts. In that sense, it is fair to say that the de-carbonization of energy sources and that of vehicles go hand in hand. Progress in one begets progress in the other, with solar-powered vehicles on the horizon. When it comes to airplanes, there has recently been a global agreement (in October 2016), under the auspices of the International Civil

Aviation Organization, to cap GHG emissions on international flights by encouraging airlines to acquire more fuel-efficient airplanes, replace jet fuel with cleaner-burning biofuels, and offset emissions on individual flights by acquiring credits from certified emission-reduction projects (e.g., reforestation, renewables).[15]

Of course, we can personally do our share to help cut carbon emissions. Apart from buying electric or hybrid cars, we can simply drive less or use public transport more often. We can also fly less, with just one long-haul international flight less per year likely to cut emissions more than all the fairly easy day-to-day adjustments (turning off the lights, eat less meat, etc.) combined. Such behavioral changes, however, are not easy to get motivated for, even harder to sustain. They can certainly be helped by good public policy providing the right kind of incentives. For instance, city governments may wish to restrict inner-city traffic by means of regulations or fees, as we have witnessed with London's successful congestion charging scheme. Or they may encourage carpooling, a useful practice made in addition organizationally easier in recent years because of smart-phone apps that help organize joint rides.

This discussion gets us to the much broader topic of what cities can do with regard to improving urban sustainability. A majority of key cities across the planet have already in one way or another become exposed to the dangers of climate change, an exposure made strategically ever more important by rapid ongoing urbanization especially in emerging-market countries such as China, Brazil, or India. City governments, closer to their citizens and relatively nimble in comparison with their regional or national counterparts, can respond more quickly and decisively to concrete urban problems arising in the wake of climate change. And they are the most appropriate level of government to innovate with new policies and approaches. They can also learn well from each other in terms of what works and what does not, as exemplified by the fruitful knowledge-sharing collaboration among the world's leading megacities grouped together in the C40 Cities Climate Leadership Group (c40.org). Urban climate-change mitigation and adaptation projects may involve decentralized distribution of solar power and other alternative energy supplies (e.g., co-generation), enhanced park lands, incentives for urban farming, more environmentally friendly public transport, better protections against flooding and surges, building prefabricated housing facilities for inhabitants of shanty towns, low-carbon building designs, retrofitting buildings for better insulation and heating/cooling, recycling facilities and zero-waste strategies, and many more.[16] We have to ask ourselves what makes Barcelona and Atlanta, two cities of similar population size and comparable living standards, so different from each other in terms of per capita GHG emissions, with the former having one tenth of the latter's carbon footprint per person. This huge gap is a matter of very large differences between the two in terms of access to low-carbon public transport, urban sprawl, availability of parks and trees, nature of buildings, waste processing, energy mix, and emphasis on "green" public policy. Another key lesson to take away from this comparison is the crucial importance of urban planning, as crystallized around how those two cities dealt, respectively, with hosting the Olympic Summer Games. Barcelona managed to remake its city, Atlanta did not!

Agriculture is a strategic battleground in our fight against climate change. This is an intrinsically difficult area of policy intervention and also for innovation, because most actors in this sector are isolated in rural areas, relatively poor (especially as concerns the majority of farmers concentrated in developing countries), and hard to reach. Then, of course, there is also large-scale agribusiness, and that segment is more easily transformable in its structure and practices—especially with regard to raising livestock whence we get a lot of the methane discharged into the air. There are several priorities to pursue in this area. Most crucial is the preservation and restoration of forests, because these act as crucial carbon sinks which absorb a portion of our greenhouse gases and so neutralize their global warming effect. In this regard, the goal is to stop the deforestation, whether motivated by slash and burn, intensive agriculture, cattle ranching, or timber extraction, as Brazil has been able to but Indonesia has not. Brazil has even started significant reforestation projects, and these should be undertaken as well in other tropical zones with wet climates. It also makes sense to plant new trees in abandoned farmland. Existing forests need to be managed better, together with cropland as well as grassland. Another crucial carbon sink is soil carbon sequestration, and this can be promoted through land restoration programs using a variety of abatement strategies (e.g., crop rotation, cover cropping, and no-till farming). Such agroecological practices, as green manuring, composting, mulching, and conversion of biomass into "biochar" agricultural charcoal, are all effective for increasing soil carbon. We should also consider promoting photosynthesis-based carbon capture by plant vegetation. These changes, aimed at increasing soil carbon to offer the world a highly effective carbon sink, point to a sustainable agriculture whose practices stand in stark contrast to the high-emission, soil-degrading strategies of agribusiness. Stranded assets here would include chemical fertilizers and fungicides both of which we are better off phasing out. Finally, farmers will have to reduce how much methane livestock, in particular cows, release by adjusting how and what these animals eat.

Industrial processes are yet another area ripe and ready for abatement. A major culprit here in terms of GHG emissions is the production of cement which we can make less pollutant by sharply lowering its clinker content and/or burning biomass rather than fossil fuels in the cement kiln. Iron and steelmaking could also become less inclined to discharge greenhouse gases to the extent that their producers replace coke with biomass in the furnaces, use gas-fired furnaces, adopt more efficient direct-casting and smelt-reduction techniques, and reduce energy demand through co-generation. A third industrial sector with lots of abatement potential is chemicals. Here, we can use more efficient motor systems, less polluting fuels (biomass, gas), improved catalysts, combined heat and power (CHP) techniques, or better ethylene cracking. In general, technological advances in the production and analysis of chemicals are crucial in our struggle with climate change, as we have already alluded to earlier in our discussion of ammonia. Chemistry is after all at the heart of the problem as well as the catalyst for many of its potentially counteracting solutions.[17]

Toward a New Ecologically Embedded Accumulation Regime ("Eco-Capitalism")

When looking at the tasks ahead in the face of the climate-change challenge, we are talking about having to transform many of the forces at the center of our capitalist economy which have brought about this problem in the first place. We will need to redo our energy mix completely, abandoning our century-long nearly exclusive reliance on fossil fuels. It is going to be far more important to emphasize energy-efficiency gains reducing costs rather than pushing larger markets, meaning more consumption, for greater revenue. And this means not least figuring out how to move away from the one-person gas-guzzling car-driving transport system, we have come to rely on and in the process reimagine how to move people around more effectively. Our agriculture has to determine not only how to grow food but also what we eat and how we eat it, while at the same time, stopping the destruction of forests in the quest for (only temporarily available) arable land. We cannot let unfettered urbanization continue to give us over-sized, polluted, perennially congested cities that are far too overpopulated for the infrastructure or housing stock they have and whose correspondingly overwhelmed local governments are powerless to deal with these problems. All this will cost a bundle, trillions of dollars, which we have yet to mobilize, while stranded assets create large losses at the same time. Much of the needed technology has yet to go beyond the drawing board or pilot-project stage. Firms, motivated by the dictate of shareholder value maximization into short-term thinking about next trimester's bottom line, will have to grapple with the long run when addressing climate change. And finance, which at this point does not even have a reliable price for carbon to work with, will nonetheless have to mobilize the massive funding of low-carbon investments. All these examples show us, simply put, that we have to change the whole system that got us there to the edge of the disaster it helped create—its priorities, its incentives, and its decision-making processes. There is a great transformation at hand. Will we succeed in carrying it out sufficiently before it is too late?

My feeling is that it is indeed possible for us to meet this challenge collectively—as enormous as it seems now. One reason is that it might very well be in the endless cumbersome and costly to carry out the needed changes in energy, transportation, agriculture, or industry, especially if we consider the possibilities of rapid technological progress, massive efficiency gains, and significant job creation all helping to reduce transition costs as we move from the current regime to the new one. And there have indeed been lately a number of fairly optimistic projections by renowned economists specializing in the question of what it would take to de-carbonize our economy thoroughly, such as William Nordhaus or Nicholas Stern, who argue that steps needed to de-carbonize our economy are both feasible and on balance beneficial for our economy.[18]

Yet there is an even deeper reason for guarded optimism. More than a decade ago, the world economy went through a major global crisis and then found itself stuck far below its potential on a slow-growth path—a post-crisis pattern a number of economists (e.g., Lawrence Summers, Robert Gordon) have come to characterize as "secular stagnation." Notwithstanding welcome improvements in economic growth and job creation in the USA, Europe, and key emerging-market economies (e.g., India) during 2017, the world economy remains stuck significantly below pre-crisis investment levels. A climate-induced investment boom remaking our energy, transportation, and urban infrastructure may in the context be exactly what the patient needs to recover. And because worldwide growth has remained so slow ever since the onset of the 2007/2008 financial crisis, even long-term interest rates continue to hover near zero if not altogether in negative territory—a historically unprecedented situation. There is rarely a better time to borrow for productive infrastructure investment projects than now! The argument that governments have a post-crisis problem of high debt/GDP ratios and therefore should not add to their deficits ignores the likely boost to growth from such an infrastructure spending boom which actually may very well end up helping governments to reduce their relative debt burdens (as the GDP denominator rises more rapidly to bring down the ratio).

But the argument about using our struggle with climate change as an opportunity to leave a major crisis behind has ultimately an even deeper context. When looking at the historic evolution of our capitalist system since its inception in the wake of the industrial revolution two centuries ago, we can see that system having moved through long waves. Also known as "Kondratiev cycles," such long waves typically have a sustained period of rapid growth (e.g., 1852–1873, 1896–1929, 1947–1973, and 1982–2007) culminate in major financial crises that usher in a decade or more of distinctly slower growth (1873–1896, 1929–1939, 1973–1982, 2007–?).[19] What motivates this long-wave pattern of boom, systemic crisis, and stagnation? One driving force, stressed by Joseph Schumpeter (1939), may be technological to the extent that technical change typically comes in growth-boosting clusters before exhausting itself. Another, pointed out by Hyman Minsky (1964), may be financial inasmuch as booms end up encouraging debtors to take on progressively more debt and higher risks which leave them ultimately more exposed to major bouts of financial instability.

While it is obviously a worthy effort to identify how a period of sustained expansion may exhaust itself and give way to slower growth, it should be of equal interest to figure out how stagnant economies may recover to move back onto a faster growth path—a situation of acute relevance today. There is nothing automatic or mechanical about this, and a bad situation may as easily get worse before getting better because of policy mistakes, political instability building up after years of stagnation, heightened social conflict, even war. But as the Marxist long-wave specialist Ernest Mandel (1980) pointed out, crisis triggers its own self-correcting forces such as elimination of excess capacities by plant closures and bankruptcies, liquidation of failed firms leaving survivors with bigger market shares, deleveraging bringing down debt levels, higher rates of workers' exploitation boosting profit rates, and the combination of cheapening of costs and more intense competition setting the stage for renewed acceleration of technological progress. Resumption of the upswing phase of fast growth over a couple of decades is greatly facilitated by rapid technological change especially in production and/or in terms of fostering new growth industries. And finally, stagnation may give way to booms in the wake of successful policy reforms. All these recovery forces should work in tandem if and when they arise in a propitious constellation to boost each other. For example, World War II gave a huge boost to the adoption of mass-production technologies centered on assembly lines and the adoption of productivity-boosting "scientific management" techniques of piecework. On its own, this enhanced productive capacity would surely create overproduction conditions. It was only with the landmark collective bargaining agreement between General Motors and the United Automobile Workers in 1948, tying real wage increases to productivity gains, that the industrial nations found a formula better-balancing aggregate supply and aggregate demand.

It is this need for reforms as well as transformations, and their composition in terms of mutually reinforcing interactions, that is at the heart of capitalism's capacity for sustained recovery at a fast pace. Rather than thinking, as mainstream economists tend to do, about macroeconomic equilibrium, we need to analyze capitalism's self-regulation in terms of the institutional configuration guiding the forces that drive its production volumes and spending levels. That method is at the heart of an alternative French approach known as "la théorie de la régulation" which emerged in the late 1970s to analyze the post-war boom in the industrial world and its subsequent disintegration into a new type of structural crisis known as "stagflation." Its key protagonists—Michel Aglietta, Robert Boyer, Alain Lipietz, and others (myself included)—have since then grounded this unique historical-institutional approach to economics into a broader analysis of the long-wave dynamics of capitalism. What they have termed "régulation" refers to the modalities of the system's internal self-balancing reproduction over time. The Régulationists look at the evolution of capitalism in terms of distinct "accumulation regimes," each with its own unique "mode of regulation."[20] Long waves then become how capitalism reorganizes itself during a period of structural crisis, coinciding with the downswing phase of such a long wave, to give rise to a new accumulation regime that, if and when it starts to work amid mutually reinforcing interaction of its growth drivers, yields a sustained period of fast-paced expansion until that initially propitious constellation exhausts itself and/or breaks down to throw the system into a yet another structural crisis.

We can see that crisis-induced reorganization promoting the emergence of a new accumulation regime play out in the last quarter of the nineteenth century when capitalism, following a major worldwide crisis (1873–1879), moved from an essentially competitive system with flexible prices and wages in both directions to a more monopolistic one thanks to the emergence of large trusts dominating key sectors, their close ties to banks, formation of industrial unions, and worldwide adoption of the British-led gold standard. Similarly, we can see the global interwar depression (1929–1939) set the stage for a series of reforms in the wake of Roosevelt's New Deal to match aggregate demand better with the economy's enhanced productive capacity and, following World War II, provide an international context with the Bretton Woods Agreement of July 1944 to give those reforms space to take root in Western Europe and Japan as well. That so-called Fordist accumulation regime fell apart amid worldwide stagflation conditions (1973–1982) at the end of which a series of pro-market reforms known as "Reaganomics" set the stage for a new, finance-dominated accumulation regime which I have elsewhere termed "finance-led capitalism" (see Guttmann 2008, 2016).[21] The "subprime" crisis of 2007/2008, which started out in a fairly obscure corner of the US bond market to ricochet around the globe until the world economy faced calamity, can in this context be seen as the onset of that regime's structural crisis and the reason for the "secular" stagnation observed since then. If we are to accept the validity of this account of long-wave patterns, then it stands to reason that the time has come to contemplate the emergence of a new accumulation regime, one centered on an ecologically embedded and more socially oriented type of capitalism which provides humankind with a chance to combat climate change effectively and so sustain our planet as inhabitable—an accumulation regime I wish to refer to here as *eco-capitalism* (for "ecological capitalism").

The birth of eco-capitalism, as is typical for the emergence of new accumulation regimes, is tied to path-breaking policy reform, in this case, the Paris Climate Agreement of December 2015. Here, we have a prescription for a fundamental revamping of our energy mix, industrial apparatus, and land-use practices while providing the governance structure to push this transformation forward in globally coordinated fashion. In its wake, we will have to remake globalization, moving us from the transnational regulation of finance-led capitalism to a *supranationally coordinated* regulation engulfing *all* nation-states in a collective effort to reduce greenhouse gas emissions rapidly to a sustainable level where those gases get naturally absorbed by our environment without further heating up the planet. We have to make that goal the organizing principle of how we run our economy from now on, if we want to have a chance to succeed.

This low-carbon imperative will change very much how firms operate, moving them away from their shareholder value maximization dictate and toward a more stakeholder-oriented set of goals beyond short-term profitability to take account of their contribution to the environment and societal well-being. Producers, while still able to earn profit, will probably have to coordinate their other objectives beyond profit with civil-society representatives while also engaging with the public sector and government agencies in cooperative rather than adversarial fashion in pursuit of the public interest. The Paris Accord's prescription for "financial-flow consistency" also implies a very different financial system. This new *climate finance* will have to be capable of funding inherently uncertain projects stretching over the very long run, socializing massive losses arising from the imposed obsolescence of our fossil-fueled economy, transferring capital globally on a massive scale toward the environmentally most fragile regions, and promoting rapid technological change. We will in that context also have to move away from the current system of privatized know-how, rooted in intellectual property rights endowing innovators with monopoly rights, and toward a system of widely shared and rapidly disseminated knowledge. That emphasis on knowledge diffusion, a prerequisite for the kind of technological change we need in pursuit of our global zero-emission objective, requires government to go beyond its current support for research and development and remake itself into an *entrepreneurial state* equipped to launch technologies, even new growth sectors, where the private sector cannot or will not do so.[22] Here, we have a chance not only to push capitalism toward a more environmentally friendly posture, but also to remake its basic tenets into a socially better balanced system. Pushing a progressive vision of eco-capitalism should revitalize a tired and often futile political debate over policy priorities while providing the needed antidote to the worldwide resurgence of nationalistic demagoguery which threatens to destroy whatever is left of our leaders' ability and willingness to coordinate in the face of global threats.

We are, of course, still nowhere near such a profound transformation. At this point, a couple of years into the new climate-accord regime, most countries are just starting to put into place their first-round climate-policy actions in line with their initial commitments made in the wake of the Paris Agreement of 2015. While most of these early initiatives are still quite modest, we can already see some examples of ambitious and innovative policies pointing the way forward to where we all will need to go. It is worth taking a closer look at these path-setting examples to learn from them, as they set themselves apart from the broad range of still rather modest measures and goals.

NOTES

1. G. Wagner and M. Weitzman (2015) provide a thorough analysis of the special systemic uncertainties associated with climate change. See for more on this also M. Wolf (2015a).
2. Climate deniers, such as H. Jenkins (2016), like to question the climate models used by the IPCCC and other groups of concerned scientists in their forward projections, especially as concerns their assumptions about gradually intensifying negative feedback effects. But a number of recent studies by leading climate scientists, including L. Cheng et al. (2016), P. Gleckler et al. (2016), and K. Trenberth et al. (2016), confirm that the major climate models in use over the last few decades have been remarkably accurate.
3. For more details on the implications of a 450 ppm limit, see International Energy Agency (2014) detailing its so-called 450 Scenario. As to the notion of carbon budgets, the IPCCC's AR5 specifies carbon budgets for projected temperature increases of 1.5°, 2°, and 3°, three per specific hike depending on the targeted level of probability achieving this particular increase. We should note that the IPCCC's carbon budget is expressed in terms of units of carbon whereas the carbon budgets of the IEA as well as those offered by Carbon Tracker are expressed in terms of carbon dioxide, for an equivalence of 1 GtC = 3.67 GtCO$_2$. Carbon Tracker (2013) discusses the methodological and modeling differences between those three carbon-budget measures we use.
4. These estimates include state entities, such as Saudi Arabia's huge Aramco. If we are just looking at the private sector, in this case the hundred largest coal companies and the hundred largest oil or gas

companies, their reserves amount to just three times the carbon budget. Coal reserves absorb two-thirds of total fossil-fuel reserves. See M. Wolf (2014a) for a more detailed discussion about Carbon Tracker's carbon budget implications for fossil-fuel reserves.

5. The global market shares of the different energy sources are taken from the excellent Web site The Shift Project Data Portal (tsp-data-portal.org) which offers access to a large array of energy and climate data.

6. R. Molla (2014) provides a useful comparison of different energy sources in terms of their respective efficiency, keeping in mind energy efficiency is composed of several criteria.

7. See A. Mooney (2016) and G. Tett (2016) for details on how the oil giants face growing pressure from institutional investors and government regulators to take better account of their exposure to climate-change regulation and uncertainties associated with the prospect of major changes in the energy mix over coming decades in its wake.

8. Global fossil-fuel subsidies are discussed in D. Coady et al. (2015), including the IMF estimates of pollution externalities as an implicit type of subsidy. Global subsidy data are collected by IEA (2015).

9. A comprehensive study of coal's renaissance during the last couple of decades can be found in J. C. Steckel et al. (2015). In an equally troubling study, A. Pfeiffer et al. (2016) discuss in great detail the implications of keeping new power plant construction within the two-degree carbon budget, concluding that we will have to commit ourselves to zero-emission energy sources very soon if we are to keep global warming trends within the Paris objectives. The high-powered Global Commission on the Economy and Climate (2014) comes to a very similar conclusion.

10. The US government under Obama (see U.S. Energy Information Administration 2016) projected that coal will hit a ceiling and be largely surpassed by natural gas over the next couple of decades while renewables will show the fastest growth rate so that all three energy sources will have roughly equal (28–29%) market shares by 2040. Someone will need to tell this to Trump and some of his key appointees at the EPA, Department of Energy, or Department of the Interior.

11. The International Renewable Energy Agency (IRENA) provides annual reviews of the number of jobs provided by the different renewables, as in IRENA (2015).

12. For more information on the increasingly competitive performance and cost structure of renewables, especially wind and solar, when compared to fossil fuels see J. Trancik (2014), M. Wolf (2014b), T. Randall (2016), and P. Clark (2017).

13. See the discussion by P. Clark (2015) on the reasons behind this apparent decoupling of economic growth and GHG emissions.

14. For the details of the various mitigation and adaptation measures in the different sectors discussed here as follows, I relied principally on the aforementioned McKinsey Report (McKinsey 2013) as well as United Nations Environment Programme (2016).

15. The deal, to take effect after 2020 and on a voluntary basis for the first six years, has too many loopholes to move the airline industry fully towards carbon neutrality over the next decade. See H. Fountain (2016) for more details on the deal and its limitations.

16. M. Scott (2013) describes how the "greening" of buildings in terms of using alternative LED lightbulbs, improving insulation, upgrading appliances, retrofitting heating as well as cooling systems, and other energy-efficiency improvements basically pay for themselves in a fairly short period of time while cutting GHG emissions substantially.

17. Speaking of chemicals! Less than a month after the global airline deal (still in October 2016) over 170 countries concluded a binding global agreement in Kigali (Rwanda) to phase out so-called hydrofluorocarbons (HFCs) used in air conditioners and refrigerators all over the world. These HFCs are very powerful heat trappers, and with the rapid spread of air-conditioners in China and India, there was a risk that their accelerating spread into the atmosphere could raise the temperature by 0.5 degrees alone. The deal allows the emerging-market economies more time to phase out HFCs in the hope that more benign substitutes will by then have come down in price.

18. See in this context the optimistic arguments by W. Nordhaus (2013), F. Green (2015), N. Stern (2015), or R. Pollin (2015) all of whom consider early action on climate-change mitigation both something that can be done and on balance beneficial to our economy.

19. Nikolai Kondratiev (1925/1984, 1926/1936) ran the Soviet Union's post-revolutionary data-collection and statistics office until Stalin banned him to the gulag for stressing the internal survivability dynamic of capitalism. His sin was to have identified a long-wave pattern among key capitalist economies on the basis of longitudinal studies of strategic price movements and credit aggregates which implied that capitalism could revive itself following a major crisis. But the statistician-economist never gave us underlying explanations for this phenomenon, only its empirically verifiable pattern.

20. Among key works by "la théorie de la regulation" published in English are the founding text by M. Aglietta (1979), A. Lipietz (1985, 1987), R. Boyer (2000), as well as R. Boyer and Y. Saillard (2001).

21. A Régulationist perspective on the stagflation crisis of the 1970s and early 1980s triggering restructuring efforts and policy reforms ("Reaganomics") which moved us to a new, more finance-dominated accumulation regime, can also be found in R. Guttmann (1994).

22. See in this context the highly relevant contribution of M. Mazzucato (2013) stressing the ability of the state apparatus, the public sector, to

play a leading entrepreneurial role in the promotion of new technologies, growth industries, and production practices.

REFERENCES

Aglietta, M. (1979). *A Theory of Capitalist Regulation: The US Experience.* London: Verso. First Published as *Régulation et crises du capitalisme* (Paris: Calmann-Levy), 1976.

Boyer, R. (2000). Is a Finance-Led Growth Regime a Viable Alternative to Fordism? A Preliminary Analysis. *Economy and Society, 29*(1), 111–145. http://dx.doi.org/10.1080/030851400360587.

Boyer, R., & Saillard, Y. (2001). *Regulation Theory: The State of Art.* London: Routledge. First Published as *Théorie de le régulation: l'état de saviors* (Paris: La Decouverte), 1995.

Carbon Tracker. (2011). *Unburnable Carbon—Are the World's Financial Markets Carrying a Carbon Bubble?* http://www.carbontracker.org/wp-content/uploads/2014/09/Unburnable-Carbon-Full-rev2-1.pdf. Accessed 23 Apr 2015.

Carbon Tracker. (2013). *Things to Look Out for When Using Carbon Budgets!* (Working Papers, 08-2014). http://www.carbontracker.org/wp-content/uploads/2014/08/Carbon-budget-checklist-FINAL-1.pdf. Accessed 12 Oct 2016.

Cheng, L., Trenberth, K., Palmer, M., et al. (2016). Observed and Simulated Full-Depth Ocean Heat-Content Changes for 1970–2005. *Ocean Science, 12*, 925–935. http://www.ocean-sci.net/12/925/2016/. Accessed 13 Oct 2016.

Clark, P. (2015, March 15). Global Carbon Emissions Stall in 2015. *Financial Times.*

Clark, P. (2017, May 18). The Big Green Bang: How Renewable Energy Became Unstoppable. *Financial Times.*

Coady, D., Parry, I., Sears, L., et al. (2015). *How Large Are Global Energy Subsidies?* (IMF Working Papers, WP/15/105). Washington, DC: International Monetary Fund. http://www.imf.org/external/pubs/ft/wp/2015/wp15105.pdf. Accessed 14 Oct 2016.

Fountain, H. (2016, October 6). Over 190 Countries Adopt Plan to Offset Air Travel Emissions. *New York Times.* http://www.nytimes.com/2016/10/07/science/190-countries-adopt-plan-to-offset-jet-emissions.html?_r=0. Accessed 7 Oct 2016.

Gleckler, P., Durack, P., Staufferii, R., et al. (2016). Industrial-Era Global Ocean Heat Uptake Doubles in Industrial Era. *Nature Climate Change, 6*, 394–398.

Global Commission on Economy and Climate. (2014). *Better Growth, Better Climate* (The New Climate Economy Project—2014 Report). http://newclimateeconomy.report.

Green, F. (2015). *Nationally Self-Interested Climate Change Mitigation: A Unified Conceptual Framework* (Working Paper No. 224). Centre for Climate Change Economics and Policy.

Guttmann, R. (1994). *How Credit-Money Shapes the Economy: The United States in a Global System.* Armonk, NY: M. E. Sharpe.

Guttmann, R. (2008, December). *A Primer on Finance-Led Capitalism and Its Crisis* (Revue de la Régulation, No. 3/4). regulation.revues.org/document5843.html.

Guttmann, R. (2016). *Finance-Led Capitalism: Shadow Banking, Re-regulation, and the Future of Global Markets.* New York: Palgrave Macmillan.

International Energy Agency. (2014). *450 Scenario: Method and Policy Framework.* Paris: IEA. http://www.worldenergyoutlook.org/media/weowebsite/2014/Methodologyfor450Scenario.pdf.

International Energy Agency. (2015). *World Energy Outlook 2015 Factsheet.* Paris: IEA. http://www.worldenergyoutlook.org/media/weowebsite/2015/WEO2015_Factsheets.pdf.

IRENA. (2015). *Renewable Energy and Jobs—Annual Review 2015.* http://www.irena.org/menu/index.aspx?mnu=Subcat&PriMenuID=36&CatID=141&SubcatID=585.

Jenkins, H. (2016, June 28). Climate Denial Finally Pays Off. *Wall Street Journal.* http://www.wsj.com/articles/climate-denial-finally-pays-off-1467151625. Accessed 12 Oct 2016.

King, D., Browne, J., Layard, R., et al. (2015). *A Global Apollo Programme to Combat Climate Change.* http://cep.lse.ac.uk/pubs/download/special/Global_Apollo_Programme_Report.pdf.

Kondratiev, N. (1925/1984). *The Major Economic Cycles* (in Russian), Moscow. Translated as *The Long Wave Cycle.* New York: Richardson & Snyder.

Kondratiev, N. (1926/1936). Die langen Wellen der Konjunktur. *Archiv für Sozialwissenschaft und Sozialpolitik, 56,* 573–609. Published in English as "The Long Waves in Economic Life." *Review of Economic Statistics, 17*(6), 105–115.

Lipietz, A. (1985). *The Enchanted World: Inflation, Credit and the World Crisis.* London: Verso. First Published as *Le monde enchanté: De la valeur a l'envol inflationniste* (Paris: F. Maspero), 1983.

Lipietz, A. (1987). *Mirages and Miracles: The Crisis of Global Fordism.* London: Verso. First Published as *Mirages et Miracles* (Paris: La Decouverte), 1985.

Mandel, E. (1980). *Long Waves of Capitalist Development: A Marxist Interpretation.* Cambridge, UK: Cambridge University Press.

Mazzucato, M. (2013). *The Entrepreneurial State: Debunking Public vs. Private Sector Myths.* London: Anthem Press.

McKinsey. (2013, September). *Pathways to a Low-Carbon Economy: Version 2 of the Global Greenhouse Gas Abatement Cost Curve.* Report. http://www.mckinsey.com/business-functions/sustainability-and-resource-productivity/our-insights/pathways-to-a-low-carbon-economy.

Minsky, H. (1964). Longer Waves in Financial Relations: Financial Factors in the More Severe Depressions. *American Economic Review, 54*(3), 324–335.

Molla, R. (2014). What Is the Most Efficient Source of Energy? *Wall Street Journal* (September 15). http://blogs.wsj.com/numbers/what-is-the-most-efficient-source-of-electricity-1754/.

Mooney, A. (2016, May 11). Academics Back Exxon and Chevron Climate Openness Vote. *Financial Times*.

Nordhaus, W. (2013). *The Climate Casino: Risk, Uncertainty, and Economics for a Warming World*. New Haven, CT: Yale University Press.

Pfeiffer, A., Millar, R., Hepburn, C., et al. (2016). The '2°C Capital Stock' for Electricity Generation: Committed Cumulative Carbon Emissions from the Electricity Generation Sector and the Transition to a Green Economy. *Applied Energy, 179*(October 1), 1395–1408. http://dx.doi.org/10.1016/j.apenergy.2016.02.093. Accessed 14 Oct 2016.

Pollin, R. (2015). *Greening the Global Economy*. Cambridge, MA: MIT Press.

Randall, T. (2016, April 6). Wind and Solar Are Crushing Fossil Fuels. *Bloomberg News*. http://www.bloomberg.com/news/articles/2016-04-06/wind-and-solar-are-crushing-fossil-fuels. Accessed 15 Oct 2016.

Schumpeter, J. (1939). *Business Cycles*. New York and London: McGraw-Hill.

Scott, M. (2013, March 17). Market for Green Buildings Warms Up. *Financial Times*.

Steckel, J. C., Edenhofer, O., & Jacob, M. (2015). Drivers for the Renaissance of Coal. *PNAS (Proceedings of the National Academy of Sciences of the United States of America), 112*(29), E3775–E3781. http://www.pnas.org/content/112/29/E3775.abstract. Accessed 14 Oct 2016.

Stern, N. (2015). *Why Are We Waiting? The Logic, Urgency, and Promise of Tackling Climate Change*. Cambridge, MA: MIT Press.

Tett, G. (2016, September 22). Energy Companies Must Act to Avoid Banks' Mistakes. *Financial Times*.

Trancik, J. (2014). Renewable Energy: Back the Renewables Boom. *Nature, 507*(7492). http://www.nature.com/news/renewable-energy-back-the-renewables-boom-1.14873. Accessed 2 Oct 2016.

Trenberth, K., Fasullo, J. T., Von Schuckmann, K., et al. (2016). Insights into Earth's Energy Imbalance from Multiple Sources. *Journal of Climate, 29*(20), 7495–7505. http://dx.doi.org/10.1175/JCLI-D-16-0339.1. Accessed 13 Oct 2016.

United Nations Environment Programme. (2016). *Climate Change*. http://www.unep.org/climatechange/. Accessed 15 Oct 2016.

U.S. Energy Information Administration. (2016). *Annual Energy Outlook 2016 with Projections to 2040*. https://www.eia.gov/outlooks/aeo/pdf/0383(2016).pdf.

Wagner, G., & Weitzman, M. (2015). *Climate Shock: The Economic Consequences of a Hotter Planet*. Princeton, NJ: Princeton University Press.

Wolf, M. (2014a, June 14). A Climate Fix Would Ruin Investors. *Financial Times*.

Wolf, M. (2014b, September 23). Clean Growth Is a Safe Bet in the Climate Casino. *Financial Times*.

Wolf, M. (2015a, June 9). Why Climate Uncertainty Justifies Actions. *Financial Times*.

Wolf. M. (2015b, March 3). The Riches and Perils of the Fossil-Fuel Age. *Financial Times*.

CHAPTER 3

The Global Emergence of Climate Policy

Getting all countries of the world (with the exception of war-torn Syria and Nicaragua, led by radical Sandinista leader Daniel Ortega) to agree to a global governance mechanism for sustained reductions in GHG emissions is a huge diplomatic achievement. Equally impressive was how rapidly this treaty, the Paris Climate Agreement, managed to get ratified, within a year of its initial passage. It is as if the whole world acted before climate denier Donald Trump had a chance to get elected. Now that he is the US president and has made the fateful decision to take his country out of the Paris Agreement, the treaty's speedy ratification protects its integrity inasmuch as America's earliest possible withdrawal will have to await the conclusion of Trump's first term in November 2020—and there are legitimate doubts that he can last even that long in power. But while the drama surrounding America's (lack of) participation in the Paris Agreement is obviously of great importance, it is only a temporary aberration. I am quite certain that the USA, under almost any other leader to come, will rejoin the global effort to cope with climate change. Nor do I have much doubt that in the meantime the rest of the world will remain committed to the letter and spirit of the Paris Agreement so that this historic accord continues to be the main organizing mechanism for the years ahead—with or without the USA returning to its fold.

The deeper question then is not whether the Paris Agreement will survive the shenanigans of an ill-informed and prejudiced demagogue willing to throw away the leadership position of the country he rules, but whether that accord can be made to work—as implied—in progressively

© The Author(s) 2018
R. Guttmann, *Eco-Capitalism*,
https://doi.org/10.1007/978-3-319-92357-4_3

more ambitious fashion over time. There is a profound contradiction embedded in that treaty between its overall goal—keeping the cumulative rise in the global average temperature below 2 °C—and the means chosen toward that objective—the 189 national climate pledges known as Nationally Determined Contributions with which to address climate change in terms of mitigation and adaptation. When taken together, these NDCs do not suffice to meet the ≤ 2 °C goal. Instead, if we assume that the emission-reduction targets laid out in those pledges are actually met within the prescribed timetable, they will keep the cumulative temperature hike to somewhere between 2.6 and 3.1 °C. This raises the question how much those national commitments can be rendered more ambitious, and when. The needed push for higher emission-reduction targets will undoubtedly center on the crucial issue of imposing a sufficiently high (and gradually rising) carbon price with which to provide adequately strong incentives for our mitigation efforts. We know we need such a carbon price, yet have been unable so far to make one stick. We shall see later (in Chapter 5) that this is a matter of political will, in the face of challenging trade-offs and administrative complexities.

The Wide Range of National Climate Pledges

Whatever we may think of the NDCs of this country or that, they are in the aggregate a crucial statement by the world community how to face global warming. Each country has been obliged to think through its objectives concerning GHG emissions over the coming decade, detail its plans for action in pursuit of those targets, and publish precisely structured documents describing its goals and policy commitments. These so-called Nationally Determined Contributions are to be reviewed and revised every five years. They are all recorded in a Public Registry maintained by the United Nations Climate Change Secretariat.[1] Not legally binding, these NDCs are more like statements of intent. Governments cannot be forced to abide by their goals, and there is only moral suasion as a form of public pressure shaming governments into compliance. The fragility of that strictly voluntary process was demonstrated when Trump's election victory in November 2016 basically threw America's climate-action pledges by the Obama Administration out the window, making it potentially that much easier for other reluctant countries (e.g., Russia, Saudi Arabia) to renege on their commitments in turn. Yet when Trump finally made good on his election-campaign threat of withdrawing the USA from the Paris Agreement in June 2017, that irresponsible act provoked a globally unanimous response reaffirming the rest of the world's redoubled effort to make that accord work. And this strong reaction showed a resilience which the community of nations will need to build on until the USA rejoins the worldwide effort against global warming.

When looking at the 150+ NDCs, their most striking feature is the remarkable degree of diversity among them. While they all follow more or less the same structure, they vary greatly in terms of their respective contents.[2] The USA, for example, committed itself under Obama to cut emissions up to 28% from 2005 levels by 2025. By comparison, the 28-member European Union proposed a cut of 40% from 1990 levels by 2030. China, by contrast, did not propose any cut at all. Instead it specified emissions peaking in 2030 while aiming for substantial improvements in energy efficiency to lower emissions per unit of GDP by two-thirds in the meantime. More generally, NDCs may have different base years or use the "business-as-usual" trend line as their starting point. Apart from widely divergent target levels of cuts, what gets cut may focus just on CO_2 or use a wider base of GHG emissions. Some countries frame their goal in terms of efficiency gains rather than emission reductions, like China did. While most countries define targets in terms of their economy as a whole, some just look at specific sectors. Where the NDCs differ the most is with regard to their respective action plans—what they contain, how detailed they are, whether they just focus on mitigation or include adaptation measures as well, lastly to what extent they also include the provision of carbon sinks pertaining to land use (known under the acronym LULUCF) and forest management (referred to as REDD+).[3] Many developing countries, comprising 78% of all NDCs, offered to pursue significantly more ambitious (so-called conditional) emission-reduction targets if they can receive adequate financial resources or technical support from the richer countries. Such far-reaching transfer of resources is thus bound to make a large difference in our global fight against climate change.

An independent consortium of research organizations known as Climate Action Tracker (CAT), accessible on its Web site climateaction-tracker.org, has been evaluating thirty-two NDCs covering together about 80% of global emissions. CAT has judged thirteen of those NDCs to be "inadequate," including Argentina, Australia, Canada, Japan, Russia, and Saudi Arabia, not only because their reduction targets are arguably insufficient, but also because climate policies currently in place there are not likely strong enough to meet even those modest targets. A number of countries with NDCs judged "medium" by CAT have recently actually weakened their climate-action policies, including Brazil's resumption of deforestation under new leader Michel Temer, Theresa May's downgrading of Britain's climate-change policy priorities following the Brexit vote in the summer of 2016, the Philippines' populist leader Rodrigo Duterte scaling back that country's traditional leadership role in the global struggle against global warming, and finally Donald Trump's systematic dismantling of Obama's initiatives against climate change. Noteworthy as well is how little regard countries heavily dependent on fossil fuels, such as Russia or Saudi Arabia, have for climate-change mitigation even though Norway has proven admirably that a major oil producer can at the same time also be a leader in the fight against climate change. Of the handful of countries with NDCs deemed "sufficient" by CAT, Bhutan's and Costa Rica's emphasis on preserving their forests as carbon sinks stand out, as do Ethiopia's comprehensive Climate-Resilient Green Economy (CRGE) Strategy or Morocco's aggressive pursuit of solar power. What these assessments make abundantly clear is the importance of political leadership and the need for a multifaceted green-economy strategy covering all the major sectors of a nation's economy.[4]

AMERICA'S CLIMATE WAR

The world's largest per capita emitter of greenhouse gases, the USA, is gripped by deep internal divisions about its climate-related priorities. This conflict does not just crystallize at the top, in Trump's systematic dismantling of Obama's climate policy which also includes rejection of the Paris Agreement. It goes through the entire body politic, setting apart the two political parties, separating the coastal regions from the vast middle of the country, pitting cities against suburbs and rural areas, and engaging industrial sectors in a clash of armies of lobbyists. There is a politically overrepresented coal region, centered on Kentucky and West Virginia but stretching into Illinois and Pennsylvania, which is fighting that sector's long and slow decline and has turned into a key political-power base for Trump. Oil and gas production, covering more or less the entire nation, has over the last decade and a half been boosted by the oil-shale boom known as "fracking" which has economically revived certain regions (e.g., North Dakota, Oklahoma), albeit at great environ-mental cost, and liberated the USA from its dependence on oil imports. Integrating the USA with Canada's oil-sands supply chain, tapping huge reserves in the Alaskan Arctic, and expanding offshore production along all three American coasts remain priority projects of the oil majors and the local politicians on their payroll. The Trump Administration has already moved on all three fronts during its first year in office. Oil is also on the mind of car manufacturers whose profit margins depend dispro-portionately on sales of large-sized "Sports Utilities Vehicles" (SUVs) consuming a lot of gas. Given widespread neglect of the country's infrastructure, it is also intrinsically very difficult and costly, if not alto-gether impossible, to affect needed changes such as building a network of super-fast trains or upgrade electricity grids for better integration of renewables.

Yet the ultimate barrier to change are the Americans themselves and their social consumption norms. They are deeply wedded to the wrong priorities, be it the national love affair with over-sized, gas-guzzling cars commuting every day over long distances, a fast-food culture in perfect alignment with farming-intense agro-business, excessive heating and cooling of structures, and a propensity for overbuilding anchored in the absence of meaningful zoning restrictions. As if this was not bad enough, the "American Way" has taken its celebration of individual rights and freedoms to absurd levels as clearly evident in the right to bear arms, a freedom punctuated by regular mass shootings and an epidemic of sui-cides. Most Americans like to keep the public sector small and on a tight shoestring, resist taxes whereby socially beneficial goods or activities may be funded adequately, and for the most part do not consider matters of communal organization (such as recycling) all that important. Instant gratification crowds out long-term planning. Politics is transactional and devoid of vision. What goes on in the rest of the world is not of much concern. All of these biases, grounded in their hard-core individualism, leave Americans ill prepared for the challenges of climate change.

America's political structure does not help. The federal government is relatively small, compared to other advanced capitalist nations, and rather badly funded inasmuch as both the rich and the corporate sector often manage to avoid paying their fair share in taxes. More than three-quarters of federal spending goes to the military, a couple of universal programs for the elderly (Social Security, Medicare), and interest on the debt, leaving very little for all the other functions of the state. Political parties have little power and play only a minor role in funding politicians. Yet political election campaigns are very expensive and last forever. This leaves politicians dependent on private funding and hence exposed to the influence of lobbies who play a disproportionately large role in shaping the country's policies. Much of America's legislative process boils down to how the lobbies relate to each other in fashioning the laws to be passed. Politicians less beholden to the power of lobbies make their mark as mavericks, less so as leaders "getting stuff done." The Senate gives disproportionate power to senators from small, land-bound, more rural states which are typically rather conservative. This reality becomes an especially important constraint when it comes to America's international treaty engagements which have to pass the US Senate to become legally binding, a hurdle that has already proven highly problematic in global climate negotiations. The House of Representatives no longer has a political center willing and able to fashion pragmatic bipartisan compromises, since most representatives have locked in incumbency thanks to a demographically manipulative process of redesigning electoral districts known as "gerrymandering" and are only worried about facing challenges from more extreme members of their own party in the intra-party primaries.[5]

It is against this unfavorable political backdrop that Trump's ascent to the US presidency has had an especially profound impact as he decided from his first days in office onward to dismantle all of Obama's policy-making legacy. While his efforts in that direction have not always been equally successful as became evident with the Obamacare repeal debacle, Trump has been highly effective targeting the environment. In the absence of any legislative breakthroughs that had escaped him amidst a deeply divided US Congress, Obama had to build his climate policy around executive decrees and directives to regulatory agencies capable of withstanding judicial attack by state attorney generals in Republican-controlled states and conservative judges populating the federal bench. These foundations are far weaker than laws of the land and hence more easily undone. Trump has already rolled back Obama's ambitious fuel-economy standards and repealed his Clean Power Plan, a set of regulations governing power plants which was at the heart of the USA' NDC commitment for GHG reductions. While these reversals will have to be approved by the courts in light of the Environmental Protection Agency's broad mandate to regulate greenhouse gases, the Trump Administration has sought to make sure that the EPA would end up so weakened that it had no longer much regulatory punch. Its current director Scott Pruitt is a fanatical climate denier, as are other Trump appointees to agencies with jurisdiction over climate policy (e.g., Jim Bridenstine at NASA, Rick Perry heading the Energy Department). Severe budget cuts, intimidation of key staff members, censorship, leaving strategic agency positions vacant, and new restraints on rule-making are all meant to paralyze the EPA beyond repair. Having taken his country out of the Paris Agreement and thereby freed it from any of his predecessor's commitments to reduce GHG emissions, Trump has instead sought to revive coal production and eased access to oil reserves for their exploitation.

Curious here is that, with the possible exception of "Coal Country" along the Appalachian mountain range from Alabama to Pennsylvania, this radical climate-policy reversal under Trump lacks broad-based political support. Not even the mainstream of the oil and gas executives or among the owners of mining companies, all of whom stand to benefit in the short run from Trump's encouragement of fossil fuels at home, would go as far as he has gone in obliterating any shred of governmental concern for the climate. A majority of Americans (seven out of ten) accept that climate change is occurring even though only about half believe this phenomenon is primarily caused by human activity. Most would not want to sacrifice much to combat that problem, with half not willing to pay more than one dollar per month on their electricity bills to address this issue.[6] Much of that lack of enthusiasm is rooted in fairly widespread ignorance about the precise nature of the problem, which leaves many Americans currently unaware of the changing climate's possible consequences and ill prepared to contemplate available solutions. Public sentiment is likely to firm up as the implications of the problem become clearer and discussions around those intensify.

Trump's rejectionist stance has already met large-scale opposition, with many states, cities, and businesses in the USA reaffirming their commitment to climate-change mitigation. Their stance goes beyond the purely political, framed instead often as one designed to remain competitive when the rest of the world surges ahead in the key areas of renewable energy, smart grids and other dimensions of energy efficiency, electric vehicles, public transportation, building designs, and urban planning. At the center of this "green" American coalition is California, which on its own would be the sixth largest economy of the world. While this crucial state has been in the forefront on ecological issues ever since the start of

the environmentalist movement in the late 1960s, it has greatly intensified its climate-action plan since Jerry Brown regained the governorship
in 2011. Governor Brown has even gone so far as entering into international agreements with China and the European Union with regard
to joint pursuit of strategies for emission reductions. And he is planning
a Climate Action Summit in San Francisco for September 2018 around
which to mobilize a broad US coalition in favor of remaining committed
to the Paris Agreement.

California intends to rely for a good portion of its GHG emission
cuts—up to a third of the total—on trading of emissions permits.
To that effect it has recently (July 2017) strengthened its existing
cap-and-trade program and extended its duration for another decade
until 2030. A year earlier, in September 2016, California's legislature
passed sweeping climate legislation which significantly intensified the
state's efforts to reclaim its traditional position as an environmental
leader. To begin with, it replaced its previous goal of hitting 1990
emission levels by 2020 with a much more ambitious 40% cut in GHG
emissions below 1990 levels by 2030. That is one of the world's steepest
reduction paths over the next decade and a half, thus turning Jerry
Brown's California into a true bellwether. To have a chance of slashing
GHG emissions to such a degree the state will have to push several
strategies aggressively at the same time. It will aim for much greater use
of renewables so that 60% of its electricity will come from such clean-
energy sources as solar, wind, biomass, renewable gas, or hydroelectricity
by 2030 and 100% by 2045. Current levels exceed already 30%. But
reaching so ambitious a clean-energy goal also necessitates greater energy
storage capacity and installation of an all-renewable electricity grid, both
rather expensive undertakings that might trigger possibly unpopular
hikes in electricity bills in years to come. Especially interesting here is
the push for bio-methane (also known as "renewable natural gas" or
"biogas") drawn from organic waste materials which is pipeline ready
and has the potential of transforming waste management.

Californians are famously dependent on and wedded to their cars.
The state has been able to set more stringent fuel-efficient standards
for cars than the rest of America, thanks to a waiver from the Clean Air
Act of 1970, which a number of other states follow as well. Recently
the California Air Resources Board (CARB), the state's clean-air regulator, has implemented much higher fuel-efficiency standards for cars
and trucks for 2017–2025 models, targeting an average fuel economy

of 54.5 miles per gallon for 2025, just when the Trump Administration is going in the opposite direction of relaxing such standards on a federal level. Under its Advanced Clean Cars program CARB also wants a quarter of the state's cars to be zero-emissions vehicles by 2025, whether electric, plug-in hybrid or hydrogen fuel cell. In support of this ambitious objective there are significant subsidies of up to $7500 per car. And car manufacturers gain credits, a kind of regulatory currency, for each ZEV they can sell. These incentives notwithstanding, demand for such ZEVs is insufficient and will continue to be so as long as the infrastructure needed for such cars (e.g., charging stations) is not adequately available. Besides focusing on energy and cars, California is also aiming its regulatory reach of climate policy at its cities and huge agriculture sector. Developers will have to abide by new zoning rules seeking to reduce urban sprawl and instead opt for denser communities connected to mass transit. And emissions of methane will have to be slashed from landfills and dairy production.[7]

The 2016 climate legislation discussed in the preceding paragraph also included passage of the Assembly Bill 197 which made significant changes to the governance dimension of climate policy. Previous governor Arnold Schwarzenegger's Global Warming Solutions Act of 2006 introduced a cap-and-trade scheme which CARB was allowed to organize and administer in highly flexible fashion without much political oversight by the state's legislature. The AB 197 provision changed that situation by adding political representatives from both Senate and Assembly to the leadership structure of CARB and shortening the terms of the board members which gives politicians an added tool of control when they have a vote whether or not to approve CARB's board members. Much of the motivation behind the increased political control over CARB came from pressure by environmental-justice activists who dislike the cap-and-trade approach for giving polluters too much leeway to continue emitting greenhouse gases provided they pay the price. An added concern was that the current decade-old structure of California's climate policy too often ignored the priorities of lower-income communities where environmental degradation is typically concentrated the most. For both of those reasons, AB 197 also directed CARB to prioritize (and impose) direct emissions reductions on large polluters (power stations, manufacturing plants, etc.) in such a way as to "protect the interests of the state's most impacted and disadvantaged communities" and "consider the social costs of the emissions of greenhouse gases."[8] This change from a market-oriented strategy for climate-change mitigation to a regulatory-directive approach, which we see unfold here, demonstrates a key point concerning the formulation of climate policy: It is an intensely political subject touching on governance issues and as such a work in progress. We will surely see this point play out everywhere over many years as the world tries to figure out how to deal with a global challenge materializing locally in every corner of the planet, affecting all the nooks and crannies of our economy.

While it is obviously encouraging to see California and other states with a Democratic majority (e.g., New York) charge ahead on the climate, their collective action does not compensate for the absence of national leadership under Trump. Climate policy has to be affected at all the levels, from the individual household or firm to local community all the way to the federal level, especially when it concerns the largest economy and greatest per capita emitter of greenhouse gases in the world. Not having the federal government engaged in climate policy at all leaves a vacuum that cannot be filled by increased action at lower levels like state or municipality. There are things that only the federal government can do, starting with enforcing minimum standards of action or performance across the entire nation and designing a nationwide policy that applies everywhere. Even if we can agree that Trump's radical stance of climate denial is bound to be just a temporary aberration likely reversed when he leaves office, another zag in the American story of political zigzags sequencing opposite presidential profiles (Bush Jr -> Obama -> Trump), the damage is already done. Effective and gradually intensifying climate policy depends not least on having an interactive clustering of highly capable scientists, urban planners, regulators, engineers, and auditors working together in interdisciplinary fashion across many different governmental layers and institutional boundaries while motivating businesses and households to change. By silencing the scientists and pushing out civil servants Trump and his "hatchet job" appointees will have destroyed the needed collective of talent to leave the US government weakened for a long time. The damage extends to the quality of the regulatory agencies, above all the EPA. The US government lacks the infrastructure other national governments take for granted, such as having a powerful Ministry (or Department) of the Environment coordinating national policy on environmental matters. Doing so much damage to its relatively limited institutional setup concerned with climate change is hence likely to have a disproportionately grave impact on policy-making capacity.[9]

More generally, the US government is not set up adequately to put effective climate policy in place. It lacks a sensible, pragmatic, yet reform-oriented political center capable of passing effective legislation. Obama found out how much the absence of sufficient support in Congress can hurt when he could not pass his legislative proposal for a nationwide cap-and-trade market even though the Democrats had a majority in both chambers of Congress during his first two years. Most laws affecting the environment take a supermajority of sixty (out of a hundred) votes to pass in the Senate, and that is extremely difficult to get. Obama tried to pass key elements of climate policy through budget resolutions (i.e., his stimulus package of 2009) requiring only simple majorities, such as subsidies for solar panels or funds for high-speed trains along the nation's most traveled corridors on either coast. But here he had to limit the amounts available in order to make space for other Congressional priorities, notably tax cuts and a variety of pet projects pushed by key legislators as a price for their support. Key policy initiatives made by executive action or through the regulatory process were subject to lawsuits by hostile lobbies or right-wing state governors. All this combines to make even relatively limited policy initiatives slow, time consuming, and costly to implement while they can be destroyed or blocked much more easily, a devastating asymmetry greatly constraining US climate policy. Add to this the fact that America's participation in legally binding international treaties requires ratification by the Senate which in its current composition is nigh impossible to get done, hence restricting greatly what the international community can do in coordinated fashion about climate change. Under these conditions, the USA cannot hope to be a global leader when the time comes, all the valiant efforts of daring governors like Jerry Brown notwithstanding. Leadership will move elsewhere, a process of passing on the baton which seems already under way.

CHINA AND FRANCE: TWO EXAMPLES OF CLIMATE-POLICY LEADERSHIP

Whereas California provides a good counterweight to Trump's reactionary policies and a model for future policy reversals on the federal level, other countries are already charging ahead to set new standards in climate policy for the rest to follow. Take, for example, China, the world's most populous nation and largest GHG emitter. Long resistant to any restrictions which could undermine the super-rapid growth enjoyed by the country's economy for two decades (1994–2014), China's leaders also had until recently rejected climate-related constraints on the grounds that rich countries never had to restrain themselves during a century and a half of unfettered industrialization. But then their attitude toward climate began to change a decade ago (in 2007) when they realized how vulnerable several of China's megacities along the coast might be amidst rising sea levels or how much more fragile the country's already stressed water supplies could become. But the most convincing reason to change its policy posture was the unbearable smog settling over China's cities for many weeks each year in the wake of the country's huge expansion of coal consumption. Besides becoming an increasingly thorny political problem as frustration spreads among its city dwellers, such air pollution posits a major health challenge, hampers productivity, and hurts tourism. China's current leader Xi Jinping has clearly understood air pollution to merit being treated as a political priority which happens to fuse neatly with a more activist stance toward climate change in his quest for global leadership.

While China has a vibrant capitalist economy, albeit one still dominated by large state enterprises in key sectors, its politics are still shaped by near-total control of the seven-member Politburo Standing Committee of the Communist Party of China. This centralized regime, while lacking basic democratic representation and depriving citizens of many individual rights, renders possible highly directive policy-making which China's leaders have from time to time made good use of to move their country forward by means of targeted and well-implemented reforms. China's War on Pollution, and its fusion with climate policy, exemplifies this centralized, all-encompassing policy formulation capacity of China's leaders for all the obvious reasons. China's climate policy involves both the state apparatus (e.g., key ministries) and the party machine. It includes the most important state enterprises and representatives of strategic industries. It makes extensive use of the country's top universities and research institutes in a finely tuned division of tasks for which the scientists and academics are given sufficient funding to yield productive results. There is an ongoing effort to integrate the steps needed for the transition to a low-carbon economy with the country's five-year planning exercise. The center's policy decisions are transmitted down to the lower levels of provincial as well as municipal governments. And the fight against climate change has been placed under the auspices of the powerful macroeconomic management agency known as the National Development and Reform Commission (NDRC) which formulates China's policies for economic and social development. All this proves the importance which China's current (Xi-Li) leadership has afforded climate-change mitigation by means of a multifaceted approach.

Already a world leader in clean coal technology, China has decreased coal consumption each year since 2013. Weaning itself gradually away from this dirty source of energy, China's current five-year plan (2016–2020) caps coal's share at 58% of the nation's total primary energy consumption by 2020, bans new coal-fired power plants until at least 2018, and phases out existing coal-fired power plants wherever they can be easily replaced with gas power plants (e.g., as in Beijing in 2017). At the same time, China has made a huge investment in the promotion of solar and wind power, becoming a global leader in the production of both while targeting a 20% share for renewables in total energy by 2030 and aiming to install an amazing 800 gigawatts of non-fossil capacity by 2020 (including 340 GW in new hydropower capacity and 58 GW in nuclear capacity amidst hopes of becoming a global leader in nuclear fission technology thanks to experimenting with fast neutron reactors). It is also aiming at a 10% share for gas by 2020, in line with its scaling back of coal. By 2030, half of its electricity generation should come from renewables, but this also presumes much better grid integration to reduce China's high rate of curtailment wasting a lot of that wind- or solar-powered electricity generation. Since China's NDC has been framed in terms of improved energy efficiency rather than outright emission cuts, it stands to reason that policy-makers there have launched aggressive initiatives to motivate large energy savings among its leading firms and imposed more stringent efficiency standards for appliances, buildings, and cars. With its rapidly rising middle class switching massively to cars, China wants to become a global leader in electric vehicles and so propel some of its hitherto strictly domestic car manufacturers into international markets (e.g., BAIC Group, BYD, Geely). This includes also a concerted industrial-policy push to become a global leader in lithium-ion battery production. While it has already the largest EV market in the world, it is still not clear what will happen to demand once state subsidies for this type of vehicle will have been permanently removed. There are some initiatives in China to lower non-CO_2 emissions, notably HFC and other F-gases used in refrigeration and air conditioning. China also committed to a sizeable expansion of its forest stock volume by 4.5 billion cubic meters by 2030, compared to 2005 levels, as it seeks to complete its 2009 pledge to enlarge its forest cover by 40 million hectares over the coming decade. Finally, China has just launched a nationwide emissions-trading scheme in July 2017 which, even though less ambitious compared to what was originally planned, may still become the world's biggest carbon market soon.[10]

The largest emissions-trading scheme up to now has been that of the European Union, in place since 2005, which covers over 11,000 installations (factors, power stations, etc.) in 31 countries. More broadly speaking, the EU has been a committed pioneer in the fight against climate change for a quarter of century now. Its ambitious NDC targets reflect that—a 20% emissions cut from 1990 levels by 2020, a 40% cut by 2030, and an 80–95% cut by 2050 (as part of the EU's 2030 Climate & Energy Framework adopted in October 2014 in the run-up to the COP21 meeting in Paris). While the EU specifies targets for the entire 28-member bloc, it allows individual countries to adopt more ambitious goals and introduce their own policy initiatives in pursuit of those national objectives. France, the host country of the landmark Paris Agreement, has traditionally taken a leadership role in climate policy. Even though it makes up 4.2% of the world's GDP, it only emits 1.2% of all greenhouse gases—thanks not least to its successful nuclear power program which supplies the country with cheap electricity at low CO_2 emissions. Beyond that France sought already early on to reduce GHG emissions, managing a 7% cut from 1990 levels by 2007 under the Kyoto Protocol. Those efforts intensified during the presidency of François Hollande (2012–2017), culminating in the adoption of the Energy Transition for Green Growth Act (known as "la loi relative à la transition énergétique pour la croissance verte" or LTECV) in August 2015 whereby the center-left government sought to boost its environmental credentials just as it intensified its global diplomacy push for the COP21 negotiations leading to the Paris Agreement a few months later.[11]

The LTECV sets in motion a significant change in France's energy mix, capping nuclear energy at 50% of electricity production by 2025 while increasing the share of renewables to 40% by 2030 and reducing its consumption of fossil fuels by 30% over the same period. France also aims for major energy-efficiency gains, halving final energy consumption by 2050 (from 2012 levels). In order to help achieve this ambitious program, the government foresees a target carbon price of €56 per ton by 2020 and €100 by 2023—both comparatively high levels when considering what is going on today with regard to carbon pricing, as discussed further in Chapter 5. And the law also targets specific sectors for major improvements in energy efficiency—buildings and homes (with tax credits and interest-free loans for energy refurbishment projects), clean transport (with incentives for purchases of electric vehicles and installation of recharging points at homes, as well as ambitious fleet-renewal targets pertaining to local and national governments), a zero-waste goal through recycling waste products into materials while halving waste going into landfills by 2050, incentives for the use of alternative renewables such as biomass and marine energy, and promoting the installation of smart meters.

Key to successful implementation of the LTECV's ambitious goals is the so-called National Low-Carbon Strategy (SNBC) which the Hollande government adopted just at the opening of the COP21 Conference in Paris. This strategy centers on successive carbon budgets for the 2015–2028 period (2015–2018, 2019–2023, 2024–2028), which set GHG emissions caps for each period at progressively lower levels to achieve the aforementioned reduction targets by 2030.[12] These budgets are further broken down into sector-specific tranches, each one with a specific emission-reduction target by 2028 for industry (24%), agriculture (12%), energy and waste (33%), transport (29%), and housing (54%). The SNBC then provides a road map for each sector how to get there. Besides improving industry's energy efficiency (which is also a matter of competitiveness), the French have put strong emphasis on moving from the standard wasteful linear-economy model of "extract, produce, consume, discard" to what they consider a circular-economy model which seeks to promote recycling and reduce the amount of waste. As to managing waste better, the SNBC aims to limit food waste, promote eco-design giving special consideration to the environmental impacts of a product during its whole life cycle, fight planned obsolescence, improve waste recovery efforts, and push reuse. In 2012, France launched a highly ambitious and innovative agroecology project to make farming more sustainable, a project involving a good deal of education among farmers as well as research and development in order to promote organic farming, better crop diversification, alternative sources of fertilizers (e.g., biomass, manure, compost), and a host of other environmentally friendly practices. With regard to transport we have already mentioned the need to switch to "clean" vehicles (biofuel, electric, etc.). Also noteworthy is the push for more energy-efficient engines with a target of 2 liters per 100 kilometers. Finally, as concerns construction, we need to stress the widespread use of smart meters as well as France's unique leadership role in eco-design. The latter centers on a variety of techniques (insulation, lighting, heating, etc.) to accelerate the spread of low-energy buildings and even energy-plus buildings that can produce more energy thanks to micro-generation techniques using renewables than they import from external sources.

Another dimension of the LTECV is how it intends to advance the formation of climate finance, notably its much-discussed Section 173 which sets financial-disclosure rules for GHG emissions and pushes firms to give serious consideration to environmental, social, and governance (ESG) criteria beyond the short-term profit motive (see Chapter 6 for more details). But obviously the LTECV has many other provisions worth taking note of. Complementing its Section 173, the government has also introduced two certification labels for appropriate financial products, applying mostly to investment funds or asset managers—the IRS ("Socially Responsible Investment") label for those emphasizing ESG criteria, and the TEEC ("Energy and Ecological Transition for Climate") label denoting either financing of the "green economy" (i.e., projects with measurably positive environmental impact in such sectors as renewable energy, public transport, energy efficiency, waste management) or corporations reporting the "green share" of their activities above a threshold. Both of these labels will be awarded by designated labelling organizations approved by the French Accreditation Committee (COFRAC). Furthermore, the law provides additional funds to the state-run development bank Caisse des Dépôts for investment projects in the public interest supporting the low-carbon transition. The LTECV also boosted the scope of the Environment and Energy Management Agency (ADEME) with the help of broadening a tax on waste materials ("taxe générale sur les activités polluantes" or TGAP) as a channel of public financing of various SNBC initiatives such as the "circular economy" mentioned above. Finally, the French government reaffirmed its intention to fight fuel poverty by providing low-income families with a direct subsidy toward their electricity bills.

In 2017, a new government took office in France, led by Emmanuel Macron whose surprising election victory transcended the highly polarized left-right divide in French politics by creating for the first time in six decades a dominant political center. Macron's reformist path has included continued emphasis on France's leading role in the low-carbon transition as best illustrated by his nomination of legendary environmentalist activist Nicolas Hulot to head a newly reorganized Ministry for the Ecological and Inclusive Transition. It did not take long for Hulot to put his own imprint on the country's low-carbon transition strategy. In July 2017, he offered his so-called Climate Program ("Plan Climat"). Its key features consist of phasing out coal-fired power plants in France by 2022, banning the sale of gasoline and diesel cars in 2040, and targeting (net) carbon neutrality by 2050. He also announced significant increases in public funds against fuel poverty and, more generally, for low-carbon investments. There will also be a new program to give low-income consumers a subsidy if they replaced their pre-1997 diesel cars or their pre-2001 gasoline cars with cleaner cars. Finally, Hulot recommitted his government to pursuit of green diplomacy especially with regard to setting up the Green Climate Fund and putting into place Macron's June 2017 proposal for a Global Environmental Treaty ("pacte mondial pour l'environnement"), both under the auspices of the United Nations. That latter treaty would introduce legally binding rights and duties pertaining to environmental protection.

France's ambitious climate policy includes several sector-specific initiatives capable of serving as models for other nations to follow, notably the circular-economy concept dealing with waste and recycling, the eco-design concept taking account of a product's environmental impact over its entire life cycle, the transformation of farming practices toward sustainable agriculture implied by agroecology, or the emphasis on environmentally friendly buildings further refining France's already superb public-building architecture. All of these initiatives presume nationwide mobilization of relevant professions involved (which tend to have strong and centralized representation in France), with additional input solicited on dedicated online platforms and through regular meetings from other stakeholders such as local municipalities, non-governmental organizations representing citizen or consumer interests, research institutes (of which France has a comparatively large variety), and specialized enterprises. Such multilayered structures of engagement have been fostered specifically for environmental policy ever since launch of a national dialogue on "greening" France's economy known as "Grenelle d'environnement" in 2007 which brought together the government, local authorities, business leaders, trade unions, professional associations, and non-governmental organizations to formulate in consensual fashion public policy initiatives pertaining to ecological and sustainable development over five-year periods. The decision in 2012 to work toward a low-carbon transition law (the LTECV discussed above) and the Plan Hulot announced in 2017 are both results of these five-year policy horizon roundtables.

Key in the process is the ministry responsible for the environment, both because it mobilizes this national dialogue and then has to coordinate policy implementation with other affected ministries in what is typically a very politicized structure of competing governmental leaders. This is why it is crucially important who the environment minister is and where he or she ranks among ministers. With Segolène Royal during the second half of the Hollande Administration and now Nicolas Hulot in the Macron Administration climate policy has lately been in the hands of powerful, charismatic, and fairly competent political figures which has made a big difference in how this policy has taken shape as a centerpiece of both governments. But given the polarized nature of public policy debates, the infighting that typically prevails within French governments of both Left and Right, and the power of pressure groups or lobbies, it has been a long-standing practice for this rather strong and interventionist state apparatus to spawn lots of new government agencies responsible for the pragmatic execution of key policy aspects beyond the highly politicized body politic of the French State. And climate policy has been no exception in this regard, as evidenced by the Conseil National de la Transition Écologique (CNTE) mobilizing the national dialogue started in 2007 or the aforementioned Agence de l'Environnement et de la Maîtrise de l'Énergie (ADEME) putting in place the concertation needed among involved parties for effective implementation of public policy.[13]

A New Role for the State

It is ultimately not surprising to see China or France taking global leadership in developing comprehensive climate-policy strategies covering their entire national economy which is what will have to be done everywhere eventually. Both of these countries, each in their own unique way, have evolved as mixed economies combining a thriving private sector with a fairly large public sector shaped by a state apparatus inclined to be what the French like to call *dirigiste*—in other words, a system in which the state takes an actively directive role over investment rather than just a regulatory one. When we look at the challenges ahead in mitigating climate change or adapting to its impact, just in terms of what California, China, or France are all attempting to do, it becomes clear that the state will have an important role to play in redirecting resources, helping to launch new technologies, providing (dis)incentives to alter deep-rooted behavior, and defining targets to move the national economy through the different phases of the transition toward a low-carbon economy. Of course, the public sector has always been important in getting new energy sources off the ground and shaping a country's energy mix, and the push into renewables will be no exception in this regard. But the scope of state intervention will be much larger this time around, because we have to transform an entire economy in a relatively short period of time if we want to keep the cumulative temperature increase within tolerable limits. It is perhaps precisely this prospect of much greater state intervention on the horizon which prompts many Americans, so wedded to the idea that "markets know best" and "government cannot do anything right," to oppose even fairly modest climate-policy initiatives as they did under Obama.

Americans will have to get over it! There is no way around the likely prospect that the transition to a low-carbon economy will unfold with a significantly greater role for the government in the economy. The key question becomes how can that enhanced role be made to work well. Governments all over the world will have to insist soon on preparing the world for a phaseout of fossil fuels, a prospect made more manageable if and when those same governments will have succeeded in pushing renewables as alternative. Of all our climate-policy objectives this one is probably the most decisive, since already proven oil and gas reserves still in the ground would, if ever allowed to be used commercially in full, exceed the Paris Agreement's permissible carbon budget by a factor of five! And coal is even worse. There will also have to be lots of publicly funded infrastructure spending to provide for environmentally sound public transport such as high-speed trains, cleaner power plants, smarter electricity grids, and so forth. As we push for the Paris objective of net zero emissions by mid-century, we will need to rethink fundamentally how we move around, what kind of food we grow and how we wish to consume it, what it will take to build energy-efficient structures, how to make cities compact and sustainable in the face of environmental pressures, how best to recycle waste, and which industrial processes to target for cleaner alternatives. If we let the private sector figure out these overhauls on its own, we will get results that are too unevenly applied, may take too long to materialize, could possibly end up distorted by the profit motive with built-in biases favoring privileged parties, and depend too much on competition at the expense of due consideration of the public good. We thus have to involve the state to assure that all these process transformations do happen on a large enough scale, with sufficient speed, and in adequately balanced fashion so as to assure large societal benefits from such reforms. The state apparatus may then have to go beyond its supervisory capacity and regulatory force as source of funding or even as producer. This is especially so when it comes to research and development related to climate change.

In Chapter 2, we made already brief reference to the idea of launching a global decade-long $150 billion initiative, the so-called Global Apollo Programme, to fund research and development aimed at the promotion of renewables, including smart electricity grids with which to address the special integration and distribution challenges this kind of energy source poses. Unlike fossil fuels whose costs depend largely on fairly inelastic supplies of resources subject to often volatile prices, the cost structure of renewables is driven by technology. They can be rendered progressively cheaper by technological improvements, thus are bound to benefit from targeted research and development efforts. This potential has already been demonstrated by the spectacular cost reductions of solar or wind power over the last decade. Climate change mitigation and adaptation will, however, require a possibly much larger research and development effort as we seek radical transformation of transportation, agriculture, industrial processes, waste management, architecture, urban planning, and so forth. Such profound remaking of key dimensions of our economy, an inescapable necessity to get done quickly if we want to take the net zero-emissions goal seriously, does not just require lots of research in how to do things differently. It also involves new ways of diffusing thus gained knowledge. Rather than privatizing the gains from innovation by according innovators intellectual property rights, we need to provide open access to the results of basic or applied research so as to accelerate the development of climate-related mitigation or adaptation projects. We will need a global network of interdisciplinary research teams working together on the entire range of issues raised by a steadily warming planet and then propagating the results of their efforts to everybody likely to benefit from such new knowledge. Technology transfer to more vulnerable and/or poorer areas of the world will have to become standard practice while also fostering differentiation in response to local specificities.

In many countries, if not most, the public sector is already deeply developed in the formation of knowledge—setting up research laboratories in public universities and maintaining the pipeline (i.e., masters and doctoral programs) feeding those with scientists and researchers, publicly funded research grants, governmental contracts to nourish new applications to their take-off stage, etc. Much of this state-funded activity, involving basic research and early stages of applied research, would not be undertaken sufficiently by the private sector because of its innately uncertain outcome, with a high probability of failure, and also its distance to obvious commercial applications for profit. However, once such research activity yields sufficiently promising results to warrant further effort along its life cycle, profit-seeking companies get into the act seeking to exploit those commercially. There is thus an implied public–private partnership in research and development where the state socializes risks while the rewards get privatized. The innovation economist Mariana Mazzucato (2013) has argued that this "partnership" needs to be reframed for better results, notably when it comes to the impending green revolution. There needs to be greater recognition of the crucial role the public sector plays in the formation and diffusion of knowledge, instead of always attacking the state as this gigantic bureaucracy stifling initiative in order to make sure it stays as much confined as possible.[14] If we thus come to appreciate the entrepreneurial dimension of the state, we can rethink this partnership and involve the state more in the funding and development of applications, diffusion of knowledge to protect its public-good dimension, and reaping the benefits of innovation directly as shareholder for the good of society (e.g., using its share of innovation rewards to invest in better public schools). This may involve also setting up public development banks, such as Brazil's strategically placed National Bank for Economic and Social Development (BNDES), and/or public-sector innovation funds, such as the British Technology Group (BTG) in the 1980s and 1990s. In this context, it is also worth mentioning the work of the Schumpeterian long-wave theoretician specializing in technological revolutions Carlota Perez (2003) whose important concept of *techno-economic paradigm shift* applies very well to the notion of "green growth" simultaneously reorganizing business strategy and government policy around an expansive notion of sustainability.[15]

TRACKING CARBON

One of the many interesting insights of Perez's studies of technological revolution is the importance of new infrastructure extensions around which the new growth industries could cluster and the older sectors reorganize. This was true for the acceleration of industrialization in the late nineteenth century through electrification and the spread of railroads just as our current information revolution has been driven by the Internet. The upcoming "green" revolution will require, among other things, a new worldwide measurement infrastructure to track the emission and atmospheric flow of carbon as well as the other greenhouse gases. We are talking here about what is commonly referred to as the "carbon footprint." Of course, we would want to include other GHG gases as well whose volume estimates we have up to now typically translated into carbon dioxide equivalents for simpler aggregation.

We will not be able to enforce the tight carbon budget implied by the Paris Agreement's ambitious $\leq 2°$ target unless we have a global GHG measurement system in place that is complete and accurate. Currently we have technological means, such as eddy covariance systems combining ultrasonic anemometer and infrared gas analyzer, to measure GHG concentrations in the atmosphere via satellites and on top of towers. We can also apply optical remote sensing techniques to measure GHG plumes locally. But much of the emission data is still primarily derived in indirect fashion from estimates based on specific models for each major emission source—cars, household appliances, food, and so forth—into which you can plug your user data for an emission output measure. Those models are widely available on the Internet as CO_2 calculators. While those models have steadily improved as scientists in recent years have come to understand better the precise nature of GHG emissions per key source, they are still only estimates with tangible potential for significant error. Carbon is actually intrinsically difficult to track, because most of it comes from indirect sources removed from the producer or consumer of the good. We are now putting into place guidelines how companies ought to measure their GHG emissions by product, across the company, and along their entire supply chain—the so-called *Greenhouse Gas Protocol*.[16] This accounting and measurement system for GHG emissions, comprising different standards for a variety of GHG-related activities, has gained international recognition by multiple stakeholders involved. It offers us a framework within which we can track greenhouse gas emissions at the level of individual firms or assess the effectiveness of mitigation projects to lower emissions. Such a protocol is crucial if we want to have a reliably accurate, complete, and verifiable measurement system in place without which it makes no sense that a carbon budget be followed.

The GHG Protocol has subdivided emissions into three separate kinds as a way to avoid double-counting. So-called "Scope 1" emissions are "emissions from sources that are owned or controlled by the organization." Those "Direct GHG" arise from stationary combustion (natural gas, fuel oil, propane, etc.) of fossil fuels for comfort heating and other industrial applications, mobile combustion of fossil fuels used in the operation of motor vehicles and other forms of mobile transportation (e.g., gasoline, diesel), process emissions released during manufacturing activity in specific industrial sectors, and unintentional fugitive emissions from a variety of sources such as natural gas distribution or refrigerant systems. "Scope 2" emissions, also known as "Energy Indirect GHG," arise when consuming purchased electricity or other sources of energy generated upstream from the organization. Finally, "Scope 3" emissions, referred to alternatively as "Other Indirect GHG," occur along a corporation's entire supply chain as "emissions that are consequence of the operations of an organization, but are not directly owned or controlled by the organization." Upstream activities generating "Scope 3" emissions include production of purchased goods and services, waste generated from operations, capital goods, commuting by its workforce, business travel, and leased assets. Scope 3 emissions downstream may arise from third-party distribution and logistics, the use and disposal of sold products, franchises, or investments. Proper recognition of this last emissions category is crucial, since for most industries beyond energy or transportations it constitutes by far the largest source of GHG emissions—for instance, 90% for agribusinesses (68% upstream, 22% downstream), 94% for construction (42% upstream, 52% downstream), and 97% in the automobile industry (23% upstream, 74% downstream, mostly from the fuel consumption of cars sold).[17]

This discussion presumes that we can actually identify individual sources of greenhouse gases on a micro-level and then measure their emissions output accurately. We are really only at the very beginning of building such a capacity, relying still mostly on model estimations derived from indirect economic measures (e.g., number of vehicle miles traveled, tons of fossil fuels burned) and atmospheric observations from satellites, aircraft, or towers dotting the Earth's surface yielding macro-level measures of gas concentrations. Such organizations, as the US National Institute of Standards and Technology's (NIST) Greenhouse Gas Measurements Program or Britain's Centre for Carbon Measurement at the National Physical Laboratory (NPL), are developing direct GHG measurement techniques. Crucially important will be portable spectrometers using optical fiber technology which allow us to focus on local sources and conduct small-scale measurements directly at the source. There is also a lot of progress with infrared and laser-based spectrometry. We can imagine connecting such spectrometers to the Internet-of-things sensors for sequential emissions readings and then embed those in blockchains for verification by interested parties.[18] Within the next decade, we should have a global measurement infrastructure in place which will allow us to treat different greenhouse gases separately, distinguish between natural and anthropogenic emissions, identify local "hotspots" and then trace how these concentrated emissions disperse in the atmosphere, provide for both production-based and consumption-based GHG accounting, and match our macro-level inventory-cum-atmospheric-observations approach with micro-level measures of specific-source emissions we aggregate. This is going to be a major effort to construct, but we cannot avoid it!

The European Union, often a leader in climate policy worldwide, has recently taken a significant first step toward the eventual creation of what will have to be a worldwide, comprehensive carbon-tracking system with passage of its Monitoring, Reporting, and Verification (MRV) Regulation in 2015 for the maritime sector applying to ships in excess of 5000 gross tonnage (GT) using EU ports. That regulatory framework provides for EU approval of monitoring plans by shipping companies, followed by the detailed reporting of fuel consumption, CO_2 emissions, and transport work of individual ships, and concluding with the verification of the information by accredited independent third-party verifiers. Starting with the Bali Action Plan adopted at COP13 in 2007, the United Nations Framework Convention on Climate Change (UNFCCC) overseeing the implementation of international climate treaties like Kyoto and Paris has worked on implementing a comprehensive MRV framework with which to evaluate and validate member countries' national climate-related mitigation actions. Somewhere between the micro-level of the individual product (EV, ships) and the macro-level of national climate-policy commitments monitored by the United Nations we will have to develop a meso-level MRV infrastructure for firms, sectors, cities, and regions where the real battle against global warming will be fought by committed actors concretely able to do something about it. This is surely also going to be one of those areas where the necessary integration of different (measurement) technologies and (micro-meso-macro-meta-) levels of analysis stands to benefit greatly from rapid advances we are currently making with machine-learning artificial intelligence, big-data analysis, and the Internet of things. There will be an army of people in charge of collecting, analyzing, and interpreting the vast amount of MRV data as our collective push toward zero net emissions intensifies.

ILLUSTRATIONS OF THE NEEDED PARADIGM SHIFT

How important it is to measure emissions accurately can be illustrated by the example of the electric car (EV), the iconic symbol of the pending "green" revolution which climate policy-makers are pushing hard and car manufacturers are obliging to switch into. We assume simply without detailed verification that EV must be better in terms of climate change than traditional cars relying on internal combustion engines. And if you are just looking at tailpipe emissions for comparison, you would be correct. But EV use batteries containing rare-earth metals such as cobalt or lithium which are very carbon-intense and mined in just a few, often politically fraught areas of the world such as the Congo. As electric-car producers, such as Tesla, push for bigger EV with longer driving range to appeal to a larger customer base, the batteries get progressively bigger. And those large vehicles also end up needing more electricity, so that the aggregate environmental impact of EV depends then not least on what energy source powers the electricity in that particular region. If it is still mostly dirty fossil fuel, then the EV will be less environmentally friendly. What materials the car is made of and whether those can be recycled also makes a big difference. Regulators should therefore adopt a life-cycle approach which takes into account how the car gets produced, including the sourcing of rare materials going into the battery, the electricity needed to power the EV, and the recycling of its components. If they were to adopt such a holistic approach to their environmental-impact assessment, they would realize that in some regions small combustion-engine cars emit smaller amounts of carbon over their entire life cycle than would large electric vehicles.[19] This seemingly surprising conclusion goes to show how radically different the state's regulatory policy

will have to be in the age of climate change, because right now governments everywhere are pushing the phaseout of conventional cars in favor
of EV without any consideration of life-cycle emission aggregates.

Another imminent transformation pertaining to motor vehicles is the
self-driving car, tangible illustration of artificial intelligence's vast potential to transform the interaction between humans and machines. In step
with rapid recent advances in AI we have progressed to the point where
driverless cars actually do their job reliably so that we can expect their
takeoff in a few years. Many companies, including firms not known to
produce cars but in need of large mobility networks moving things or
people (e.g., Uber, Amazon, Google), are jumping into the fray which
indicates already now, at the very birth of this technology's life cycle,
how it will transform the car industry. It will not just do that, but more
importantly perhaps render possible an entirely different way of moving
about. Currently we have societies of car owners using fuel-driven vehicles to drive to work, usually alone, everyone pretty much at the same
time. We end up with huge traffic jams during rush hour, and the rest
of the time cars are just sitting around in vast parking lots dotting the
landscape everywhere. This is a wasteful and environmentally harmful
way of moving people. Fast-growing emerging-market economies just
mimic rich societies, going within a generation or two from bicycles via
mopeds and motorcycles to cars the size of which grows over time while
the number of people in them at any point declines. We are talking here
about a deeply rooted carbon-intense consumption pattern on a global
scale. But the imminent prospect of driverless electric cars allows us to
imagine having those self-driven cars pick up passengers and drop them
off in a continuous flow throughout the day, helped by large centers
where those cars can recharge their batteries, take a break for servicing,
and even pick up or drop off people opting for standard center-to-center
trips rather than individualized taxi-like trips. It makes perfect sense to
have fewer (non-polluting) cars more or less in constant motion and
recover very large amounts of public space now used up by parked cars.

Implementing such a new motorized people-moving system also
illustrates the importance of a point made earlier, transforming the state
into a progressive facilitator of GHG-saving innovation. Government-
run laboratories can help foster technological advances in the way cars
are built and operate, focusing in particular on transforming battery
technology away from reliance on rare metals, increasing usage of environmentally friendly, recyclable materials, and helping make small cars

safer and more comfortable. Discriminatory pricing in favor of those social-use cars and against privately owned cars will help anchor people's shifting preferences concerning how they travel to and from work. Local government can also impose much higher parking fees and congestion charges for entering city centers in individual-use cars. Zoning regulations and building standards can promote sensible design of those aforementioned car-and-people processing centers. It might also help to give people greater flexibility in terms of when and where they get their work done so that we no longer have rush hours where large numbers of people go to work or return home at the same time. Instead we end up with a constant flow of people which those driverless cars can pick up or drop off no matter what the hour of the day. We would also expect traffic regulation to be adjusted for the staggered flow of multi-passenger travel throughout the day, assuming people indeed have more flexible working hours and places. And, of course, the state can greatly facilitate recovering public spaces now used for parking to be redeployed for alternate uses.

How effectively we fight climate change will in the end depend not least on how willing and able firms are to make their GHG emissions a central focus of their operations and reorganize accordingly. Take, for illustrative example, Microsoft. Already in 2012 the world's leading software company, whose co-founder Bill Gates has long been an advocate for sustainable development goals and strong climate action, adopted a so-called Global Carbon Fee to help advance its goal of making its worldwide operations carbon neutral as soon as possible. This fee is paid for like a tax by all of its subsidiaries across the globe for their respective emissions and then collected in centralized fashion to fund investments in support of the company's carbon-neutrality goal. In the process, Microsoft has been able to transform its corporate culture so as to prioritize reducing its own carbon emissions, recycling e-waste, making its buildings more efficient, and promoting green power. It also funds carbon-offset community investments across all continents, including improved cookstoves, rainforest preservation, or bio-digesters for production of methane fuel and fertilizers, which compensate for GHG emissions the company cannot eliminate such as the air travel of its employees. Thanks to this pace-setting carbon-fee program Microsoft has over the last five years been able to eliminate 9.5 million metric tons of GHG emissions, purchase 14 billion kilowatt of green power (including even geothermal and biomass), and benefitted 7 million people with its carbon-offset community projects. This remarkable climate-policy model by one of America's leading high-tech companies points to what a sufficiently high carbon tax might be able to do for climate policy if properly priced, collected, and re-allocated, a key topic we shall return to later in greater detail in Chapter 5.[20]

Finally, we can envisage major initiatives across the planet aimed at radically redesigning and rebuilding cities. A forerunner of what may be possible in that area of climate policy emerged in October 2017 when Toronto, Canada's largest city and member of the C40 coalition of cities (see www.c40.org), announced an agreement with Alphabet's (i.e., Google's) urban-innovation platform Sidewalk Labs to develop a "smart city" parcel named Quayside on the currently empty east side of the harbor. Having just rendered this run-down neighborhood better secured by a $1.2 billion flood-protection initiative, the 12-acre pilot project should lead into a much larger waterfront project at a later date. Sidewalk Labs, a New York-based interdisciplinary incubator for new city-based technologies, has thus gained its first big project integrating its major innovations in architecture, construction, and urban planning to demonstrate how data-driven technology may improve the quality of inner-city life.

The area under development on Toronto's waterfront would ban private cars and instead rely on autonomous vehicles carrying people while industrial robots move trash and recyclables in underground tunnels. Traffic gets regulated by intelligent signal systems. A digital layer composed of data-collecting sensors and cameras equipped with artificial-intelligence capacity measures the movement of people, traffic patterns, noise levels, or air quality as well as the performance of trash bins and electricity. Included here would also be hyper-local weather sensors, designed to encourage more comfortable microclimates rendered possible by how building get spatially distributed in relation to each other (e.g., for better protection against wind). The collection of all that data, rendered widely accessible as a matter of good governance and to encourage further innovation, makes it possible to relax zoning restrictions and building codes whose objectives are now handled by data flows. The easing of these regulations in turn permits radical innovations in architecture and construction methods which Sidewalk Labs hopes to turn into a model for future urbanization, from modular buildings with easily rearranged interior-design arrangements and pop-up retail to allow rotation of stores to a much more varied mix of interconnected multi-use structures within essentially pedestrianized neighborhoods. Quayside's buildings would have design features and contain materials to make them cheaper and faster to build while offering an unprecedented degree of flexibility for rearrangement of space. The project will also emphasize energy-efficiency features which should lower per capita consumption of electricity substantially (estimated up to 95%) while constantly keeping track of energy consumption patterns.[21]

It remains to be seen whether a profit-driven company specializing in the collection and monetization of private data can build a livable urban space from scratch that feels like a city. But there is no doubt that Google's push into "smart city" technology, covering every facet of urban life integrated into a whole larger than the sum of its parts, marks a major leap forward in rethinking cities for the twenty-first century, and by extension also in preparation for climate change. The great innovation economist Carlota Perez, mentioned already in the preceding section, was quite right when she embedded the concept of technological revolution into the broader notion of techno-economic paradigm shift (see also Perez 2009) which implies a radical rethinking of how we do things, what our priorities are, and which tools we want to use for set tasks. From the few examples introduced in this chapter—the climate plans of outliers California, China, and France, the implications of major rethinking with regard to (electric, driverless) cars, Microsoft's push for carbon neutrality centered around a "global carbon fee," and Google's vision of a "smart city"—it should already become clear that climate change forces us to remake our economic system fundamentally from the bottom up and in the process rethink the socio-political-economic nexus within which we live and work. The question now is whether economists, to whom the great British economist John Maynard Keynes (1936) already ascribed potentially great power in terms of their influence on political leaders and the public discourse, are getting ready for this paradigm shift in the making.[22]

NOTES

1. The Secretariat, headquartered in Bonn (Germany), serves the Convention of the Parties as well as its associated bodies in implementing the provisions of the Paris Agreement and subsequent COP initiatives. Its (interim) Public Registry, detailing each country's NDC, is accessible at United Nations Framework Convention of Climate Change (2016a).

2. Analyses of how the NDCs differ from each other, and also what they
share in common, can be found in World Bank (2016) and United
Nations Framework Convention on Climate Change (2016b).
3. LULUCF stands for land use, land-use change, and forestry whereas
REDD+, which is the acronym for "reducing emissions from deforestation and forest degradation" with the "plus" sign implying emphasis on
the sustainability of forests, refers to a variety of agreements regulating
the management of forests.
4. See Climate Action Tracker (2016). On Ethiopia's green-economy strategy see also United Nations Development Programme (2011).
5. The highly reputed Pew Research Center, accessible at www.pewresearch.
org/topics/political-polarization/2017/, is a good place to go regularly for a sense of how deeply polarized American politics have become
amidst growing gulfs between Democratic and Republican voters on a
whole range of issues. For a good explanation of gerrymandering see C.
Ingraham (2015).
6. See A. Aton (2017) for a summary of recent surveys on Americans'
attitudes toward climate change and what to do about the problem.
Those surveys confirm fairly broad but also rather shallow support for climate-change mitigation measures as long as they cost individual households very little. Nearly two-thirds of those surveyed would not even
accept an additional $10 on their monthly electricity bills.
7. Summaries of Governor Brown's recent ambitious climate-policy initiatives can be found in C. Megerian and L. Dillon (2016) and H. Tabuchi
(2017).
8. The law in question specified "social costs" to mean "an estimate of the
economic damages, including, but not limited to, changes in net agricultural productivity; impacts to public health; climate adaptation impacts,
such as property damages from increased flood risk; and changes in
energy system costs, per metric ton of greenhouse gas emission per year."
See A. Carlson (2016) or C. Megerian and M. Mason (2016) for more
details on these implications of AB 197.
9. For more details on Trump's devastating strategy concerning the EPA see
S. Frostenson (2017).
10. China's climate action policies are recorded and discussed in the official government site *China Climate Change Info-Net* (http://en.ccchina.gov.cn).
11. For a summary of the French energy transition law see the official government sites *Energy Transition* (http://www.gouvernement.fr/en/
energy-transition) and *Adoption of the National Low-Carbon Strategy for
Climate* (http://www.gouvernement.fr/en/adoption-of-the-national-
low-carbon-strategy-for-climate). That ambitious law foresees €70 billion
in additional green-growth spending and creation of 100,000 new jobs.

12. While we have already discussed the idea of a finite carbon budget to burn through before achieving net carbon neutrality shortly after mid-century if we are to keep global warming to 2 °C, it stands to reason that such a strategy necessitates breaking down this aggregate into smaller budgets for countries to follow. France's strategy in this regard borrowed from the example of the UK whose Climate Change Act of 2008 introduced five-year carbon budgets on a downward path to meet that country's target of an 80% cut in all six GHG emissions by 2050 (from 1990 levels).

13. These governance issues pertaining to French environmental policy are well discussed in Organization of Economic Cooperation and Development (2016), and that holds true more generally for all OECD country reports on this particular dimension of public policy.

14. In a June 2013 TED talk (see www.ted.com) Professor Mazzucato pointed out that the "really cool stuff" making the iPhone smart were all funded by the government—the Internet and Siri by the Pentagon's Defense Advanced Research Projects Agency (DARPA), the Global Positioning System (GPS) by the US military's Navstar project of satellite networks, and the touchstone screen by grants from the Central Intelligence Agency (CIA) and the National Science Foundation (NSF) to a research team at the public University of Delaware.

15. For a closer look at Carlota Perez's fascinating historical and interdisciplinary approach to the interaction between technology and society regularly transforming each other I would recommend browsing her personal Website www.carlotaperez.org.

16. The Greenhouse Gas Protocol, introduced by the World Resources Institute (WRI) and the World Business Council for Sustainable Development (WBCSD) in 2001, has been expanded systematically ever since to include several standards (e.g., Corporate Standard, Project Protocol, Product Standard). In 2006, the International Organization for Standardization introduced the IOS 14064 standards for greenhouse gas accounting and verification. The two accounting systems are supposedly being made compatible by the three sponsoring organizations.

17. I got these sectoral Source 3 estimates from Alain Grandjean, head of Carbone4, a Paris-based company specializing in GHG emission measurements (see www.carbone4.com) which has done pioneering work identifying and estimating Source 3 emissions of various kinds. He also coined the phrase "carbon impact ratio" referred to below.

18. Blockchain (also called "distributed ledger") technology, which emerged first with the cryptocurrency Bitcoin, will revolutionize how networks authenticate specified interactions among them in decentralized fashion and thus transform the online operations of organized groups. See I. Kaminska (2017) for a foretaste of this technology's possible reach.

19. For more details on how small combustion-engine cars may compare favorably to large electric cars in terms of carbon emissions see the fascinating article by P. McGee (2017).
20. These figures are reported in the United Nations Framework Convention on Climate Change (2017) and also in A. Nikolova and T. Phung (2017).
21. See A. Bozikovic (2017) and A. Marshall (2017) for comprehensive discussions of the fascinating new urban-planning elements of Google's Quayside project in Toronto.
22. In the last paragraph of his *General Theory* Keynes (1936) wrote: "The ideas of economists and political philosophers, both when they are right and when they are wrong, are more powerful than is commonly understood. Indeed, the world is ruled by little else. Practical men, who believe themselves to be quite exempt from any intellectual influences, are usually slaves of some defunct economist."

References

Aton, A. (2017, October 3). Most Americans Want Climate Change Policies. *Scientific American.* https://www.scientificamerican.com/article/most-americans-want-climate-change-policies/.

Bozikovic, A. (2017, October 17). Google's Sidewalk Labs Signs Deal for 'Smart City' Makeover of Toronto's Waterfront. *The Globe and Mail.* https://www.wired.com/story/google-sidewalk-labs-toronto-quayside/. Accessed 15 Nov 2017.

Carlson, A. (2016, August 24). Does AB 197 Mean the End of Cap and Trade in California? *LegalPlanet.* http://legal-planet.org/2016/08/24/does-ab-197-mean-the-end-of-cap-and-trade-in-california/. Accessed 17 Aug 2017.

Climate Action Tracker. (2016). *Tracking (I)NDCs: Assessment of Mitigation Contributions to the Paris Agreement.* http://climateactiontracker.org/indcs.html. Accessed 4 Aug 2017.

Frostenson, S. (2017, April 4). We Knew Trump Wanted to Gut the EPA. A Leaked Plan Shows How It Would Be Done. *Vox.* https://www.vox.com/energy-and-environment/2017/4/4/15161156/new-budget-documents-trump-gut-epa. Accessed 11 Nov 2017.

Ingraham, C. (2015, March 1). How to Steal an Election: A Visual Guide. *Washington Post.* https://www.washingtonpost.com/news/wonk/wp/2015/03/01/this-is-the-best-explanation-of-gerrymandering-you-will-ever-see/?utm_term=.3263f0af3fcd. Accessed 8 Nov 2017.

Kaminska, I. (2017, June 14). Blockchain's Governance Paradox. *Financial Times.*

Keynes, J. M. (1936). *A General Theory of Employment, Interest and Money.* London: Macmillan.

Marshall, A. (2017, October 19). Alphabet Is Trying to Reinvent the City, Starting with Toronto. *Wired*. https://www.wired.com/story/google-sidewalk-labs-toronto-quayside/.

Mazzucato, M. (2013). *The Entrepreneurial State: Debunking Public vs. Private Sector Myths*. London: Anthem Press.

McGee, P. (2017, November 8). Electric Cars' Green Image Blackens Beneath the Bonnet. *Financial Times*.

Megerian, C., & Dillon, L. (2016, September 8). Gov. Brown Signs Sweeping Legislation to Combat Climate Change. *Los Angeles Times*.

Megerian, C., & Mason, M. (2016, August 22). 'An Exercise in Threading the Needle': Lawmakers Perform Balancing Act to Move Climate Legislation Forward. *Los Angeles Times*.

Nikolova, A., & Phung, T. (2017, February 19). Microsoft's Carbon Fee: Going Beyond Carbon Neutral. *Our Stories*. Yale Center for Business and the Environment. http://cbey.yale.edu/our-stories/microsoft's-carbon-fee-going-beyond-carbon-neutral. Accessed 10 May 2017.

Organization of Economic Development and Cooperation. (2016). *OECD Environmental Performance Reviews—France 2016*. Paris: OECD. http://dx.doi.org/10.1781/9789264252714-en. Accessed 16 Aug 2017.

Perez, C. (2003). *Technological Revolutions and Financial Capital: The Dynamics of Bubbles and Golden Ages*. Cheltenham, UK: Edward Elgar.

Perez, C. (2009). *Technological Revolutions and Techno-Economic Paradigms* (Working Papers in Technology Governance and Economic Dynamics, No. 20). The Other Canon Foundation. http://edu.hioa.no/pdf/technological_revolutions.pdf. Accessed 15 Nov 2017.

Tabuchi, H. (2017, March 24). California Upholds Auto Emissions Standards, Setting Up Face-Off with Trump. *New York Times*.

United Nations Development Programme. (2011). *Ethiopia's Climate-Resilient Green Economy*. http://www.undp.org/content/dam/ethiopia/docs/Ethiopia%20CRGE.pdf.

United Nations Framework Convention on Climate Change. (2016a). *The Interim NDC Registry*. http://unfccc.int/focus/ndc_registry/items/9433.php.

United Nations Framework Convention on Climate Change. (2016b). *Aggregate Effect of the Intended Nationally Determined Contributions: An Update*. http://unfccc.int/resource/docs/2016/cop22/eng/02.pdf.

United Nations Framework Convention on Climate Change. (2017, January 23). *Microsoft's Climate Fee—A Shining Example of Corporate Climate Action*. http://newsroom.unfccc.int/climate-action/how-microsofts-carbon-fee-is-driving-climate-action-forward/. Accessed 4 Oct 2017.

World Bank. (2016). *The NDC Platform: A Comprehensive Resource on National Climate Targets and Action*. http://www.worldbank.org/en/topic/climatechange/brief/the-ndc-platform-a-comprehensive-resource-on-national-climate-targets-and-action.

CHAPTER 4

Rethinking Growth

Looking at what economists might have to say about climate change, one of their major concerns in this context is bound to be economic growth. This connection gets typically framed around two different, yet obviously interrelated causalities. Will measures to mitigate the warming of the planet likely depress growth or boost it? And to what extent will the likely negative consequences of global warming end up undermining the growth capacity of countries damaged by those effects over the long haul? Depending how you answer these two questions will determine how much you will commit now to taking on the burden of climate-change mitigation as opposed to taking a wait-and-see attitude until you have a better idea of the environment's actual degradation and how much of an impact this will have on hampering growth. Unfortunately, this debate plays out against a background of recent historical trends which complicate matters greatly. On the one hand, we have had a leap of globalization which fostered rapid growth worldwide for a decade and a half (1992–2007), especially in many so-called emerging-market economies experiencing rapid urbanization on the back of export-led growth. This boom period moved the world onto a steeper growth path of GHG emissions, thus accelerating global warming. And then the world economy experienced a sudden and serious financial crisis in the late 2000s from which it has only slowly recovered, putting us on a slow-growth pattern which has only sharpened the disagreements over climate change between those worried about further hampering growth versus those viewing the climate challenge as a major opportunity to

© The Author(s) 2018
R. Guttmann, *Eco-Capitalism*,
https://doi.org/10.1007/978-3-319-92357-4_4

rekindle faster growth. Set against this polarized background, we may argue in accordance with the French Régulation Theory (see Chapter 2) that the challenge of climate change necessitates a profound structural transformation toward a different type of capitalism, one we may term *eco-capitalism*, in which the very nature of the growth process gets qualitatively transformed.

A decade after the Great Recession, a global systemic financial crisis of unprecedented virulence, we are still stuck worldwide in a slow-growth pattern which constrains policy-makers, feeds political instability, and makes any proposal containing significant up-front costs a non-starter. In this environment, it has proven very difficult to push for the kind of measures described in the preceding two chapters as regards our transition to a low-carbon economy. The initial costs involved, from carbon taxes and higher energy prices to massive amounts of stranded assets in strategically important sectors of our economy, are just too much to ask for. And so too many are holding off on the needed measures, hoping for better days when to start this difficult transition. In the meantime, we are losing very valuable time, not realizing how much our current inaction will make the challenge of this transition more difficult in the end. What we need here rather urgently is a new vision of growth that incorporates the measures required for climate-change mitigation while also addressing associated challenges of income inequality, job insecurity, excessive concentration of power, and uneven development—a politically attractive alternative to our current system which has left so many of us exhausted, dispirited, and existentially threatened. Otherwise, we not only run the risk of unacceptably long delays in action, but—even worse—face the prospect of letting a mounting populist-nationalist backlash move us actually in the wrong direction, as Brexit and Trump have already shown the potential of doing.

FROM ASSET BUBBLES TO SECULAR STAGNATION

The ongoing acceleration of global warming, with average worldwide temperatures having increased by about 0.75 °C over the last four decades and rising at a noticeably faster pace since 2000, is clearly related to a burst of greenhouse gas emissions from 1970 onward. This trend acceleration has been especially concentrated among two groups of countries, lower-middle-income countries whose GHG emissions in gigatonnes of CO_2 equivalents rose from 3.4 $GtCO_2$ in 1970 to 7.9 $GtCO_2$ in

2010, and upper-middle-income countries facing an even more impressive rise from 5.9 $GtCO_2$ in 1970 to 18.3 $GtCO_2$ in 2010. By contrast, neither low-income countries nor high-income countries saw any significant increase in emissions during that period (from 3.2 $GtCO_2$ to 3.4 $GtCO_2$ and 14.4 $GtCO_2$ to 18.7 $GtCO_2$, respectively). Much of those spectacular increases in both groups of middle-income countries came from energy and also industry.[1] In other words, the intensification of global warming in recent decades is directly related to changes in the world economy which have seen accelerating industrialization and rapidly rising energy output in a significant number of countries we either characterize as lower-middle-income countries (which the World Bank defines as those with a gross national income per capita between $1026 and $4035) or as higher-middle-income countries (with per capita GNI below $12,475). The former includes such populous nations as Bangladesh, Egypt, India, Indonesia, Nigeria, Pakistan, Philippines, Ukraine, and Vietnam; the latter Brazil, China, Iran, Malaysia, Mexico, Russia, South Africa, and Turkey.

Stepping back, we can see that our planet is now facing the environmental consequences of having brought three billion people into the world economy within the span of just one generation. In the decades following World War II, these countries mentioned above were more or less closed economies, either run by communist parties as centrally planned command economies (e.g., China, Russia, and Ukraine) or pursuing an import-substitution strategy of industrialization behind protectionist walls (e.g., Mexico, Brazil, and Indonesia). Either way, single-party states favored domestic monopolies as "national champions" while a large government sector also provided infrastructure, maintained income levels, and regulated prices. Initially capable of spurring accelerated industrialization and so foster satisfactory growth in the wake of high investment shares, this model was bound to exhaust itself thanks to built-in inefficiencies that became worse over time. Its stagnation intensified when the advanced capitalist economies emerged during the early 1980s from their decade-long stagflation crisis with renewed vigor, pushing technological advances, redeploying resources into knowledge-intense services, and benefitting from market-opening reforms espoused by a conservative counterrevolution known as "Reaganomics."[2]

We all know about the progressive disintegration of the Soviet Empire or the profound changes sought by the successors of Mao for the Chinese economy in the late 1980s and early 1990s. Less well remembered is the so-called *LDC Debt Crisis* of the 1980s during which many developing countries reneged on their sovereign debts owed to Western banks in the wake of which they had to undertake profound policy reforms in return for continued financial assistance. This "deal" was enforced by the International Monetary Fund's structural-adjustment programs whose acceptance by troubled debtor countries provided the "green light" for continued access to more bank loans and eventual restructuring of their foreign debts.[3] Of vital importance here were wide-ranging neo-liberal policy reforms pushed by the IMF, World Bank, and US Treasury (hence known as the *Washington Consensus*) which centered on fiscal austerity, privatization, and deregulation. While initially destabilizing and difficult to administer politically, these pro-market reforms eventually transformed those developing countries into "emerging-market economies" as they were increasingly integrated into the world economy through much-increased trade and cross-border investment flows. That process was further deepened in 1994, a year of crucial policy reforms in China and Brazil, and then again in the late 1990s following another major international debt and currency crisis starting in Eastern Asia during the summer of 1997 and moving via Russia all the way to Latin America a couple of years later.

Gradually, from the early 1990s onwards, these middle-income countries benefited from substantial capital flows from the US-dominated center to the periphery which were motivated by transferring manufacturing capacity to low-wage pools, outsourcing amid creation of global supply chains, capacity expansion in booming natural resource sectors, and global portfolio diversification of various investment pools (mutual funds, pension funds, and hedge funds) taking advantage of burgeoning local financial markets as well as a new worldwide market for sovereign bonds. One key aspect in the transition from being a developing country to becoming a emerging-market economy, besides offering attractively high interest rates on local financial investments, was the pegging of exchange rates at competitive levels. When such pegs came under attack in the great currency crises of the 1990s (see the breakdown of Europe's Exchange-Rate Mechanism in 1992/1993, Mexico's "Tequila" crisis of 1994, the "Asian" crisis of 1997–1998), the newly industrializing countries thus affected basically all decided to push accumulation of foreign-exchange reserves in order to shield themselves better against sudden reversals of capital flows and speculative attacks on their currencies. Those currency crises also served as useful trigger for downward adjustments in exchange rates to even more competitive levels. It is thus fair to say that emerging-market economies turned into such with the help of export-led growth strategies yielding large current-account surpluses that translated into steadily growing foreign-exchange reserve hoards. This neo-mercantilist strategy could achieve rapid growth rates even though domestic wages levels (and hence also consumption) were kept in check, thus making sure that billions of people in the Southern hemisphere entered the world economy first as cheap labor pools before turning into enthusiastic middle-class consumers with pent-up demand.[4]

For all of those economies to pursue current-account surpluses at the same time, in addition to the chronic ones driving the German and Japanese economies, someone had to accept large trade deficits in turn— and this surplus-absorbing deficit country was the USA. That counterbalancing process started in the early 1980s when Reagan's tax cuts and spending hikes yielded large US budget deficits, which in turn triggered widening US trade deficits amid larger capital imports and significant dollar appreciation. While the Reagan Administration tried to rebalance its twin deficits a bit through international (interest- and exchange-rate) coordination efforts during the second half of the 1980s, such adjustment could only go so far as the rest of the world was becoming more dependent on Americans absorbing its trade surpluses as the world's "buyers of the last resort."[5] And Americans could play this role because of the world-money status of their currency. With US dollars serving as principal international medium of exchange which countries use to settle their payments obligations with each other or keep their reserves in, the USA supplied these dollars to the rest of world by running balance-of-payments deficits that mobilized the needed net outflows of dollars from their country of issue into the hands of foreigners. To the extent that the rest of the world absorbed those dollars in international circulation for cross-border payments or as reserves, it automatically financed US deficits (as, for instance, when central banks parked their $-reserves in US Treasuries), and the USA in turn was able to borrow from other countries in its own currency. This is a profoundly privileged debtor position. Whenever its dollar-denominated foreign-debt instruments (e.g., Treasuries) held abroad come due, the USA can just create new dollars and issue new debt to replace the maturing claims.[6] The USA is thus in a position to run deficits like no other country, and in the absence of any external constraint, its economy can operate at a higher level of activity than the rest of the world subject to the normal financing constraints. Gradually, from the early 1990s to the late 2000s, the US economy was thus spending progressively more than it produced, with its current-account deficit in excess of 7% of GDP at the onset of the subprime crisis of 2007 absorbing in effect three-quarters of the rest of the world's surpluses.

Fueling such American excess spending and its international embedding in an explosion of trade, cross-border capital flows, and steadily rising commodity prices, thereby underpinning export-led growth in strategically important industrial nations and emerging-market economies, was the super-fast expansion of finance on a global scale. This evolution too emanated from the center, the USA, to the periphery, notably Europe and Eastern Asia. Following America's systematic deregulation of money and banking during the 1970s and early 1980s, financial institutions and markets took off from 1982 onwards to reshape the modus operandi of capitalism amid relentless financial globalization. The combination of deregulation, innovation, and computerization transformed finance, with (mutual, pension, and hedge) funds replacing traditional bank deposits as principal channel for savings and securities crowding out loans as primary form of credit. The result was a growing dependency by an ever-widening circle of economic actors—corporations, households, even governments, and of course financial institutions themselves—on debt financing while at the same time also making the pursuit of financial income sources—interest, dividends, capital gains, etc.—an increasingly important objective.

On the micro-level, this evolution translated into financial assets and liabilities becoming increasingly dominant in the balance sheets and motives of individual actors, a process which has been characterized as *financialization*.[7] On the macro-level, this same structural change toward the centrality of finance fostered a "debt economy," in which growth depended on credit and, tied to it, endogenous money creation for more or less automatic debt monetization, as well as a "bubble economy," which powered the economy forward at greater speed through recurrent asset bubbles and their requisite stimulative wealth effects. On the global level, three consecutive US bubbles—the stock-market mania of the "corporate raiders" 1984–1987, the dot-com euphoria of 1997–2000, and the housing boom of 2003–2007—provided the institutional context for capital gains fueling Americans' excess spending and so accommodating the rest of the world's push for exports while at the same time, supplying the latter with the financial resources to support thereby accelerated growth.

This US-centered, finance-mediated growth dynamic, signaling the upswing phase of a new long wave starting in the early 1980s and as such perhaps best characterized in Régulationist terms as a historically bound new accumulation regime I have elsewhere characterized as *finance-led capitalism* (see Guttmann 2016), also led other regions— Europe, Eastern Asia, Latin America—to rely more on debt, financial assets, and endogenous money creation for their own expansion around trade surpluses and rising commodity prices. Asset inflation and commodity inflation went hand in hand while prices of manufactured goods and increasingly tradable services (see Internet) remained stable or even fell. The rapid, finance-driven growth of this long-wave upswing, having become a worldwide phenomenon by the early 1990s, required extensive investments in expanding energy production not least to accommodate the accelerating industrialization of the emerging-market economies. Most of that growth in energy output relied on fossil fuels, specifically lots of coal in very populous economies (India, China), a remarkable surge in natural gas supplies, and obviously also oil now expanding into new areas of extraction (offshore, oil sands, and oil shale).[8] The newly industrializing countries of the Southern hemisphere have also witnessed rapid urbanization. Of the thirty-one megacities with more than 10 million inhabitants in 2016, twenty-four are located in the less developed regions of the "global South," nine of those in China or India alone. Of the ten additional urban agglomerations expected to become megacities by 2030, each one of them is located in developing countries. Both of these trends—rapid energy growth centered on fossil fuels and the massive expansion of cities—have contributed greatly to the increase in GHG emissions warming the planet.

This global growth pattern, driven by US deficit spending and export-led growth in much of the rest of the world, broke down when the bursting of America's housing bubble triggered a freeze in global money markets (following the collapse of Lehman Brothers in September 2008) and a deep recession from which the world economy has only recovered very slowly over the past decade. That systemic crisis of 2007–2008 basically forced an adjustment to lower chronic external imbalances and thus, by necessity, a one-time downward shift in the global level of economic activity which has grown more slowly and stayed below its pre-crisis potential ever since. Not only has the American propensity toward asset bubbles subsided, but the global commodity super-cycle has turned down to end up depressing prices of many raw materials or intermediate products which in turn left their producers at lower output levels. Whereas the G-20 countries had early on, in the London summit of April 2009, committed to maintain free trade, they ended up practicing a form of monetary protectionism instead. In November 2008, the Federal Reserve introduced an unorthodox monetary policy known as "quantitative easing," involving massive purchases of government bonds to flood the banking system with liquidity, which had the effect of pushing the dollar down and so help boost the competitive position of American producers both at home and abroad. This squeezed the euro-zone as the euro sharply appreciated to make euro-denominated products more expensive for everyone else while foreign products were rendered cheaper inside the euro-zone. That region then went into its own crisis in October 2009 which in turn triggered "quantitative easing" by the European Central Bank and massive depreciation of the euro. Apart from feeding back to the USA, the euro's depreciation hit above all China and other emerging-market economies. This "currency war," a phrase coined by Brazil's finance minister Guido Mantega in September 2010, was one of the key transmission mechanisms by which the US-initiated "subprime crisis" of 2007–2008 spread throughout all four corners of the globe. The world economy was thus subjected to three consecutive shocks—the subprime crisis of 2007–2008, the euro-zone crisis of 2009–2012, and the collapse of commodities prices in 2013–2014. Add to this widespread fiscal austerity by many governments who had seen the crisis dramatically push up their country's debt-to-GDP ratios, and the end result of all of these forces was systemically inadequate aggregate demand across most regions of the world.

With demand globally six to eight percent below the bubble-fed pre-crisis trend line for nearly a decade (2008–2017), it is not surprising that several well-reputed economists have characterized this lasting slow-growth scenario as "secular stagnation."[9] We should add here that such a pattern is quite typical in the aftermath of systemic crises signaling the turning point of long waves into their downswing phase, as we could also witness from 1873 to 1896, during the 1930s, and from the early 1970s on all the way into the early 1980s. In the most recent case, the global growth dynamic simply lost its two key engines—American asset bubbles and the intra-firm trade of global supply chains combining with a commodities boom to fuel the rapid expansion of emerging-market economies. Globally depressed aggregate demand has translated into significant excess capacities across a large number of sectors whose production capacity had been boosted greatly during the boom years of the 1990s and 2000s to levels that ultimately proved unsustainable. Such overproduction conditions, which depress margins, trigger plant closures, and drive out the weakest firms, have been intensified by technological changes, mostly automation of production and the Internet rendering a lot of traditional brick-and-mortar production methods obsolete.

These wrenching adjustments in many manufacturing and trade-dependent service sectors have made it difficult for millions of workers to get over the trauma of the systemic 2007–2008 crisis, as they have had to cope with persistent retrenchment pressures throughout the decade following that brutal downturn. Amid such uncertainty, there has been a political backlash among once-privileged blue-collar workers in industrial nations against those forces they perceive to have been most responsible for their plight—globalization, irresponsive elites looking only out for their own advantage, cheap foreign labor causing "unfair" trade, immigrants. That hostile reaction has already reshuffled the political scenery in a growing number of countries, crushing traditional party alignments especially concerning the political Left (e.g., America's Democrats, the Socialists in Spain or France, and Britain's Labor Party) amid an ethnocentric, nationalist backlash that has moved Britain to leave the European Union and led to the election of Donald Trump in the USA. One aspect of this rightward shift, part of its nostalgic factory fetishism, is an exaggerated allegiance to energy independence centered around traditional fossil-fuel sources and opposition to globally coordinated climate-change mitigation efforts which the anti-globalists view as an infringement on national sovereignty. To the extent that these angry working-class voters view scientific evidence with suspicion as just another expression of elite power, they may even reject evidence of a warming planet. Just when we need to mobilize majority opinion in favor of meaningful climate-change mitigation measures in the wake of the Paris Agreement, we find ourselves having to face a resurgence of climate doubters or deniers rejecting overwhelming evidence to the contrary.[10] Both the Trump Administration and the British cabinet formed after the Brexit referendum include many climate deniers, thus having rendered two crucial countries unable to lead the world in this struggle. This makes a new vision of economic growth, anchored in a rearticulated context of sustainable development, a matter of great urgency.

GOING BEYOND THE VISION OF A "STEADY STATE"

One way to elaborate a more climate-friendly vision of economic growth is to draw from the debates among economists concerned with the environment. In the face of growing consciousness of our economy using up non-renewable natural resources and damaging our environment (with pollution and waste), ecological economists such as Kenneth Boulding (1966), Nicholas Georgescu-Roegen (1971), and Fritz Schumacher (1973) began to look for alternatives to the standard neoclassical growth paradigm just when the environmental movement was about to take off in the late 1960s and early 1970s. Culminating in the notion of a "steady-state economy" put forth by Herman Daly (1980), this group of environmentally conscious economists pushed for limits on growth which they often argued by linking economic activity to the laws of physics.[11] This linkage, their argument goes, shows us how the transformation of resources into marketable goods absorbs finite resources and generates waste both of which impose physical limits on how far we can push this process. Technological improvements rendering such resource transformation more efficient help temporarily, but are ultimately bound to be outpaced by further growth and thus do not allow us in the end to escape the entropy that comes with the economy's expansion. We can identify many avenues of environmental degradation which those ecological economists highlighted even before climate change became an issue during the 1990s—depletion of non-renewable resources, excessive use of renewable resources threatening their reproduction (e.g., deforestation), pollution, the endangering of species, loss of biodiversity, human overpopulation, and urbanization. All these negative consequences, of which global warming is only the latest if not the most severe threat, are a counterweight to the material riches we seek to create and so need to be subtracted from that wealth. As they accumulate, their relative burdens increase. And their intensification ultimately puts into question the quest for faster growth. For instance, what were China's double-digit growth rates during much of the 1990s and 2000s worth to its urban population if this breakneck expansion left its cities' air so polluted that breathing became difficult there? Should we really allow the unrestrained expansion of coastal cities, as happens these days all over the world, if the destruction of wetlands and buildup of shoreline structures threaten to unleash highly damaging floods?

It is this balancing of economic growth and environmental degradation which has prompted Herman Daly and other ecological economists to call for the implementation of a "steady-state economy." This notion refers to maintaining a given stock of physical capital and population size, something viewed as desirable inasmuch as such constancy of our labor and capital resources alleviates the environmental stressors arising from economic activity. The classical political economists, from Adam Smith (1776) to David Ricardo (1817) and John Stuart Mill (1848), all viewed the capitalist economy eventually coming to rest in a steady state as the accumulation of capital would exhaust profitable investment opportunities when the decline in the rate of profit approached its zero bound. Later on, John Maynard Keynes (1930/1933) foresaw the saturation of capital accumulation over the coming century in similarly optimistic fashion as Mill had done, in terms of reaching such a state of abundance that the daily struggle for survival, which he termed the "economic problem," would give way to a life of leisure and cultural pursuits. Keynes (1936) welcomed in this context in particular the "euthanasia" of the rentier living off interest (or more generally purely financial) income, a prediction obviously contradicted by today's dominance of all sorts of capital income in finance-led capitalism.

The ecological argument in favor of a steady state is quite different. It derives not from abundance of capital, but in connection with the gradual erosion of the environment within which the economy is embedded. If the economy destroys that which nourishes it, such destruction needs to be stopped. Ecological economists, notably Herman Daly, thus call for an imposition of a steady-state economy by government action. His policy proposals included issue and selling of depletion quotas which restricted quantitatively the throughput of resource flows needed to maintain a stable stock of productive capital, population control so that birth rates would match death rates by distributing transferable reproduction licenses to all fertile women, and massive income redistribution with the help of caps on income and wealth. These ideas still reverberate in the climate debate today, notably among environmentalists favoring the slowdown of population growth as a crucial mitigation measure.[12] Daly (2015) himself continues to push for a zero-growth scenario which he justifies nowadays with the notion of a "full world." In the wake of decades of rapid global economic expansion, we have become sufficiently wealthy, but are at the same time also approaching the physical limits of our environment's qualitative degradation. Some of his colleagues are going even further, calling for "de-growth" as in the case of Peter Victor (2012). Such calls for a contraction of the economy are meant to counteract our propensity for overconsumption which these more radical ecological economists view as the primary cause behind both environmental degradation as well as social inequalities.

Daly's vision of a steady-state economy has been criticized as too restrictive. There are in essence two counterarguments to his pessimistic prescription. For one, it might be argued reasonably that over time, our economy is getting increasingly dematerialized as we move away from agriculture and manufacturing of goods to the provision of services which increasingly take place in cyberspace. Whereas the infrastructure underpinning the Internet and its breakneck expansion across the globe requires a lot of energy, the activities organized online do not. And those can increase exponentially with barely any additional costs to the environment. This trend toward the immaterial alters the physical relationship between the economy and environment by rendering production less energy-intense and hence less entropic overall. Based on past evidence, we can also argue with good reason that technological progress can overcome environmentally imposed limits to growth. That second objection to Daly's vision of a steady-state economy is a question of profound pertinence today in the face of climate change, as we have witnessed for instance over the past decade with the spectacular cheapening of wind and solar power making it possible to envisage the kind of radical remake of our energy mix we need.

We thus face a clash of visions between environmental pessimists and technological optimists. It does not help that limits-to-growth scenarios pushed by environmental economists are disliked to the extent that they imply stagnant incomes and inadequate job creation. They may even be less popular now that the global economy seemed to have entered a sustained period of stagnation as is typical in the aftermath of major systemic crises such as the one we experienced in 2007/2008. After almost a decade of slow growth, our concerns with job security and income erosion have intensified greatly, making us far less inclined to accept arguments in favor of more or less permanent stagnation. Much greater inequality in the distribution of income and wealth has fueled more frustration with the status quo, as prospects for the future and social mobility have deteriorated. A permanent zero-growth proposal is sure to be a difficult sell under those unfavorable conditions. Still, at the same time, we cannot just rely on the automaticity of technological progress which may or may not come about sufficiently or rapidly enough. Even if we can see already indications of technology's potential to help us cut down GHG emissions, we have no way of knowing for sure that these changes will go far enough to alter the very growth dynamic of our global economic system so much that it assures sustainable development into the foreseeable future. Yet this debate is too important for the future of our planet to be left unresolved. We need to address it by searching for a synthesis which puts concern for the environment, specifically rapid and sustained reductions in GHG emissions, at the center of our growth model going forward. We will have to ask ourselves frankly what it will take to make this goal feasible.

HARTWICK'S RULE AND "WEAK" SUSTAINABILITY

To make climate-change mitigation central to how our economy operates without accepting the environmentalists' zero- or slow-growth prescription as the only alternative path forward, we have to go back to the standard economic growth model and restore an appropriate role for the consideration of nature in it. Right now that role does not exist, to the extent that mainstream economists have ended up completely sidelining considerations of the environment in their growth models. This omission has its roots way back in Classical Political Economy of the nineteenth century which treated nature as land and then collapsed land into productive capital.

Today's standard growth models operate typically with just two factors of production—capital and labor. There used to be a third factor given much consideration, namely land. At the very inception of modern economics, during the eighteenth century when our pre-industrial societies were still dominated by agriculture, land was actually at the center of economic analysis. The French Physiocrats, for example, viewed land as the only source of surplus and hence crucial to the growth process. Classical economists, from Adam Smith (1776) to Jean-Baptiste Say (1803) and David Ricardo (1817), started downplaying the role of land by arguing that capital and labor too created value and thus contributed to a national economy's surplus as a source of growth. With regard to land, those thinkers were primarily concerned with the relation between the produce derived from land and the rent accruing to the landlord. But we already see in their writings a clear subjugation of land's productive contribution to that of capital, a bias that a century later led such economists as Alfred Marshall (1890) or Philip Wicksteed (1910) to argue that land should simply be considered part of capital. From there, it was only a small step to dropping land altogether as an independent factor of production and subsume it under the notion of capital, either in terms of the limited use of land by industrial firms putting up their structures (e.g., factories) for production or in terms of drawing raw materials from the land for further processing and profitable use (e.g., mining, agriculture, and lumber).

The progressive marginalization of land in economic analysis reflected both the diminishing share of agriculture in advanced capitalist economies and also how little land use came to matter as a cost factor to industrial firms. Once land became thus absorbed into the factor capital, it was also no longer considered a factor constrained by fixed supply. The use of land could be augmented by clearing it, by irrigation, through use of fertilizers, with the help of land reclamation projects if needed, and other techniques of exploration so that it would simply not constitute a barrier to growth. But when we look at land in such fashion solely as a productive resource, as part of capital which we are constantly seeking to augment, then we ignore that our economy is embedded in nature. Neoclassical growth models thus imply an economy separated from society and entirely disconnected from nature. In the age of a warming globe, this restrictive view of how our economy operates will no longer do. We need to set the economy back into its societal context and reconnect it with nature so that we can save it from its own self-destruction amid anthropogenic climate change.

It has been an important achievement of environmental economics to have revived the idea that humans depend on what nature has to provide. Toward that objective Fritz Schumacher (1973) introduced the crucial notion of "natural capital." Going beyond the mainstream economists' limited concepts of "land" and man-made "capital," Schumacher's "natural capital" recognizes the fact that non-human life produces essential resources such as a river supplying fish or an apple tree generating an edible fruit of high nutritional value. His concept refers to natural resource inputs (e.g., water and soil) and ecosystems services (e.g., decomposition of wastes and crop pollination) for economic production. Natural capital can be improved or degraded by human action, and there is growing agreement that ecological health is essential to how well our economy performs for its participants.

Neoclassical economists have integrated this notion of "natural capital" in terms of a trade-off with other types of (man-made) capital, be it human or physical ("productive") capital. They see the two as interchangeable substitutes for each other. In other words, natural capital is allowed to decline as long as this reduction is matched by sufficient increases in human or physical capital. Known as *Hartwick's Rule*, as formulated by John Hartwick (1977) and further refined by Robert Solow (1986), the key here is that such substitution of natural capital by man-made capital leaves the total amount of capital at least constant over time from one generation to the next.[13] We are thus allowed to deplete some natural resources as long as that process benefits human capital or physical capital commensurately. A pertinent example would be mining coal to produce electricity which in turn fosters activities, such as heating, refrigeration, or lighting, that improve our quality of life. Hartwick's Rule corresponds to a "weak" definition of sustainability, in contrast to a "strong sustainability" definition according to which "natural capital" and "human (or physical) capital" are complementary, but not interchangeable.

Neoclassical economists, for whom industrialization and rapid economic growth are overriding priorities, find the capital substitution logic underpinning the weak-sustainability concept irresistible, since it protects the primacy of private manufacturing capital and its profit-seeking activities even in the face of environmental degradation. While production-induced damage to the environment should be deducted from economic growth to get a more accurate picture of how much the economy has truly improved, mainstream economists usually fail to make such adjustments. Instead, they simply assume that economic growth is an end in itself, with its output and income gains automatically trickling down to improve our standard of living and assure sustainability. In reality, however, we cannot assume such automatic trickling down. If economic growth causes pollution undermining our health or congestion wasting a lot of our time stuck in traffic, then it may not improve our quality of life and/or may prove unsustainable over the long run. Hence, we need to treat the issues of economic growth, living standards, and sustainability as interrelated, but nonetheless separate objectives. They are arguably all of equal importance.

In the weak-sustainability framework of standard economic theory, such issues as pollution or traffic congestion are treated as externalities, inasmuch as the social costs arising from such spillover effects on third parties fall outside the private cost–benefit calculations of both producers and consumers causing these effects in the first place. Their market transaction simply ignores how others are impacted. Such unaccounted social costs borne by others reduce the total social welfare below the private benefits accruing to both sides in that transaction. To the extent that the market mechanism fails to consider social costs imposed on others, it will produce too much of that negative externality. Such a market failure needs correcting by obliging the private parties responsible for creating these negative spillover effects to "internalize" their costs. Other than having governments impose regulatory limits on negative externalities, these social costs could also be captured either through taxes or by means of well-defined property rights encouraging the parties involved to deal with the problems they cause others.[14] Adequately set taxes or correctly defined property rights should give producers sufficient incentives to develop resource-saving technologies and new capital substitutes. We have seen either one of these approaches emerge in the debate over climate change with proposals for a carbon tax and introduction of tradable pollution permits in so-called cap and trade schemes, both of which we shall discuss in the following Chapter 5.

Suffice it to say here that climate change cannot be reduced to an externality even though it is a form of pollution not taken into account by the producers and consumers of GHG-emitting goods and services. It is simply much too vast a problem, amounting in its scale to a systemic threat for the entire global economy. We have in effect a two-step reduction of the problem by mainstream economists. By subsuming nature under the notion of capital so as to allow its substitutability with human or physical capital, they effectively accept environmental degradation as the price to pay for increased wealth. In addition, economists reduce climate change to a problem of pollution that can be treated as a negative externality, proposing market-friendly solutions, like a market for emissions permits ("cap and trade"), which cannot do justice to the scale and complexity of this systemic threat to our planet's survivability. We see this double marginalization of climate change also inhabit the macroeconomic models economists use to predict growth and related features such as inflation, job creation, budget deficits, or a country's balance of payments.

MACROECONOMIC MODELING OF CLIMATE CHANGE

This is not the place to give the reader a primer in macroeconomic modeling. Suffice it to say that climate change itself and the measures we may take to deal with this issue will both have important consequences for the functioning of our economy. They thus deserve to be studied within the framework of macroeconomic models. To the extent that those models project forward on the basis of past relationships and trends, there is always the question how transformative developments, like a changing climate systemically altering the environment within which the economy is embedded, can be meaningfully factored into such models.

This question is especially challenging to address with regard to climate change because of the tremendous uncertainties associated with that process playing out over a very long time horizon. We are uncertain about the stock of greenhouse gases building up in the atmosphere, depending on our collective mitigation efforts aimed at lowering emissions. We are furthermore not sure by how much that stock, whatever it ends up being before getting eventually stabilized, will raise temperatures. It is not just a matter of the average temperature hike, but also of the temperature's variability in time and space around that mean. We know little about either, yet both are determinant in their respective impacts on the weather. We need to understand a lot more about the relationship between temperature trends, spatial as well as temporal deviation ranges concerning temperature, and weather patterns so as to prepare ourselves for a range of extreme weather events becoming a lot more common, notably droughts and storms. How do we factor extreme-weather events into our meteorological and macroeconomic models? And, of course, we also face a lot of uncertainty about the impact of climate change on levels and patterns of economic activity. While we can gradually improve our estimates of costs and benefits of mitigation measures as they unfold, we know far less in advance about how best to adapt to the consequences of climate change and what such adaptation will cost us. What can we model and how much are these models worth in the face of these four different sources of climate-related uncertainty? The answer lies in "trial and error" as well as "learning by doing." We have been doing exactly that so far. Our models keep getting better, but they are still quite far from being good enough.

Macroeconomic modeling of climate change entered the world's consciousness with somewhat of a bang in October 2006 when the UK government published the so-called Stern Review, named after environ-mental economist Nicholas Stern who chairs the Grantham Research Institute on Climate Change and the Environment at the London School of Economics.[15] This comprehensive review of the impact of cli-mate change on the world economy comes to the conclusion that cli-mate change represents an extremely serious threat to our collective well-being, through its varied and potentially severe impact on our envi-ronment, health, food supplies, and water resources, the worst of which could still be averted if decisive international collective action were taken at once. Stern aims to stabilize the GHG stock at 500–550 ppm CO_2e which would presumably translate into a temperature hike of about 3 °C. Getting to that level he estimated to cost 1% of global GDP, updated a couple of years later to 2% in response to belated realization how much climate change seemed to have accelerated in the late 2000s. By comparison, not doing anything might push temperatures up by 5–6 °C at which point the world economy might lose at a minimum 5% of its GDP permanently, if not a much greater loss (potentially reaching 20% of GDP). While the Stern Review has been criticized from many different angles, as part of a broader discussion it provoked, most critics of the report accept its basic conclusion about the urgency of mitigating action on a global scale and its relative cost efficiency compared to what happens if we fail to take such proactive action and continue instead with business as usual.

Stern used for his estimates an integrated assessment model (IAM) known as Policy Analysis of the Greenhouse Effect (PAGE2002) developed by Cambridge University Professor Christopher Hope (2006, 2013) who subsequently improved and updated it into the PAGE09 model yielding significantly higher estimates for climate-re-lated damage impact and mitigation benefits. IAMs are ready-made for the analysis of climate policy, since they synthesize diverse knowledge, data, methods and perspectives into an overarching framework to address complex problems crossing the boundaries of different disci-plines, in this case, biophysics and economics. There is a second major integrated assessment model widely used for estimating long-term climate-change impact and comparing it to the costs of mitigation, known as Dynamic Integrated model of Climate and the Economy (DICE) and developed by Yale University Professor William Nordhaus (1992, 2017) which has been refined and updated regularly since first introduced in 1989.

Apart from its interdisciplinary orientation, the IAM approach is advantageous inasmuch as its models are relatively simple and sufficiently flexible to look at several multilayered interactions between geophysical variables, such as GHG emissions or temperature trends, and economic variables, such as per capita GDP or population growth, from different angles while allowing new driving variables to enter the calculation. It can also factor in conditions of uncertainty rather well by running Monte Carlo simulations many times, based on random inputs (set within realistic limits, of course) so as to identify a range of possible outcomes in different scenarios and analyze the impact of risks on forecasting targeted variables in terms of probability distributions. But IAM has also some serious drawbacks. Its results depend to a very large degree on the quality of the data inputs, which necessitates continuous searching for improvements of those and then running the model again to upgrade. It is fair to say that IAM requires ongoing work to keep relevant. Its use of GDP equivalent to express the results of its cost-benefit analysis may be too narrow inasmuch as the actual damage from climate change is likely to go beyond just a loss of income and involve loss of life or irrecoverable destruction of the environment. Only directly observable impacts carrying market prices, which can be measured accurately, are entered reliably into the equations of the model. Non-market damage, as for instance on health, ecosystems, or biodiversity, is typically excluded because such impact cannot be reliably quantified in monetary terms. Nor can the IAM take into consideration cross-sectoral effects, as is bound to happen quite substantially in our quest for net carbon neutrality (shift in energy mix impacting the provision of electricity, greater reliance on trains squeezing the car industry, etc.). And finally, IAM also ignores cross-border spillovers affecting trade, financial flows, migration patterns, or so-called "socially contingent events" referring to social and political turmoil arising in the wake of climate change. This is arguably a lot of ignoring. Many aspects of climate-related damage cannot be properly captured by IAM modeling.[16]

Climate change has a deep and transformative impact on a limited number of sectors, notably energy (with a mix of alternative energy sources) and electricity (comprising various power plant technologies and electricity grid integration). Bottom-up sectoral models can provide lots of insights into how specific sectors can achieve successful transitions to low-carbon paths. These models are typically engineering-based rather than economy-based, and as such interested in identifying a menu

of technical choices or abatement opportunities based on precise spec-
ifications of capital equipment being used. Such modeling aims, in the
end, to measure how specific actions or technologies might impact car-
bon-emitting equipment, such as power plants or car engines. Their
focus is on justifying mitigation efforts in terms of the financial and economic benefits they yield in terms of developing new technologies and
adopting the best technologies available. Bottom-up models can be
usefully plugged into top-down macroeconomic models, such as IAM,
for higher-quality provision and better calibration of data while at the
same time providing an alternative measure of GHG emission-reduction
potential.[17]

Economists have also been using computable general equilibrium (CGE) models to estimate an economy's reaction to either climate-change damage or to mitigation policies aimed at lower GHG
emissions. We favor the CGE approach when we have good actual data
sets for the present to fit into the model's equation set but lack the kind
of long time-series data typically used in econometric forecasting models.
The beauty of the CGE approach is its focus on cross-sectoral interaction, using input-output tables or social-accounting matrices to trace the
interdependence of markets/sectors. Both climate change and climate
policy are intersectoral phenomena containing lots of spillover effects
worth identifying, hence fitting rather well for this kind of multisector
approach to macroeconomic modeling. Policy instruments to encourage
mitigation and adaptation, such as taxes and subsidies, directly change
relative prices across a large number of sectors. At the center of CGE
models are behavioral equations which are typically framed in terms of
elasticities initiated by a change in price(s). This is arguably a rather narrow way of framing the analysis of policy effects, but also one that can
be easily standardized. CGE models are comparative-static in nature,
comparing situations "before" and "after," say, a policy change. They
contain usually more variables than equations, making thus some of the
former exogenous. This is typically the case for technological change. In
reality, however, it would be more appropriate to look at climate-related
technological change as an endogenous process motivated either by climate-related damage altering relative prices and motivating the search
for technical solutions or by climate-policy inducing actors to change
their production methods and products. We should therefore push more
in the direction of endogenous-growth models which are able to capture the positive spillover effects of emission-reduction technologies in

terms of, say, less pollution or improved total factor productivity or the virtuous-cycle dynamic of greening the economy in the accelerating transition to a low- or zero-carbon (net) economy. Be that as it may, it might be precisely because of the need to trace cumulative processes—both in terms of climate-related damage and successful policy inducements—that we should go beyond the CGE's static approach and instead opt for a dynamic model, such as the dynamic stochastic general equilibrium (DSGE) model. These are more difficult to use, to the extent that all markets must clear at the same time, but have the advantage of taking explicit account of uncertainty.[18]

Another significant improvement in macroeconomic modeling techniques consists of using agent-based models (ABM) which address the crucial question of micro-foundations more meaningfully than either CGE or DSGE models have been able to. The latter base their macro-level variables on aggregating a "representative agent," be it a household or a firm, under the neoclassical premise that it is a rational optimizer. More realistic, however, is the presumption that in the face of such a high degree of uncertainty as posited by climate change individual actors opt instead for satisficing behavior aiming at a range of acceptable outcomes. Moreover, both the systemic nature of climate change as well as its multiple sources of uncertainty give rise to a much wider variety of behavioral propensities and expectations among actors. Both of these characteristics, satisficing behavior and heterogeneity among actors, are at the core of what ABM allows us to track. This type of computational microscale model replaces the mainstream models' representative agent with rule-based agents whose diverse, dynamic, and interdependent behavior, located in specific time and space within networks that tie them to others, can produce complex and non-linear patterns of interaction even if they are based on simple behavioral rules. Those agents may be differentiated by social status, pursue different interests as long as those can be justified as bounded rational, use heuristic or simple decision-making rules, and are capable of learning. Their social interactions can be simulated to result in herding, behavioral shifts, and all kinds of complex behavior which make ABM capable of capturing a lot more dynamic macro-level outcomes than the equilibrium-constrained IAM, CGE, and DSGE models tend to depict. Apart from the rapidly improving availability and quality of computational software for ABM, this approach has the advantage of also cutting across many disciplines of relevance for climate policy, with applications in biology, business, finance, technology, network theory, organizational theory, urban planning, and economics. Possible interdisciplinary integration using ABM may in the future vastly improve our understanding of systemic impacts on socioeconomic networks.[19]

In recent years, there have been increased efforts among Post-Keynesian economists, in their search for an alternative to mainstream equilibrium models, to merge agent-based models on the micro-level with a new macro-scale approach known as stock-flow consistent (SFC) models as first espoused by W. Godley and M. Lavoie (2007). While very promising in terms of providing a better integration of micro-foundations and macro-dynamics toward a more realistic depiction of capitalist economies in forward motion than standard CGE and DSGE models can offer, we are still only at the very beginning of this effort. An impetus for such AB-SFC integration has been the financial crisis of 2007/2008 which mainstream macroeconomists could not foresee because of their marginalization of finance linked to their treatment of money as an exogenous stock variable. The AB-SFC framework, by contrast, is built around the assumption of money's endogeneity from which follows the imperative of properly integrating the real and financial sides of the economy as an inseparably intertwined nexus of monetary circuits.[20]

Whereas the ABM part may tell us more realistically what is happening at the micro-level with groups of individual actors and yield realistic value parameters for data, the SFC model consists of a rigorous, self-enclosed, and complete accounting framework for all the stocks and flows of the economy in question. To the extent that we live in a fully monetized economy of cash flows, someone's assets are somebody else's liabilities just as much as someone's income is somebody else's expense. You can therefore depict any such economy in terms of sector-aggregated balance sheets whose interdependent stocks (assets and liabilities) get changed over time by the flows of funds between the groupings making up the structure of that economy. The SFC model accomplishes this task by starting out with aggregate balance sheets of initial stocks, adding a matrix of transaction flows between the different sectors, and then integrating those intersectoral flows for revaluation of initial stocks (in a revaluation matrix). You thus get to show an economy in motion over distinct time intervals, with all of its fund transfers covered. How it moves depends on a set of behavioral assumptions applied to the various sectors we want to study in greater detail as well as the parameter values we put into the model for calibration.[21]

There are many attractive features to SFC models. They cover the entire system in a logically consistent and comprehensive manner, with all funds going somewhere or coming from somewhere clearly identified. There are no unidentified parts of the system left unexplained, and this makes SFC models more complete than their various neoclassical counterparts (at the expense of typically requiring lots of equations). SFC models are also intrinsically very flexible and can therefore be used for lots of different purposes and situations. As long as there are distinct flows between the different sectors to trace, we can basically design any configuration of sectors and transactions by means of simple balance-sheet aggregation and identification of intersectoral flows. The economy in SFC modeling can be closed or open vis-à-vis the rest of the world, can have a financial sector of whatever complexity we deem desirable, can be run under different behavioral assumptions yielding distinct response patterns for comparison, and can be made to play out over different time horizons. This type of model is especially useful in predicting a system's adjustment dynamic to a shock or in identifying the buildup of ultimately unsustainable balances. SFC models can even predict the timing of a tipping point when the growth dynamic under investigation shifts to a new pattern. With all of these advantages, we expect SFC models to be especially helpful in addressing issues related to the transition to a low-carbon economy, and their applicability may further improve when wedded with agent-based models depicting different behavioral responses or biases among heterodox agents (e.g., three different scenarios how oil companies deal with reserves and their valuation in stock prices). We have already seen AB-SFC model's potential for fairly accurate predictions.[22]

MEASUREMENT CHALLENGES

No matter how much we can and will improve macroeconomic modeling of climate change, the usefulness of all these models ultimately depends on how good the data are that we are feeding them with. And here we are facing endemic biases distorting our results in favor of too little action too late. It is, as we have noted already, highly problematic to restrict your cost estimates just to income or output losses. Climate change goes far beyond that inasmuch as it may cost lives, destroy natural capital, and damage the environment in permanent fashion—irreplaceable losses that cannot be measured in money. Just because they do not carry monetary values, the models underestimate, if not altogether ignore, these losses. Yet the point is that these losses are too big to be given monetary values, not too small.

Another measurement challenge arises from the fact that, no matter how "dynamic" the models are, they are never going to capture all the interactive feedback loops associated with climate change. Some of these—the melting of the polar ice caps, the thawing of the tundra, the rise of the oceans' sea levels—are bound to accelerate global warming trends in exponential fashion. There is thus also a cost of inaction, of waiting too long to address the issue meaningfully until it is too late to avoid tipping points where our meteorological–ecological system shifts irreversibly into a whole different gear. What we consider today to be extreme-weather events will at that point have become the new normal. None of the macroeconomic modeling techniques used today are ready to factor in the normalization of such extreme weather which remains marginalized at the tail ends of meteorological probability distributions for temperature, precipitation, or wind. Yet just in the summer of 2017, we experienced three consecutive hurricanes in the Caribbean and Gulf of Mexico each of which broke extreme-rain or extreme-wind records on its own. It is high time for us to pay much closer attention to such extreme-weather events, extending to droughts and wildfires, all of which have already greatly intensified in frequency, strength, and duration even though the cumulative temperature rise so far has been only 1 °C. Imagine what the world will be like when the average temperature will have risen 3 °C (Nicholas Stern's goal) or 5–6 °C (the "business as usual" scenario)!

Standard macroeconomic models of climate change also tend to underestimate the benefits of effective measures aimed at lowering greenhouse gas emissions to limit the process. While such mitigation entails significant up-front costs, the ensuing transition to a low-carbon economy carries two hugely advantageous sources of positive spillovers both of which the models tend to downplay, if not altogether ignore. One is that climate-change mitigation cuts down on pollution, improves how and what we eat, enhances transportation, and generally is bound to make our environment better. The other arises from mitigation-induced technological change yielding efficiency gains, not least from various scale, scope, and network economies, across the entire industrial matrix. We can expect not only a lot of new applications, but also widespread improvement in how we run our economy, from much better energy efficiency via sharp reductions in waste to higher-quality and more durable consumer goods. Remaking our economy into a low-carbon system is the equivalent of a full-scale industrial revolution, bound to generate cumulative and dynamic gains across the board.

Another challenge concerns the appropriate discount rate to be used. The benefits of mitigation typically arise later, possibly far into the future, certainly after we have incurred the costs of putting those measures into effect. We therefore have to discount those future benefits back to present value in order to make them comparable with the costs. The higher the discount rate, the more future benefits get discounted so that they end up less valuable today. Even relatively small variations in the discount rate can have a big impact on the aggregate present value of future benefits, especially when those stretch over a long period or arise far into the future. These, unfortunately, are precisely the kinds of projects the struggle against global warming entails—long-term and stretching far into the future. What's worse, those kinds of projects carry above-average risks quite apart from their long time horizons, because of the uncertainties associated with climate change while also often enough involving new and untested technologies. Higher risks have to be taken account of by adjusting the (risk-weighted) discount rate upward appropriately. If we hence apply the standard rules of financial economics to climate-related investment projects so as to end up with above-average discount rates of, say, $\geq 5\%$, we will discard the majority of mitigation projects as not worth their while. We cannot follow such standard practice here, but instead must opt for a much lower discount rate when it comes to climate change.[23] This is an ethical question inasmuch as it is ultimately a matter of how much we value the lives of our children and grandchildren relative to our own. If we want to be fair to future generations, then we need to use a much lower discount rate as a result of which future benefits are worth a lot more today than they would be if we followed a purely financial rationale in favor of the present. What the necessity of such a low discount rate implies more broadly will be discussed further below when we look at the exigencies of climate finance in Chapter 6.

Finally, I want to take issue with the use of the business-as-usual scenario in standard models as the benchmark for comparison purposes. We are typically asking ourselves in those model-driven cost–benefit calculations whether a given mitigation strategy to reduce GHG emissions is, given its costs in terms of additional investment spending or premature obsolescence of existing productive capital, worth undertaking compared to doing nothing. But we are thereby implicitly assuming that we can just continue into the indefinite future along the emissions path we are currently on. This means staying more or less with fossil fuels and burning up much of the already proven reserves, with renewables gradually rising over the next couple of decades to perhaps a quarter of the energy-supply total. This, incidentally, is what most of the oil majors, like Exxon Mobil or British Petroleum, still expect to happen. Yet our scientists already know enough about ongoing environmental and meteorological trends to view such a scenario as a recipe for disaster. So how can we in all earnestness still assume business as usual to be a viable option? Only by continuing to adhere to Hartwick's Rule underpinning the "weak" sustainability notion that the current path of environmental degradation is justified because of its capacity of wealth creation. And indeed the IAM models, like DICE and PAGE, or standard CGE models used in mainstream macroeconomics to deal with climate change all tend to project significant increases in GDP per capita between now and 2050. This completely ignores the likelihood of climate change itself impacting negatively on the growth capacity across large swaths of the world economy. If that is the case, and we must assume that there is a good chance it will be, then the business-as-usual path projected forward should indicate a flattening out of growth to come rather than assume the same average pace of growth that we have had up to now.

ECOLOGICAL MACROECONOMICS AND "STRONG" SUSTAINABILITY

Given all of these measurement challenges, we cannot just continue standard macroeconomic modeling and rely on its typical projections of costs and benefits. Instead, we need to develop a special branch of macroeconomics dedicated to treating those climate-related costs and mitigation-driven benefits more accurately and fairly. This is precisely what seems to be happening. We are indeed witnessing the birth of something called "ecological macroeconomics" of which the aforementioned SFC-based studies of low-carbon transition paths (see note 22) are representative. Apart from the authors of those two studies cited, the emerging field of ecological macroeconomics includes Tufts University Global Development and Environment Institute's Jonathan Harris (2009), Tim Jackson (2009/2017) at the University of Surrey, where he directs the interdisciplinary Centre for the Understanding of Sustainable Prosperity (CUSP), Armon Rezai (2011), who works at the Institute for Ecological

Economics of the Vienna University of Economics (Austria), and the already mentioned Peter Victor (2012) of University of Toronto.[24]

What unites these macroeconomists is their concern for the negative consequences of economic growth in terms of the environment, rooted in Nicholas Georgescu-Roegen's (1971) path-breaking work of bringing natural resource flows into economic modeling so as to pinpoint ecological constraints. Mostly inclined to be of the Post-Keynesian variety, ecological macroeconomists see great promise in merging SFC models with input–output analysis to track the feedback loops between economic growth and environmental degradation. At the center of their work, especially when looking at it from a political perspective, is their notion of post-growth which should not be confused with de-growth or zero growth or no growth. Rather does it point to the need for a different growth pattern than the one currently driven by limitless consumer needs and highly leveraged, often speculative financial operations. What post-growth emphasizes more than anything else is the idea of sustainability, of identifying socially and ecologically sustainable (post-)growth paths which the planet and society can live with from one generation to the next.

Apart from sounding more appealing to a large audience than post-growth, sustainability has the added advantage of putting the idea of securing survivability at a high level of quality of life at the center of societal concern. Not only is there a projection of caring for coming generations as much as for oneself today, addressing the crucial ethical question of intergenerational justice (since the yet-to-be-born have by definition no say in what we decide to leave them with to face). But there is also the issue of intra-societal justice built into that sustainability concept, implying a society grounded in the ideals of equality and opportunity.[25] Ecological macroeconomists thus worry as much about income inequality, for instance, as about environmental degradation. This is not just a moral issue, but may well extend to being politically pragmatic when making that connection. Just as the power structure shaping an economy bent on environmental destruction is the same gaming the system in favor of the few to the exclusion of so many, so are the forces pushing for environment preservation bound to be also likely to insist on income equality and "justice as fairness" (to use a phrase made famous by John Rawls).

Ultimately it boils down to breaking away from the aforementioned Hartwick's Rule and the presumed substitutability of natural capital with human or physical capital underpinning that rule at the heart of the mainstream "weak" sustainability notion. If we follow that rule, as we still do and standard macroeconomic models still assume, we will always accept environmental degradation, hence also climate change, as the "cost of doing business," in other words as the necessary price we pay for becoming richer and consuming more. Following this perspective, we will of course be willfully inclined to underestimate the potential damage from global warming and overstate the costs of doing something about this. Ecological macroeconomics pushes for a paradigm shift to get us out of this harmful posture and put us into a different normative context, namely that of "strong" sustainability, which acknowledges that natural capital does not get duplicated by humans and therefore cannot be replaced by manufactured goods or services. Moreover, there is natural capital of vital importance without which we cannot continue to exist (e.g., ozone layer and rain forest). We must first identify which natural capital is critical to our survival, and then decide what changes in our modus operandi are needed to protect this critical natural capital against its gradual destruction or disappearance. Environmental protection gains priority over economic growth so that the latter only occurs when and if the former is assured. Such a shift from a "weak" notion of sustainability to a "strong" one does not just open the way for a new approach to macroeconomic modeling, ecological macroeconomics; it transforms capitalism itself.[26]

Notes

1. These GHG emission-growth data are from the Technical Summary in the appendix of IPCC (2014).
2. "Reaganomics" refers to a conservative counterrevolution in the aftermath of the election victories of Margaret Thatcher in Great Britain (May 1979) and Ronald Reagan (November 1980) which triggered relatively radical economic-policy changes towards large-scale tax cuts, deregulation, privatization, and social-spending reductions in income-maintenance programs.
3. The LDC debt crisis was eventually resolved with the introduction of so-called *Brady Bonds* in 1989 giving creditors a choice of transforming their loans into tradeable bonds to reduce their exposure to LDCs while at the same time giving those hard-pressed countries debt relief. That innovation spawned a rapidly growing market for emerging-market bonds which tied those recently opened newly industrializing countries in lasting fashion to the global capital markets.

4. To the extent that the emerging-market economies, led by China, could grow rapidly by means of exports, they did not have to rely as much on domestic consumption. This enabled them to have productivity gains outpacing real-wage growth without running into overproduction crises. In turn, we now have—a generation later—1.5 billion new middle-class consumers in those countries as a potentially global source of growth in the years to come.

5. We are talking here about the Plaza Agreement of the G-5 countries (USA, Japan, Germany, France, and Great Britain) for an orderly decline of the overvalued dollar in October 1985, followed by the Louvre Agreement (G-5 plus Canada and Italy) in February 1987 to keep key exchange rates floating within regularly readjusted target zones to reduce currency-price volatility. These agreements implied coordinated currency market and interest rate moves by the central banks concerned. But the Louvre Agreement ultimately failed in 1989, as these cooperation requirements were not adequately grounded in coordinated policy-mix adjustments addressing underlying macroeconomic imbalances within and between the G-7 economies.

6. The term "exorbitant privilege" was first coined in 1964 by Valéry Giscard d'Estaing, then De Gaulle's finance minister, to refer to the benefits the USA enjoys from the world-money status of its currency. See B. Eichengreen (2011) for a broader discussion of America's privileged position as issuer of the international medium of exchange.

7. The multifaceted notion of "financialization" has become the central focus of much recent work in several heterodox economic traditions—from American Radicals (Epstein 2005) and Socio-Economists (Krippner 2005) to Post-Keynesians (Stockhammer 2004; Hein 2012) as well as Marxists (Magdoff and Bellamy Foster 2014). Elsewhere, in a chapter entitled "Financialization Revisited—A Meso-Economic Approach" (Guttmann 2016, Chapter 4) I have tried to integrate these different approaches to financialization within a broader Régulationist perspective. The implications of financialization for the global growth pattern, centered on the bubble-prone and increasingly indebted US consumer serving as "buyer of last resort" for the rest of the world, have been analyzed in R. Guttmann (2009).

8. China's coal production, for example, more than quintupled from about 600 million metric tons in 1980 to a 2012 peak of 3650 million metric tons, whereas India's rose nearly as fast during the same period from about 115–630 million metric tons. Global natural gas production more than doubled between 1980 and 2012 from 1485 billion cubic meters to 3230 billion. And worldwide oil production volume rose from 53.5 million barrels per day in 1982 to about 78 million in 2014.

9. Recent revivals of the Depression-era notion of "secular stagnation," as applied to the slow-growth pattern over the last decade, can be found in L. Summers (2016), P. Krugman (2013), and R. Gordon (2015). For an excellent summary of the different, yet overlapping views on this post-crisis phenomenon see C. Teulings and R. Baldwin (2014).

10. We have already noted the climate skepticism of the Republican voter base and Trump at the very beginning of the book. Suffice it to say that Trump's pick to head the Environmental Protection Agency, former Oklahoma Attorney General Scott Pruitt, has a long record of resisting federal climate-change mitigation measures and fought Obama's EPA tooth and nail. And when the Brexit referendum brought Teresa May to power, one of the first steps she took as new Prime Minister in July 2016 was to abolish the Department for Energy and Climate Change in favor of a new Department for Business, Energy, and Industrial Strategy.

11. The path-breaking work of N. Georgescu-Roegen (1971), regarded by many as the foundational text in ecological economics, used the second law of thermodynamics, the law of entropy, to point to the physical limits of economic growth transforming low-entropy resources into high-entropy waste when we produce material goods.

12. See, for instance, R. Engelman (2009), T. Jackson (2009), or P. Ehrlich and J. Harte (2015).

13. Hartwick's Rule, framed in terms of "intergenerational equity," requires the next generation to be left at least as much total (i.e., natural, human, and productive) capital as the preceding one had available. This also implies that non-renewable resources, such as fossil fuels, be allocated correctly over time with the help of an appropriate discount rate.

14. The idea of using a tax to internalize the social costs arising from negative externalities came to A. Pigou (1920). R. Coase (1960) proposed dealing with social costs by providing the parties creating those costs with market incentives to reduce them, with the associated deal-making requiring well-defined property rights and low transaction costs.

15. See N. Stern (2006).

16. Vivid Economics (2013), N. Stern (2016), and R. Pindyck (2017) discuss in more detail how macroeconomic models of climate change, such as IAM, drastically underestimate the costs of global warming over the long haul while at the same time also downplaying the cumulative benefits from effective mitigation on the environment and the economy's performance.

17. Often the mitigation benefit estimates of the bottom-up models exceed those of the top-down models because of the latter's implicit bias, rooted in their Walrasian perfect-competition framework, that climate-change mitigation is seen as disrupting otherwise efficient markets. See on this point also D. Van Vuuren et al. (2009).

18. Meaningful and balanced discussions of the use of CGE models for climate change and in climate-policy analysis can be found in Frontier Economics (2008) and S. Döll (2009). S. Dietz and N. Stern (2015) or M. Roos (2017) offer good examples of how endogenous growth models capturing the positive feedback loops of technological change can help us appreciate better how mitigation works and might be more beneficial than standard macroeconomic models generally portray. For illustration how DSGE modeling of climate change may apply usefully see Y. Cai et al. (2013) or C. Hambel et al. (2015).

19. An example of agent-based modeling with relevance for climate policy is V. Rai and A. D. Henry (2016). J. D. Farmer et al. (2015), looking at new "third wave" approaches to climate-change modeling in economics, highlight the potential contribution of ABM.

20. We now have a benchmark AB-SFC model we can build on, provided by A. Caiani et al. (2016), as well as a few other AB-SFC integration efforts of interest, notably P. Seppecher et al. (2017).

21. Good survey articles tracing the evolution of SFC models in their manifold applications and uses are E. Caverzasi and A. Godin (2014) or M. Nikiforos and G. Zezza (2017) who noted the potential of this modeling technique for useful contributions to ecological macroeconomics.

22. Two good early examples of how SFC models may help clarify the issues involved in the transition to a low-carbon economy are the recent papers by Y. Dafermos et al. (2017) and E. Campiglio et al. (2017).

23. In his famed aforementioned review, N. Stern (2006) estimated the appropriate climate-related discount rate to be 1.5%.

24. See in this context also the comprehensive survey of the field presented in L. Hardt and D. O'Neill (2017).

25. The dimensions of justice invoked here, as constituent elements of sustainability and hence lending that notion a deeply normative context beyond environmental preservation, relate back to the important discussions of the justice concept by political philosophers John Rawls (1971) and Michael Walzer (1983).

26. For greater detail on the juxtaposition between the "weak" and "strong" notions of sustainability, see E. Neumayer (2003) or J. Pelenc et al. (2015).

REFERENCES

Boulding, K. (1966). The Economics of the Coming Spaceship Earth. In H. Jarrett (Ed.), *Environmental Quality in a Growing Economy* (pp. 3–14). Baltimore, MD: Johns Hopkins University Press. http://dieoff.org/page160.htm. Accessed 8 Dec 2016.

Cai, Y., Judd, K., & Lontzek, T. (2013). *The Social Cost of Stochastic and Irreversible Climate Change* (NBER Working Papers, No. 18704). National Bureau of Economic Research. http://www.nber.org/papers/w18704.

Caiani, A., Godin, A., Caverzasi, E., Gallegati, M., Kinsella, S., & Stiglitz, J. (2016). Agent Based Stock-Flow Consistent Macroeconomics: Towards a New Benchmark. *Journal of Economic Dynamics and Control, 69*(1), 375–408.

Campiglio, E., Godin, A., & Kemp-Benedikt, E. (2017). *Networks of Stranded Assets: A Case for a Balance Sheet Approach* (AFD Research Papers, No. 2017-54).

Caverzasi, E., & Godin, A. (2014). Post-Keynesian Stock-Flow-Consistent Modelling: A Survey. *Cambridge Journal of Economics, 39*(1), 157–187. https://doi.org/10.1093/cje/beu021.

Coase, R. (1960). The Problem of Social Cost. *Journal of Law and Economics, 3*(1), 1–44.

Dafermos, Y., Nikolaidi, M., & Galanis, G. (2017). A Stock-Flow-Fund Ecological Macroeconomic Model. *Ecological Economics, 131*(1), 191–207.

Daly, H. (1980). *Economics, Ecology, Ethics. Essays Towards a Steady-State Economy.* San Francisco, CA: W. H. Freeman.

Daly, H. (2015). *Economics for a Full World.* Great Transition Initiative. http://www.greattransition.org/images/Daly-Economics-for-a-Full-World.pdf. Accessed 9 Dec 2016.

Dietz, S., & Stern, N. (2015). Endogenous Growth, Convexity of Damage, and Climate Risk: How Nordhaus' Framework Supports Deep Cuts in Carbon Emissions. *Economic Journal, 125*(583), 574–620. https://doi.org/10.1111/ecoj.12188.

Döll, S. (2009). *Climate Change Impacts in Computable General Equilibrium Models: An Overview* (HWWI Research Paper, No. 1-26). Hamburg Institute of International Economics. https://www.econstor.eu/bitstream/10419/48201/1/64016790X.pdf.

Ehrlich, P., & Harte, J. (2015). Biophysical Limits, Women's Rights and the Climate Encyclical. *Nature Climate Change, 5,* 904–905. https://doi.org/10.1038/nclimate2795.

Eichengreen, B. (2011). *Exorbitant Privilege: The Rise and Fall of the Dollar and the Future of the International Monetary System.* New York: Oxford University Press.

Engelman, R. (2009, June 1). Population and Sustainability: Can We Avoid Limiting the Number of People? *Scientific American.* https://www.scientificamerican.com/article/population-and-sustainability/. Accessed 9 Dec 2016.

Epstein, G. (2005). *Financialization and the World Economy.* Cheltenham, UK: Edward Elgar.

Farmer, J. D., Hepburn, C., Mealy, P., & Teytelboym, A. (2015). A Third Wave in the Economics of Climate Change. *Environmental and Resource Economics, 62*(2), 329–357.

Frontier Economics. (2008). *Modeling Climate Change Impacts Using CGE Models: A Literature Review*. http://www.garnautreview.org.au/CA25734E0016A131/WebObj/ModellingClimateChangeImpacts/$File/Modelling%20Climate%20Change%20Impacts%20-%20Frontier%20Economics.pdf.

Georgescu-Roegen, N. (1971). *The Entropy Law and the Economic Process*. Cambridge, MA: Harvard University Press.

Godley, W., & Lavoie, M. (2007). *Monetary Economics: An Integrated Approach to Credit, Money, Income, Production and Wealth*. Basingstoke, UK: Palgrave Macmillan.

Gordon, R. (2015). Secular Stagnation: A Supply-Side View. *American Economic Review, 105*(5), 54–59. https://www.aeaweb.org/articles?id=10.1257/aer.p20151102. Accessed 25 May 2017.

Guttmann, R. (2009). Asset Bubbles, Debt Deflation, and Global Imbalances. *International Journal of Political Economy, 38*(2), 46–69.

Guttmann, R. (2016). *Finance-Led Capitalism: Shadow Banking, Re-regulation, and the Future of Global Markets*. New York: Palgrave Macmillan.

Hambel, C., Kraft, H., & Schwartz, E. (2015). *Optimal Carbon Abatement in a Stochastic Equilibrium Model with Climate Change* (NBER Working Papers, No. 21044). National Bureau of Economic Research. http://www.nber.org/papers/w21044.

Hardt, L., & O' Neill, D. (2017). Ecological Macroeconomic Models: Assessing Current Developments. *Ecological Economics, 134*, 198–211. https://doi.org/10.1016/j.ecolecon.2016.12.027

Harris, J. (2009). *Twenty-First Century Macroeconomics: Responding to the Climate Challenge*. Cheltenham, UK: Edward Elgar.

Hartwick, J. (1977). Intergenerational Equity and the Investing of Rents from Exhaustible Resources. *American Economic Review, 67*(5), 972–974.

Hein, E. (2012). *The Macroeconomics of Finance-Dominated Capitalism—And Its Crisis*. Cheltenham, UK: Edward Elgar.

Hope, C. (2006). The Marginal Impact of CO_2 from PAGE2002: An Integrated Assessment Model Incorporating the IPCC's Five Reasons for Concern. *Integrated Assessment, 6*(1), 19–56.

Hope, C. (2013). Critical Issues for the Calculation of the Social Cost of CO_2: Why the Estimates from PAGE09 Are Higher Than Those from PAGE2002. *Climatic Change, 117*(3), 531–543.

IPCC. (2014). *Fifth Assessment Report: Climate Change 2014—Synthesis Report*. Cambridge, UK: Cambridge University Press. http://www.ipcc.ch/report/ar5/.

Jackson, T. (2009/2017). *Prosperity Without Growth: Economics for a Finite Planet*. London: Routledge. Published as a Much Revised Second Edition *Prosperity Without Growth: Foundations for the Economy of Tomorrow* in 2017.

Keynes, J. M. (1936). *A General Theory of Employment, Interest and Money*. London: Macmillan.

Krippner, G. (2005). The Financialization of the American Economy. *Socio-Economic Review, 3*(2), 173–208.

Krugman, P. (2013, September 25). Bubbles, Regulation, and Secular Stagnation. *New York Times.* https://krugman.blogs.nytimes.com/2013/09/25/bubbles-regulation-and-secular-stagnation/.

Magdoff, F., & Foster, J. B. (2014). Stagnation and Financialization: The Nature of the Contradiction. *Monthly Review, 66*(1), 4–25.

Marshall, Al. (1890). *Principles of Economics*. London: Macmillan.

Mill, J. S. (1848). *Principles of Political Economy*. London: John W. Parker.

Neumayer, E. (2003). *Weak Versus Strong Sustainability: Exploring the Limits of Two Opposing Paradigms*. Cheltenham, UK: Edward Elgar.

Nikiforos, M., & Zezza, G. (2017). *Stock-Flow Consistent Macroeconomic Models: A Survey* (Working Paper, No. 891). Levy Institute. http://www.levyinstitute.org/publications/stock-flow-consistent-macroeconomic-models-a-survey.

Nordhaus, W. (1992). *The DICE Model: Background and Structure* (Cowles Foundation Discussion Papers, No. 1009). https://cowles.yale.edu/sites/default/files/files/pub/d10/d1009.pdf.

Nordhaus, W. (2017). *Evolution of Modeling of the Economics of Global Warming: Changes in the DICE Model, 1992–2017* (Cowles Foundation Discussion Papers, No. 2084). https://cowles.yale.edu/sites/default/files/files/pub/d20/d2084.pdf.

Pelenc, J., Ballet, J., & Dedeurwaerdere, T. (2015). Weak Sustainability Versus Strong Sustainability. *Brief for GSDR 2015.* Global Sustainable Development Report, United Nations. https://sustainabledevelopment.un.org/content/documents/6569122-Pelenc-Weak%20Sustainability%20versus%20Strong%20Sustainability.pdf. Accessed 24 Nov 2017.

Pigou, A. (1920). *The Economics of Welfare*. London: Macmillan.

Pindyck, R. (2017). The Use and Misuse of Models for Climate Policy. *Review of Environmental Economics and Policy, 11*(1), 100–114.

Rai, V., & Henry, A. D. (2016). Agent-Based Modelling of Consumer Energy Choices. *Nature Climate Change, 6*(June), 556–562. https://doi.org/10.1038/nclimate2967.

Rawls, H. (1971). *A Theory of Justice*. Cambridge, MA: Harvard University Press.

Rezai, A. (2011). The Opportunity Cost of Climate Policy: A Question of Reference. *Scandinavian Journal of Economics, 113*(4), 885–903.

Ricardo, R. (1817). *On the Principles of Political Economy and Taxation*. London: John Murray.

Roos, M. (2017). Endogenous Economic Growth, Climate Change and Societal Values: A Conceptual Model. *Computational Economics, 50*(1), 1–34.

Say, J.-B. (1803/1821). *A Treatise on Political Economy*. London: Longman. First Published in French as *Traité d'économie politique*.

Schumacher, E. F. (1973). *Small Is Beautiful: Economics as If People Mattered.* New York: Harper & Row.

Seppecher, P., Salle, I., & Lavoie, M. (2017). *What Drives Markups? Evolutionary Pricing in an Agent-Based Stock-Flow Consistent Macroeconomic Model* (Document de travail du CEPN, No. 2017-3). Centre d'Economie de l'Université Paris Nord. https://hal.archives-ouvertes.fr/hal-01486597/document.

Solow, R. (1986). On the Intergenerational Allocation of Natural Resources. *Scandinavian Journal of Economics, 88*(1), 141–149. https://doi.org/10.2307/3440280.

Smith, A. (1776). *The Wealth of Nations.* London: W. Strahan and T. Cadell.

Stern, N. (2006). *Stern Review: The Economics of Climate Change.* London: HM Treasury. http://unionsforenergydemocracy.org/wp-content/uploads/2015/08/sternreview_report_complete.pdf. Last accessed 18 Nov 2017.

Stern, N. (2016). Current Climate Models Are Grossly Misleading: Nicholas Stern Calls on Scientists, Engineers and Economists to Help Policy-Makers by Better Modelling the Immense Risks to Future Generations, and the Potential for Action. *Nature, 530*(7591), 407–409.

Stockhammer, E. (2004). Financialisation and the Slowdown of Accumulation. *Cambridge Journal of Economics, 28*(5), 719–741.

Summers, L. (2016, March/April). Secular Stagnation: What It Is and What to Do About It. *Foreign Affairs, 95*(2). http://larrysummers.com/2016/02/17/the-age-of-secular-stagnation/. Accessed 25 May 2017.

Teulings, C. & Baldwin, R. (Eds.). (2014). *Secular Stagnation: Facts, Causes, and Cures.* London: CEPR Books. https://scholar.harvard.edu/files/farhi/files/book_chapter_secular_stagnation_nov_2014_0.pdf. Accessed 25 May 2017.

Van Vuuren, D., et al. (2009). Comparison of Top-Down and Bottom-Up Estimates of Sectoral and Regional Greenhouse Gas Emission Reduction Potentials. *Energy Policy, 37*, 5125–5139.

Victor, P. (2012). Growth, Degrowth and Climate Change: A Scenario Analysis. *Ecological Economics, 84*, 206–212. https://doi.org/10.1016/j.ecolecon.2011.04.013.

Vivid Economics. (2013, May). *The Macroeconomics of Climate Change.* A Report Commissioned by the UK Department for Environment, Food and Rural Affairs.

Walzer, M. (1983). *Spheres of Justice.* New York: Basic Books.

Wicksteed, P. (1910). *The Common Sense of Political Economy.* London: Macmillan.

CHAPTER 5

Pricing Carbon

One of the trickier issues of central importance in the macroeconomic models of climate change is the determination of an appropriate carbon price without which it is difficult, if not altogether impossible, to estimate damage-related costs or mitigation-induced benefits. That carbon price must in some fashion relate to the social cost of carbon dioxide's damage as well as the economic costs of avoiding its emission both of which we have yet to learn how to measure accurately. The fact that many governments have in recent years taken first steps in the direction of establishing a carbon price, either by setting up so-called cap-and-trade markets for emissions permits or by introducing carbon taxes, does not yet address this pressing issue adequately. Almost none of those experiments have up to now worked well enough to impose a sufficiently high and gradually rising carbon price, and there are good reasons for these disappointing results. We will have to go quite a bit further than we have so far collectively dared to go. And if we do not, we will only have ourselves to blame! The issue is urgent, since we risk getting to face this issue finally only in a moment of panic when, because the carbon price has been so low and applied so spottily for such a long time, we suddenly see it imposed everywhere and at a much higher level to push the world economy into drastic cuts of GHG emissions amidst increasingly troubling signs of potentially disastrous climate-related damage. Such a sudden U-turn scenario represents a substantial systemic risk posed by climate change which we can avoid by taking earlier and decisive action on the global pricing of carbon.

© The Author(s) 2018
R. Guttmann, *Eco-Capitalism*,
https://doi.org/10.1007/978-3-319-92357-4_5

COPING WITH A FINITE CARBON BUDGET

If we really want to be serious about meeting the ≥ 1.5 °C <2 °C objective of the Paris Agreement, we will have to learn how to live within the carbon budget it prescribed. A carbon budget refers to a finite amount of GHG emissions that can be emitted over a specified period of time. Its target amount should obviously be in line with what scientists deem to be "tolerable" levels of global warming, and there is still some argument as to what that quantity would be. The notion of a carbon budget started taking root at the beginning of this century, in 2001, with a worldwide effort sponsored by the so-called Global Carbon Project to identify all sources of GHG emissions and quantify their output. In its Fifth Assessment Report, the IPCC (2014) estimated how much we could still emit in CO_2 (as well as other greenhouse gases, the effects of which on global warming were included in more complex models) to keep cumulative temperature hikes limited to 1.5, 2, or 3 °C, respectively, with each of these limits assessed at having a 66% chance, a 50% probability, or just a 33% likelihood for each.[1] The IPCC scientists have defined an overall cumulative carbon-emission budget of 2900 $GtCO_2e$ from 1870 onward if we want to cap human-induced global warming to less than 2 °C. Of that amount we had spent 1900 $GtCO_2e$ by 2011. Given that the world emits about 40 $GtCO_2e$ per year, we have right now (mid-2017) about 785 $GtCO_2e$ left to emit before chances of keeping the cumulative temperature hike within acceptable limits start getting increasingly unlikely. So we will have to have reached carbon neutrality by the time this budget gets used up. At that point remaining GHG emissions should match counteracting carbon sinks as well as carbon capture and storage facilities able to absorb fully what still gets emitted. In other words, we need to reduce GHG emissions by about 85% from current levels by mid-century, with annual cuts averaging 2–3%. This kind of downward path is feasible, provided society gives this objective high priority and reorganizes itself accordingly.

We are nowhere near such a downward path. True, global carbon emissions seemed to have peaked in the mid-2010s despite an acceleration of economic growth in many regions of the world. Those emissions did not rise at all for three years during 2014–2016. But then they grew by 2% last year (2017) to 37 billion metric tons of CO_2, much of it due to an unexpected pickup in China's burning of coal in response to a draught hampering the country's generation of hydroelectric power amidst dwindling rivers. Add to this 4 billion metric tons from land use changes, mostly deforestation, for a total of 41 $GtCO_2$. In the meantime daily average values for carbon dioxide in the atmosphere rose for the first time steadily above 400 ppm in 2016, reaching an all-time record peak of 412.63 ppm on April 26, 2017. Arguably most worrisome has been the warming trend. Every single year since 2012 has been globally the hottest on record, with a hike of about 0.94 °C above the twentieth century average in 2016 and nearly 1.2 °C above pre-industrial levels prevailing in 1850. The pace of warming seems also to be accelerating, with a record increase of 0.13 °C in average from 2015 to 2016 in the wake of strong El Niño conditions prevailing at the beginning of that year.[2]

The declining emissions path all the way to (net) zero by 2060 envisaged by the Paris Agreement is nowhere in sight. Instead we are on an actual emissions path still sloping upward, albeit at a slower pace than in the 2000s. While that is better than the business-as-usual scenario, we have yet to curb emissions globally in sustained fashion. The just announced 2017 CO_2 hike has dashed experts' hopes (which were raised during 2015/2016 in the wake of a first year-to-year decline) that the needed stabilization can endure as of yet. That unexpected increase in emissions coincided with aggressive Trump Administration efforts to expand domestic energy production in the USA by promoting construction of oil and gas pipelines, oil exploration off-shore or in Alaska, and a revival of coal mining—all steps moving us even further in the wrong direction, compounded by systematic destruction of America's climate-policy infrastructure. We know that existing fossil-fuel reserves have to stay in the ground for the aforementioned carbon budget to have any chance of realization. Even just burning 20% of already proven reserves would exhaust the entire budget all on its own. And over the next five years, there will be a wave of replacements for the rapidly aging power-plant infrastructure in the USA and elsewhere for which we have to move massively away from oil or coal. Trump's push for fossil fuels thus arrives at a crucial time which bodes ill for the future.

What if we miss the Paris Agreement goals by a lot and then fail in addition to toughen national NDC commitments as prescribed in the early 2020s, as Trump's refusal to play by its rules ends up weakening this global accord beyond repair? What if during the same period we see very troubling signs that the negative repercussions of global warming are spreading steadily, as droughts last longer, storms intensify in frequency and strength, wildfires are reaching a new scale, sea levels rise in accelerating fashion, more ice is melting ever faster, and Paris begins to feel like living in Spain? We will then face, possibly less than a decade from now, the prospect of a very fateful and difficult choice between suffering huge amounts of climate-related damage of a permanent kind or committing instead to much more radical action aimed at pushing global GHG emissions levels down as rapidly as possible. Having suppressed taking the problem seriously, we may now go to the other extreme in the face of a climate panic and impose top-down measures to cut emissions at all cost. This would in all likelihood prove extremely disruptive to the world economy, amounting to a systemic crisis from which we cannot exit as rapidly as in the 2007–2009 crisis because of its structural nature depending on slow-moving environmental change. We should do our best to avoid such a horrible choice, and the only sure way to do that is to accelerate global mitigation efforts now so as to start pushing emissions levels down much more gradually and steadily until we have that problem reasonably under control. This will require pricing carbon so that we can get properly incentivized to want to get rid of it. And such a carbon price will have to be set to internalize what is otherwise a huge and growing externality, the emission of greenhouse gases capable of changing our climate.

THE SOCIAL COST OF CARBON

Currently we live in a high-carbon economy based on fossil fuels as our primary energy source and lots of other carbon-intense processes as well as products (e.g., cars using fuel combustion engines). We have across our entire industrial matrix, spanning the globe, lots of sources of greenhouse gases such as carbon dioxide or methane. These pollutants trap heat in our atmosphere and thus cause global warming which in turn triggers a variety of possibly damaging consequences from climate change. As it stands, those GHG sources causing the problem are emitting their pollutants without any cost, thus enjoying a huge implicit subsidy. They are allowed to ignore the nefarious social consequences of their activity while pocketing the private benefits it yields (i.e., profit) at the same time. We need to make the emitters, as well as society at large, take account of these social costs in order to create sufficiently strong incentives to reduce those damaging GHG emissions. We can only do that by putting an adequate price on carbon and the other greenhouse gases, a price that needs to be paid by making fossil fuels and other GHG sources proportionately more expensive. That price relates to the social costs these pollutants cause which it seeks to internalize after all. Hence, the strategic importance of what has come to be referred to as the *social cost of carbon* (SCC).[3]

We can then define the social cost of carbon as an estimate, measured in dollars, of the long-term economic damage done by a metric ton of carbon dioxide emissions in a given year stretched over a long period of time. That same measure also applies to the value of damage avoided from a small (i.e., one-ton) reduction of CO_2 emissions and can therefore be used to calculate the benefits of emission-reducing mitigation projects. Such damage includes changes in agricultural productivity, human health, property devaluation from increased flood risk, and energy costs (heating, air-conditioning). We should be clear that limits pertaining to still-unknown effects of climate change over the long haul, insufficient data, and still relatively unsophisticated modeling techniques oblige us to exclude certain dimensions of climate-related damage which we are not able or willing to capture at this point. For example, none of the social-cost estimates used anywhere include ocean acidification, a key factor in the bleaching of coral reefs and alarming declines of fish supplies. Nor do they include calculation of cumulative impacts from negative feedback loops such as the thawing of the Arctic permafrost. Whatever SCC estimate we are able to come up with right now, it will by its very nature of incompleteness end up giving us a price too low! Of course, as we get a better picture of climate-related damage in all of its many facets, more accurate data, and improved models over time, we are able to upgrade and improve our social-cost estimates. The Obama Administration did precisely that in 2010, 2013, and just before it left office in January 2017, much to the chagrin of the climate deniers!

I have already pointed out (in Chapter 4, for instance, when discussing the macroeconomic models) that climate change is a process fraught with several major uncertainties, hence by definition very difficult to make assessments and estimates about. Just for this reason any social-cost measure will be highly approximate and captured best as a range. Add to this additional measurement complications directly associated with the SCC formula. To begin with, there is the need to determine an appropriate time horizon stretching way into the future, a crucial question in light of the very persistent and cumulative nature of climate change. The longer the time horizon we consider, the higher the SCC estimate we get (even with discounting). It makes quite a difference whether our estimates stop at 2100 or go all the way to, say, 2300. There is also the question of whether to account for the geographically uneven impact of climate change. We are pretty sure already that climate change will have far more severe impact in poorer developing countries stretching along the equator whereas the emissions causing all that damage come mostly from richer countries located in the Northern hemisphere. Our SCC estimates should be adjusted accordingly via a redistribution of compensation values to be fair, and the more we include such equity weighting the higher our social-cost estimate. Ideally, we would use those higher estimates to finance transfers of resources from the high-emission regions to the high-damage areas as has been committed to in the Paris Agreement with creation of the so-called Green Climate Fund. A third complication relates to the question whether to use the median or the mean of calculations as the central estimate. When the distribution is skewed to the right (toward higher damage being more probable), then the mean is larger than the median and should be the measure we use. Finally, with the SCC involving such a long time horizon over which we discount future costs to present value, the choice of our discount rate makes a huge difference to calculated cost estimates. The higher the discount rate, the lower the cost estimate and the less benefit we get from mitigation. This, as we shall see, is a very controversial topic over which there exist fundamental disagreements depending on political persuasion.

The idea of a social cost of carbon emerged around 2001 when implementation of the Kyoto Protocol sparked a worldwide effort to identify and measure all GHG emissions for a better picture of the problem. In November 2007, the US Ninth Circuit Court of Appeals remanded a fuel economy rule on the grounds that the Department of Transportation had failed to consider the social cost of carbon in its cost-benefit analysis, stating that "[w]hile the record shows that there is a range of values, the value of carbon emissions reduction is certainly not zero." In 2009, the Obama Administration set up an interagency working group (IWG) comprising scientists and economic experts from various federal agencies to calculate the social cost of carbon and harmonize how that measure is used in rule-making across the US government.[4] As it stands, over 100 federal actions have been implemented using the official IWG measure of the SCC, covering regulations with more than $1 trillion in benefits. The IWG provides four values for its SCC estimates, three of which distinguished by their respectively different discount rates of 2.5, 3, and 5%. The fourth estimate assumes a high-damage, low-probability climate-change scenario at the 95th percentile of the frequency distribution of SCC estimates based on a 3% discount rate. In its latest 2016 update, these social-cost calculations came to $14 (at 5%), $46 (at 3%), $68 (at 2.5%), and $138 (for the high-impact scenario) per metric ton of CO_2 emissions in 2025 (in 2007 dollars). The central SCC estimate in 2016 dollars for right now (2017) is about $43/ton.

The IWG estimates use calculation averages from three different, widely cited and extensively peer-reviewed impact assessment models (IAM) which, as we have already mentioned in the preceding Chapter 4, link physical impacts to the economic damage of CO_2 emissions. All three of these models have a similar structure comprising a socioeconomic module projecting economic growth, population growth, and increases in GHG emissions, a climate module translating emissions increases into projected weather patterns (temperature, precipitation, sea levels, likelihood and range of extreme-weather events, etc.), a module assessing costs and benefits of climate change over a long time period (until 2300), and finally using discounting to calculate present values of these future projections. In that earlier discussion of IAM we had already made reference to two of them, Christopher Hope's PAGE model and William Nordhaus' DICE model. The third IAM is University of Sussex Professor Richard Tol's (2009) FUND model which stands for Climate Framework for Uncertainty, Negotiation and Distribution (see also fund-model.org). Tol's FUND is significantly and consistently less pessimistic than the other two IAMs. All three get regularly updated, often in response to the new assessment reports of the IPCCC, which then in turn prompts governmental recalculations of the SCC. We are therefore gradually getting improved SCC estimates as the damage functions become more precise. In addition, there was a concerted effort launched by Obama's EPA to subject social-cost estimates of the other greenhouse gases to a similarly rigorous procedure, yielding distinct dollar figures for the social costs of methane ($SC\text{-}CH_4$) and of nitrous oxide ($SC\text{-}N_2O$) as opposed to carbon dioxide ($SC\text{-}CO_2$) to replace the less accurate formula of multiplying the $SC\text{-}CO_2$ by these other gases' relative global warming potential in use ever since the Kyoto Protocol.

While there are many factors weighing on the accuracy of the SCC estimates, clearly the most important one is the choice of an appropriate discount rate. The EPA under Obama took the sensible approach of providing estimates based on a range of three different discount rates, not least to show the huge impact this choice has on the carbon price. At the same time, the agency also based its so-called central estimate on the 3% discount rate, justifying this choice as the one best applicable to consumers who stand to be impacted the most by climate change. The trouble is that even this relatively modest discount rate may be too high when it comes to climate change for two powerful reasons. One is that the problem is one of exceptionally long duration, stretching possibly over centuries given how long those greenhouse gases stay in the atmosphere and also considering its cumulative nature of possibly accelerating the warming trend. Such a time horizon justifies in and of itself a much lower discount rate than would be typically applied to corporate investment projects lasting at most a couple of decades. But in addition we have here a problem that is hurting future generations in potentially ever-worsening fashion, and we simply are morally obliged to discount those unborn lives of our grandchildren and great-grandchildren as little as possible if our actions today can help avoid the worst for them then. Ideally, with this argument in mind, we would set a discount rate of zero to assure that we take the well-being of our off-springs as seriously as our own. That is surely not going to happen, because all that money thereby induced for spending on climate-change mitigation does have some positive opportunity cost in terms of returns foregone from alternative investment options thereby not undertaken. Yet this argument drives us to the second reason justifying a much lower discount rate. We cannot argue here in conventional terms of opportunity costs which apply when talking about private money being invested for private gains. Minimizing damage from climate change through aggressive and timely pursuit of mitigation measures is a public good benefitting all, whether they pay for it or not. And if we consider mitigation such a public good, then it merits a below-market "social" rate of discount, actually way lower.

A survey (see Moritz Drupp et al. 2015) interviewing 197 economists a couple of years ago yielded an average long-term social discount rate of 2.25%. The Stern Report, making the ethical argument that puts emphasis on inter-generational fairness, proposed an even lower discount rate of 1.4%. Whereas the US and UK governments provide guidance to use discount rates of 3 and 3.5%, respectively, both accept the need to use lower discount rates for periods in excess of thirty years, approaching 1% for time periods greater than 300 years.[5] As the Organization for Economic Co-operation and Development reports (see Smith and Braathen 2015), other countries have opted similarly in favor of applying lower discount rates, ranging between 1 and 3%, for cost-benefit analysis when it comes to long-term projects.

The trend toward a more sensible approach to discounting climate change damage and its mitigation has been rudely disrupted in November 2016 by the election victory of climate denier-in-chief Donald Trump. In March 2017, his administration disbanded the IWG to take over climate policy and redirect it radically away from the Obama Administration's stance and the Paris Agreement. Having put an outspoken and long-standing critic of the Environmental Protection Administration, Scott Pruitt, at the head of that agency, it did not take long for policy reversals to kick in. Trump's key objective has been to knock out Obama's Clean Power Plan, and such a step required a new cost-benefit analysis permitting resumption of coal-fired power plants. The EPA under Pruitt's leadership has for that purpose proposed using two discount rates, keeping the 3% one (because that one is legally enshrined as the official one) and introducing a 7% rate (in lieu of 5%) while dropping the 2.5% rate. Such a 7% rate, long demanded by conservative think tanks such as the Heritage Foundation, is justified as akin to the average return shareholders earn in the stock market. It is to wonder whether climate-change mitigation benefits or the costs of environmental disaster unfolding over decades, if not centuries, deserve to be evaluated with this kind of completely private, basically short-term opportunity cost of investment. Pruitt's EPA also decided to confine the cost estimates of environmental damage from climate change to the territory of the USA alone rather than the entire globe, thus ending up with much lower numbers for cost. So now the world's largest per capita emitter of greenhouse gases presumes that such pollution stays within its geographic boundaries as if Americans lived in a different atmosphere than the rest of the world. Finally, the new EPA policy ignores improvements in emissions-related pollution from coal-fired plants as a health benefit if those arise in areas of relatively low concentrations of fine particulate matter from burning of fossil fuels.[6] These manipulations led the EPA to cut its SCC estimate to between $1 and 6 per metric ton by 2020, compared to the $45/ton estimate under Obama in 2016—a reduction of 87–97%.

In the face of these policy reversals, the USA under Trump has basically decided to abstain from climate-change mitigation. Obviously, we have already illustrated in Chapter 3 with the examples of Jerry Brown's California and Microsoft that states and corporations, in addition to cities and civil-society organizations, continue to push their struggle against climate change. But the absence of the federal government, whose apparent commitment to boost GHG emissions under Trump goes against the grain of the global mobilization, will greatly hamper that struggle. At one point, it may well dawn on American policy-makers and those electing them that their recent withdrawal from sensible climate policy, besides being irresponsible to an unconscionable degree, also hurts America's chances to maintain its leadership role while capitalism reorganizes itself globally during the course of the twenty-first century. If this reconsideration fails to materialize and the Republican Party remains in power unreconstructed as is, then the US economy will remain stuck in an obsolete growth model while others—notably China and Europe—charge ahead remaking their economies into low-carbon, energy-efficient, waste-recycling, organic-farming systems outcompeting the USA as it remains wedded to a twentieth-century high-carbon growth model. Americans will either change or forfeit their leadership. In the meantime, others are charging ahead reinforcing their climate policy launched in Paris at the end of 2015. For example, ever since the COP21 in Paris there has been a very significant increase in large multinationals, as for instance Microsoft (see Chapter 3), adopting an internal-carbon price to move their investments and operations toward low-carbon initiatives. The carbon prices they have chosen vary greatly, as illustrated by the $1.02/ton of Nestle, the $6.70 of BMW, the $11.17 of Societé Generale, the $23.92 of Ernst & Young, the $40 of Royal Dutch Shell and BP, the $86.04 of Britain's National Grid PLC, and the $100/ton set by Novartis.[7]

Governments too are eyeing adoption of carbon prices at sufficiently high levels. In September 2017, newly elected French president Emmanuel Macron, for instance, called for a carbon price of at least 25–30 euros ($30–36) per ton as a carbon floor price, coupled with introducing an EU-wide common energy market and more cross-national electricity interconnections toward eventual creation of an European super-grid extending to regions around its borders (North Africa, Turkey, Kazakhstan). Decoupling from Trump's America, the Canadian government under Justin Trudeau has recently introduced the so-called Pan-Canadian Framework on Clean Growth and Climate Change which foresees, among other elements of climate policy, the introduction of a nationwide carbon price in 2018 at C-$10/ton which will rise rapidly to C-$50/ton ($38) by 2022. Adding importantly to the building of a worldwide consensus, Trump's USA excepted, has been the High-Level Commission on Carbon Prices set up in November 2016 by France's Hollande Administration and concluding its fact-finding mission six months later with a report.[8] The commission, headed by Nobel Laureate Joseph Stiglitz and Lord Nicholas Stern, concluded in this report that aiming for the Paris Agreement objectives would require an explicit (i.e., actually and broadly charged) carbon price of at least \$40–80 per ton in 2020 and \$50–100/tCO$_2$ by 2030. These are reasonable and ultimately fairly modest targets, assuming that they are part of a broader climate-policy package which all Paris signatories adhere to. The fairly broad price range set by the Stiglitz-Stern Commission reflects current uncertainty about climate-policy implementation, differences in living standards between countries, the extent of international support, and availability of financial transfers from rich to poor countries. Depending on evolving circumstances affecting these variables, that range should be narrowed over time.

While this initiative marks an important step in the ongoing debate over carbon pricing, we have to accept that the social cost of carbon is intrinsically difficult to measure and fraught with massive amounts of uncertainty. It may therefore ultimately not be the best valuation basis for carbon prices, raising the question what other alternatives there may be available to calculate adequate price signals with which to incentivize steps necessary for the transition to a low-carbon economy. We should keep in mind that the SCC, estimating the damage caused by GHG emissions more or less over their life cycle in today's dollars, is essentially a measure of the externality caused by these pollutants. It tells us what society should be willing to pay to avoid the future damage caused by carbon emissions. That number will go up over time, as the atmosphere contains greater GHG concentrations and the extent of thereby engendered damage rises. But there is another, parallel measure to consider, the so-called marginal abatement cost (MAC) which measures the estimated cost of actually reducing those emissions rather than the damage caused by them. Specifically, the MAC measures the additional cost of reducing carbon emissions by one ton which can then be taken as the price required to yield the emissions reductions needed for us to move along a chosen pathway of atmospheric CO$_2$ concentrations. This number is also quite uncertain, depending on the evolving set of available global abatement options, technological change needed to make these options feasible, and the degree of international cooperation as well as global commitment to policies encouraging these options to be taken.

The MAC measure has two important advantages. First of all, it is linked to an explicit emissions stabilization target (e.g., the Paris Agreement's 450 ppm level). And secondly, we have already pretty good cost estimates for a large variety of mitigation initiatives that we shall undertake in the transition to a low-carbon economy (e.g., switching to LED lighting, insulation retrofits in commercial and residential buildings, electric cars, renewable energy). We can therefore construct MAC curves with a significantly lesser degree of uncertainty than the SCC, as has already been demonstrated for example by McKinsey (2013). As more mitigation actions will have already been undertaken as a result of which the world community is moving closer to its GHG stabilization target, the MAC should decrease. In other words, the MAC curve is downward sloping just as much as the SCC curve is upward sloping (if we measure atmospheric GHG levels—the atmospheric concentrations pathway—on the x-axis). Where the two curves intersect, under optimal conditions of perfect international cooperation, you get an equilibrium price of carbon where the marginal abatement costs equal the benefits of climate damage avoidance. We can think of this as the shadow price of carbon (SPC).[9]

CAP AND TRADE

While SCC and MAC are theoretical carbon-price concepts, we have already seen actual carbon-price mechanisms put in place over the last decade or so. These have their roots in the notion of externalities. In mainstream economic theory such issues as pollution are considered externalities to the extent that they represent socially costly spillover effects on third parties which those causing that problem ignore when calculating their own benefits and costs. Since those private agents only look at their own private costs and ignore the social costs affecting third parties via spillovers, they will end up producing more of that negative externality than they should have. It is then up to government to correct this market failure, either by imposing regulatory limits on such socially costly activities, or forcing these social costs to be internalized through an emissions tax, or encourage the parties involved to address the problem they cause others by negotiated settlement between them (often involving a market-based resolution).[10] We have seen either one of these approaches emerge in the debate over climate change.

The favored approach so far has been the one using a market-based negotiation mechanism rooted in using well-defined property rights as suggested in the Coase Theorem, the so-called cap-and-trade scheme. The notion of a market-based approach to cut GHG emissions dates back to before we even began to worry about climate change, notably the Clean Air Act Amendments passed by overwhelming majorities in the US Congress in 1990 which, among other major initiatives, anchored a preference for market-based principles in general while specifically foreseeing the introduction of emissions banking and trading. In the Kyoto Protocol of 1997, this preference became explicitly anchored in the so-called Flexibility Mechanisms of which there were two project-based mechanisms involving the trading of emission-reduction credits between countries: Joint Implementation (JI), exchanging credits between developed countries listed in the Annex 1 of the Kyoto Protocol; and the Clean Development Mechanism (CDM) which engages Annex 1 countries with others not listed in that annex, typically developing countries. Kyoto's third mechanism involved instead trading of emission permits for greenhouse gases, both within and between countries. That provision sparked a number of such emissions-trading initiatives across the globe in the early 2000s whereby the "cap-and-trade" idea embodied here became the primary environmental-policy tool to seek cost-efficient GHG emission reductions in lieu of taxes or command-and-control regulations. Starting in 2002 there were pilot programs in Britain and Denmark, superseded finally by the launch of the European Union's Emissions Trading System (EU ETS) in 2005 as the world's hitherto largest (multi-country) carbon market.[11]

This market-based scheme consists of qualified government units setting aggregate GHG emission limits in line with their declared national reduction targets and breaking those down to the individual level so that polluters know how much of the GHG they are allowed to emit in a given year. Participating polluters are then assigned emission permits in line with their pollution quota, with each permit equivalent to one metric ton of carbon dioxide. Those permits are tradable so that they can be sold by polluters, who end up emitting less than their cap, to others who wish to pollute more. Such trading of emission permits establishes in effect a market price for carbon, normally quoted in the local scheme's currency per ton of CO_2 or its equivalent (tCO_2e) the latter of which applies to other greenhouse gases, such as methane or nitrous oxide

(N_2O), quoted as standard multiples of CO_2 according to their global warming potential in compliance with a formula established in Kyoto. The strength of such carbon markets depends not least on an effective enforcement mechanism requiring measuring, reporting, and verification (MRV) by a centralized regulator who must also be able to impose effective fines and sanctions on violators exceeding their assigned pollution limits or lying about their actual emissions. Another key question concerns the allocation of the emission permits, either for free or by auction, with permissible "baseline" pollution quantities based on past historical performance and realistic reduction paths going forward. Some "cap and trade" schemes allow unused permits to be carried forward to the next period. To the extent that caps need to be lowered over time in line with announced emission-reduction targets, carbon markets may also have provisions for retiring emission permits, including making it possible for environmentalist NGOs to purchase them with the aim of taking them out of circulation or for polluters to obtain a tax break when donating permits to non-profit entities.

Notwithstanding a carbon market's typical transaction costs (associated with monitoring, reporting and verification of emissions, obtaining emissions permits, or evaluating abatement options), emissions trading emerged as an elegant way to internalize what up to now had been an intangible externality, the emission of greenhouse gases, by forcing it through a market mechanism for the evaluation of thereby privatized costs and gains. What serves as property right here in the Coasian sense is the permission to emit carbon dioxide or other greenhouse gases which can be traded away to another polluter willing to pay for it. Such possible exchanges establish a market-determined price for the externality, say a ton of carbon dioxide, on the basis of which emitters can evaluate in comparative fashion the costs of pollution versus abatement in order to determine their incentives for gains in pursuit of the most cost-efficient option. If the marginal cost of abatement based on available means to reduce CO_2 emissions is less than the price of the emissions permits, then it makes sense to pursue emissions reduction, accumulate thereby excess supplies of emissions permits, and then sell those off to the highest bidders for a gain. If on the other hand marginal abatement costs are above the price of emissions permits, then one would rationally prefer to continue polluting to the point of buying additional permits to do so. All depends here on having the carbon markets establish sufficiently high prices for emission permits to make the incentives work in favor of widespread pursuit of abatement strategies. Cap-and-trade schemes thus enable countries to meet GHG emission-reduction targets at the lowest possible aggregate cost of mitigation while incentivizing its polluters to innovate with regard to pollution-reduction strategies and technologies.[12]

We have seen quite a number of carbon markets introduced over the last dozen years using the cap-and-trade approach.

- The first, also up to now the largest, was the European Union Emissions Trading Scheme whose launch in 2005 coincided with the Kyoto Protocol taking effect. The EU ETS includes thirty-one countries (the EU plus Iceland, Norway, and Liechtenstein) and covers more than 11,000 installations with a net heat excess of 20 megawatts in the industrial and energy sectors which together are responsible for close to half of the area's CO_2 emissions and 40% of overall GHG emissions.
- The New Zealand Emissions Trading Scheme (NZ ETS), starting in 2008, provides partial coverage (energy and industry, as well as forests as a net sink, but not the country's large agriculture sector) capturing a bit more than half of all emissions in that country.
- That same year saw also the launch of the Swiss ETS which allowed energy-intensive entities covering about 10% of Switzerland's total GHG emissions to trade emissions permits in lieu of paying the country's CO_2 levy.
- A year later, in 2009, nine states in the Northeast of the USA (Connecticut, Delaware, Maine, Maryland, Massachusetts, New Hampshire, New York, Rhode Island, and Vermont) introduced the so-called Regional Greenhouse Gas Initiative (RGGI, pronounced "Reggie") to cover all their fossil-fueled power plants 25 MW or larger.
- Tokyo, the world's largest city, launched Asia's first mandatory emissions-trading scheme, the Tokyo Cap-and-Trade Program (TCTP), in 2010 to target energy-intense buildings for GHG emission cuts which in three years amounted to 22% from the 2009 baseline.
- In 2012 California and Quebec, both parties of the initially far larger joint American-Canadian Western Climate Initiative which subsequently fell apart especially on the US side, launched their respective cap-and-trade programs covering power plants, refineries,

and other large emitters of greenhouse gases which they subsequently linked. That linkup will in all likelihood soon be expanded to include also Ontario which initiated its own ambitious cap-and-trade scheme in 2017.

- In 2013, major fossil-fuel producer Kazakhstan introduced its own emissions-trading scheme (KAZ ETS) aimed at the country's principal oil and gas extraction firms, coal mines, and power companies. Opposition by these powerful sectors obliged the Kazakh government two years later to suspend the program, with the intent of reintroducing it under new rules in 2018.
- At the beginning of 2015, South Korea launched the then-second-largest carbon market, the Korea Emissions Trading Scheme, which covers the country's industrial heart—specifically 525 firms comprising all electricity generators, car manufacturers, steel producers, electro-mechanical firms, airlines, and petrochemical refiners. The KETS is Korea's main plank with which to achieve its planned 30% cut in GHG reductions by 2020.
- In July 2017, we saw China set up its own carbon trading scheme covering the six most important GHG-emitting sectors of the economy, an initiative that might eventually even dwarf the EU's scheme in scale once it operates fully as initially envisaged.

While those local and regional cap-and-trade schemes have typically sprouted on their own, the product of national environmental-policy dynamics pushing in the direction of emissions trading to combat climate change, there is a global dimension to these initiatives which already emerged with the Kyoto Protocol and got further refined in the Paris Agreement. Whereas GHG emissions occur in specific places, once those gases enter into the atmosphere they get evenly distributed across the globe. Hence, it does not matter from the point of view of climate-change mitigation where we cut emissions or set up compensatory carbon sinks. The impact of any emission cuts or new sinks will be felt globally no matter where they occur. This fact has encouraged mechanisms for global emission-reduction offsets between countries so that less developed countries, where rapid growth spurs faster increases in GHG emissions while less stringent regulations and lower price levels offer cheaper emission-reduction opportunities, can also benefit from emission-lowering projects even though they are not formally (yet) parties to cap-and-trade schemes. Of course, such offsets reinforce an already troubling bias of cap-and-trade schemes in favor of letting major emitters in rich countries off the hook, by providing them with more low-hanging fruit to pick from before having to make the tough climate-change mitigation choices. In addition, these transfer mechanism have also given rise to complaints of environmental imperialism and/or racism inasmuch as they have encouraged controversial projects hurting specific populations in poor countries or the environment they live in.[13]

Under the Kyoto Protocol's JI and CDM initiatives emission-reducing projects generate carbon credits, known respectively as Emission Reduction Units for JI projects or Certified Emission Reduction units for CDM projects, which host countries can sell to Annex 1 countries to count toward the latter's emission-reduction commitments. All JI and CDM projects have to be approved and verified by a complex evaluation mechanism under the auspices of the United Nations Framework Convention on Climate Change (UNFCCC) to ascertain that they would not have taken place otherwise ("additionality"), comply with stringent eligibility criteria, and actually achieve claimed emission reductions. Those ERUs and CERs are exchangeable with Kyoto's so-called Assigned Amount Units (AAUs), which every signatory (Annex 1) country gets assigned up to the amount of its permitted emissions, thereby rendering them convertible with various emission permits. That convertibility made it possible for the ERU and/or CER carbon credits to get integrated into the various cap-and-trade schemes, allowing those to offer the option of carbon-credit offsets within quantitative limits set for such substitutions. Notwithstanding administrative complexities and policy-related uncertainties depressing usage of these cross-border flexibility mechanisms, during the relevant period from 2001 to 2012 (when the Kyoto Protocol expired) 1 billion CERs were issued, of which 60% went to China, as well as 300 million ERUs of which more than three-quarters went to Russia and Ukraine for eventual re-sale mostly to the European Union.

Article 6 of the Paris Agreement phased out both JI and CDM in favor of a new mechanism, Internationally Transferred Mitigation Outcomes (ITMOs), with which to transfer emission reductions between countries. It is not yet clear to what extent ITMOs will be more actively traded than their two predecessors. On the one hand, they apply to a much larger number of countries following climate-mitigation targets than was the case under Kyoto. But on the other hand, precisely because so many developing countries now pursue their own emission-reduction targets, they may be more inclined to use emission-cutting projects toward these goals rather than letting other countries claim those. In addition, Paris contains tighter rules than Kyoto against double-counting which may further reduce selling of emission reductions between countries. Yet at the same time the Paris Agreement also provides more flexibly for a wider variety of transfer arrangements than the hitherto standard carbon-market mechanism, including bilateral (possibly even pluri-lateral) arrangements sharing mitigation outcomes involving cross-border projects or publicly funded reductions supported by multilateral finance entities such as the Green Climate Fund. We have yet to explore fully all the possible approaches opened up by the broadly worded Article 6 provisions which we are just beginning to figure out collectively in terms of their potentially wide-ranging implications.[14] Whatever specific transfer arrangements we will launch over the next decade in compliance with the Paris Agreement, it is clearly implied by this accord that ITMOs can be most easily moved between countries within internationally integrated carbon markets. There is hence a strong impetus for setting up more cap-and-trade schemes in pursuit of climate-mitigation targets laid out in various countries' NDCs and hooking those up toward gradual creation of an increasingly globalized carbon market—a goal explicitly promoted by various organizations such as the International Carbon Action Partnership, the already cited International Emissions Trading Association, or the World Bank's Networked Carbon Markets initiative.

THE TROUBLE WITH MARKET-DETERMINED CARBON PRICING

A key argument in favor of building climate policy around cap-and-trade schemes rather than carbon taxes is that the former target quantities rather than prices, the more appropriate policy tool if the principal objective focuses on (quantity) reductions of GHG emission levels. Up to now, however, cap and trade has had decidedly modest, if not altogether disappointing, emission-reduction effects. Installations covered by EU ETS managed to decrease their GHG levels by 8% in the first five years of that scheme's existence (2005–2010) and by 5% during Phase 3 of the scheme (2013–2017), according to the European Commission's own estimates. These emitters could have achieved more significant emission cuts had they spent the money vested in that market instead directly on targeted mitigation measures such as upgrading power plants. The scheme will miss its already modest 21% reduction target (from 2005 levels) set for 2020. The NZ ETS has yet to yield any emission reductions in the nine years of existence, not least because its key emitters have relied heavily on buying up cheap Kyoto carbon credits, above all CERs from China, so as to continue their emissions at presently high levels. The Swiss ETS and RGGI in the Northeast of the USA are too small in scale to have any tangible impact. In contrast, Tokyo's ETS has been largely successful by targeting large commercial building in a stringent policy framework which encouraged concrete mitigation steps on a large scale across the vast city, such as replacing lighting, installing better cooling systems, and using new energy-saving technologies. There is a lesson to be learned here, namely the need to achieve emission reduction by means of concrete mitigation measures. Emissions data from California have shown modest annual reductions of 2% from emitters covered by its cap-and-trade scheme, a pace which will have to be sped up significantly for that scheme to play its useful role in that state's ambitious climate policy.

More generally, the unacceptably low decreases in GHG emission levels tied to carbon markets up to now have much to do with the inability of these schemes to send sufficiently strong price signals to incentivize emitters in the direction of abatement rather than continued pollution. This failure has been a global one, evident in pretty much all cap-and-trade schemes and hence possibly a systemic flaw. The prices for emission permits or carbon offsets established in those markets have been too low to begin with and subsequently declining further, precisely the opposite of what should have happened for these markets to be effective.[15] Take, for example, the EU ETS, the world's oldest and largest cap-and-trade scheme. It started in 2005 with an initial price of about 24€/tCO_2 (or per EU Emission Allowance, EUA) which rose in short order to an all-time peak of about 30€ in April 2006 before collapsing amidst spreading indications of initial excess allocations of EUAs all the way down to 10 cents by September 2007. New rules for caps and EUA allocations, the so-called Phase II, pushed the carbon price back up to 29.20€ a ton in July 2008 before renewed over-allocation of permits combined with falling economic activity during the Great Recession and worsening climate-policy divisions to depress that price again to a 10–15€ range throughout much of 2009 and 2010 before the subsequent euro-zone crisis caused further price erosion to below 4€ in June 2013. Even as recently as 2016 the EU ETS' carbon price remained for much of the year below 4€ per ton. Today's (September 23, 2017) closing price was 6.50€, better but still far from adequate. Or take

California's carbon-market price which started above $22 per ton, but then fell to below $14 within a year's time in the face of over-allocation of permits. For nearly two and a half years, from the beginning of 2014 to late 2016, the price remained in a $12–13.50 range before rebounding a bit amidst new rules passed in the wake of the hookup with Quebec, greater legal certainty following state victories in a couple of crucial lawsuits and passage of market reforms in California's legislature, as well as successful implementation of a $13.57/ton price floor in emission-permit auctions. Currently the actual price is approaching $14, just high enough for the carbon market to do its job reasonably. Both the KETS and the NZ ETS have had low carbon prices throughout 2017 as their respective governments delayed crucially needed policy clarification and reform steps related to auctions of emission permits and so eroded confidence in those schemes. China's scheme is expected to yield carbon prices in the $10–15 per ton of CO_2 range, consistent with other national experiments cited above but not high enough to motivate needed abatement effort.

A key problem undermining the price levels in any of these carbon-market experiments concerns the initial allocation of emission permits. Most schemes did so for free and issued too many of them, with both biases nourished by the desire of government officials in charge to get skeptical emitters on board. The industries concerned, notably energy producers, public utilities, and key industrial manufacturers, form powerful lobbies able to push for generous allocations to their sectors as a prerequisite for their members' participation. This gets us back to the question of governance in climate policy, as established "brown" sectors responsible for much of the GHG emissions are still politically powerful while newer "green" sectors representing a cleaner future are not yet adequately represented. In countries, whose political process depends heavily on the interplay of lobbies—from the "free-market" USA to oligarchy-controlled Russia—effective climate policy is far more difficult to put into place than in countries with stronger governments and leaders capable of acting in the nation's best long-term interest as they pursue the public good. Free allocation of emission permits, in lieu of auctions, also offers polluters a source of windfall gains to the extent that these can be resold at the prevailing carbon price. On a more basic level, such free allocations only reinforce what is the most basic flaw of cap-and-trade schemes, namely to legitimate the status quo of continuous pollution as the most viable option going forward for individual firms.

As if the initial allocation procedure were not already fraught with enough problems, the use of emissions permits itself tends toward their oversupply. Governments set their caps typically for a period of four or five years against a benchmark level rooted in current or recent emissions performance, and any improvement over that period then risks circulating too many of these permits relative to newly lowered emission levels. Add to this the great sensitivity of emission levels to fluctuations in economic growth so that any slowdown immediately ends up depressing carbon prices in cap-and-trade settings. We have also seen repeatedly, as with the EU ETS, California's scheme, the KETS, or the NZ ETS, how vulnerable such schemes and their pricing mechanism are to political and legal uncertainty of which there is inevitably going to be a lot given their complex and inherently controversial nature. Not only are market-determined carbon prices low, but they are also too volatile. The prospect of major price fluctuations, already witnessed in those markets over even short time intervals of just a few weeks or months, invites inevitably lots of speculators impacting either directly on spot prices or seeking their gains indirectly via carbon emissions futures and options traded in NASDAQ Commodities or the Intercontinental Exchange.[16] Such excessive price volatility is, however, very harmful to the long-term investment and planning horizons needed in the fight against climate change where much of the investment activity involves developing new technologies and replacing durable structures that have innately long life cycles (e.g., power plants).

Administrators of the various cap-and-trade schemes, from the European Commission to CARB in California, have recognized the problem of inadequate carbon pricing and tried to address the issue with various reforms. All these market-based initiatives are designed to unfold in consecutive three- to five-year phases, not only to tighten the emissions caps but also to make adjustments in how they operate. Pretty much each of the aforementioned cap-and-trade schemes has thus had a chance to fortify its pricing mechanism—from relying more on auctions and less on free allocation of emissions permits to figuring out different ways to remove excess allocations of such permits (retiring them, putting them in reserve accounts, having them bought up by public-interest organizations for withdrawal from circulation, etc.) and, finally, to imposing minimum-price floors.[17] And our discussion above of the scheme's price fluctuations so far indicates that some of these measures in support of higher carbon prices have had limited success, as has been recently the case in California. Still, I doubt very much that cap-and-trade schemes can impose, let alone sustain, sufficiently high prices for GHG emissions to induce needed abatement efforts leading to the kind of emissions reductions the world needs to achieve urgently over the next decade or so. It would take much tighter caps, well-run auctions replacing free allocations entirely, global spread of such carbon markets and their effective linkup, ingenious transfer arrangements of ITMOs so that they become widely popular and easily used, as well as broader coverage of emission sources captured by those schemes—and we are still years away from any of these prerequisites being put in place while time is running short. Even if we assumed to have created such an effective, global carbon market, its principal purpose would still be allowing sources of GHG emissions to continue polluting provided they are willing to pay for it. True, governments could use the revenue collected from the cap-and-trade auctions on climate-related initiatives. But why not mobilize the private sector to spend all that money vested in the acquisition and trading of emissions permits instead directly on concrete climate-change mitigation steps that would lower emissions as a result? We may get faster cuts that way, no matter how high the price of carbon emissions. Ultimately that question bears cold-eyed reflection, without being stuck by one's own ideological preferences filtering the choice. The Coase Theorem applies after all only to limited externalities whose social costs are sufficiently small to be privatized and thus integrated into the actors' individual cost-benefit calculus. Greenhouse gas emissions are decidedly *not* such a limited externality; they are an existential threat covering much, if not most, of our economy's contemporary organizational structure and sectoral makeup. A problem of this scale is simply too big for cap and trade as the main policy tool, and its resolution too urgent.

What About Carbon Taxes?

Cap-and-trade schemes have up to now dominated climate policy by carrying the argument that they create a great deal of certainty with regard to how much emissions will be cut in compliance with the caps imposed. In reality, carbon trading's predominance is probably more due to its acceptability among businesses affected for whom its introduction—often with free allocation of emission permits and low prices for such permits—has not yet had much impact on the bottom line. But that is also precisely why this quantity instrument has so far failed to do its job for society's benefit, namely to achieve the needed emissions cuts. Instead it has offered greenhouse gas emitters a rather cheap option to continue polluting. Perhaps recent adjustments, such as tightening caps, relying more on auctions when issuing emissions permits, and imposing minimum floor prices for them, will have the desired effect of pushing up carbon prices and thereby inducing greater efforts at lowering emissions. But how long can we afford to wait for these reforms to bear fruit, if ever, when time presses? Given how far we have to go and how little time we have left to get there, it seems to me that cap-and-trade schemes will have to sustain carbon prices at least twice as high as current levels, operate in many more countries than is currently the case, be effectively interconnected to move in the direction of a global marketplace, and incorporate a much larger volume of (ITMO) offsets than under Kyoto within the next five years to remain credible. We should use the Paris Agreement's 2020 deadline for new NDCs with which to crank up ambition as the moment to make an honest assessment of worldwide progress concerning cap and trade. At that point we may very well decide to push carbon taxes instead.

Carbon taxes have the advantage of imposing a sufficiently high, stable, and hence predictable price on greenhouse gas emissions with which to evaluate the profitability of investment projects, assess climate-related risks, and pursue emission-reduction targets. We can see the need for such a powerful price instrument right now in the fact that a growing number of companies have already adopted internal-carbon prices many of which are significantly higher than the market price established by the relevant region's cap-and-trade scheme.[18] But at this point, in late 2017, carbon taxes still lag significantly behind cap and trade. According to World Bank (2016), we will have forty-four (national or subnational) carbon pricing initiatives implemented or under consideration by the end of 2017 which, if we include China's nationwide cap-and-trade scheme, cover almost a quarter of all GHG emissions across the globe. Less than half of those, nineteen to be precise, involve carbon taxes which together cover barely 10% of global GHG emissions. Finland, Poland, Sweden, Norway, Denmark, Latvia, Slovenia, and Estonia have carbon taxes dating back to the 1990s, while those of British Columbia, Switzerland, Iceland, Ireland, and Japan were introduced between 2008 and 2012. The UK adopted such a tax on fossil-fuel producers as a carbon price floor in 2013. France, Mexico, Portugal, South Africa, Chile, and Alberta have initiated carbon taxes only over the last three years. In many countries the tax level is very low, in line with the prices typically established in cap-and-trade schemes and ranging from less than a dollar per ton (in Poland or Mexico) to $10/tCO_2$e (Iceland). Another six have levels between $19 and 25, whereas four outliers have comparatively much higher carbon taxes of $52/ton (Norway), $60–65/ton (Finland), $86/ton (Switzerland), and $131/ton in Sweden.

The very high carbon tax in Sweden, ratcheted up gradually since 1991, has provided strong incentives for both producers and consumers to search for environmentally friendly solutions and has allowed this Nordic country of 9.5 million inhabitants to enjoy rapid economic growth while systematically cutting GHG emissions. Sweden is on schedule to reach its goal of carbon neutrality by 2050, using bioenergy for direct heating of homes, phasing out fossil fuels in favor of renewables, running vehicles increasingly on biofuels or natural gas, imposing congestion charges on inner-city traffic, and making buildings far more energy efficient. An early global leader in environmental policy, Sweden has consistently shown the rest of the world the way forward with innovative initiatives rooted in an unusually high degree of citizen engagement, research and development, as well as pace-setting policy initiatives. Swedish cities, notably Stockholm, Malmö, or Linköping, are forerunners in sustainable urban planning and have already managed to create fully carbon-neutral neighborhoods by pursuing multifaceted strategies with regard to energy efficiency, smart grids, transportation, building codes, heating and cooling strategies, or recycling of waste material. The carbon tax, together with a long-standing energy tax on fuels, carries very low administrative costs. Consecutive step-by-step increases to their currently high levels have been accompanied by compensatory reductions of income tax charges on low-income earners and cuts in labor taxes boosting employment. Where the environmental taxes bite, consumers and producers have clean alternatives available with which to reduce their tax burden. All these provisions have made the carbon tax widely accepted among Swedes as an example of good fiscal and environmental policy.[19]

Another example of a successful carbon tax is the one introduced by the Canadian province of British Columbia in 2008 at an initial level of 10 Canadian dollars per ton of carbon dioxide equivalent and rising every year before it was frozen at 30 dollars in 2012. Collected efficiently at the retail level (e.g., at the pump for gasoline or diesel), this tax applies to fossil fuels used for transportation, home heating, biofuels, and electricity (most of which consists of cheap hydroelectric power in the province). Apart from its broad base, BC's carbon tax is also revenue neutral inasmuch as it has been used to lower the province's corporate income taxes to the lowest level in Canada as well as reduce the two lowest brackets of the personal income tax in favor of low-income earners. This has reinforced widespread support for the carbon tax, with an annual tax shift in excess of 1.5 billion Canadian dollars. Most importantly, evidence shows that British Columbia's per capita consumption of all taxed fuel types has declined between 5 and 15% over the past decade while the province has consistently been among the fastest growing in all of Canada. Whatever problems may be associated with BC's carbon tax has to do with its strictly local nature, as the inhabitants of Vancouver find it easy to get cheaper gas across the border in Washington State and lower-cost (because untaxed) electricity or fuel can enter the province on a more competitively priced basis in the absence of border-tax adjustments. Particularly interesting has been the political context of BC's carbon tax which was initially introduced by the right-leaning provincial Liberal Party whose politicians have ever since successfully defended this measure as the most market-friendly climate-policy option against recurrent attacks by the left-leaning New Democrats favoring cap and trade instead.[20] That problem may be somewhat attenuated in 2018 when Prime Minister Justin Trudeau's initiative for nationwide carbon pricing takes hold across Canada, but would have been even more alleviated had Washington's voters across the border not rejected Initiative 732 in 2016 to introduce the nation's first carbon tax.

Even though majorities of Americans worry about climate change and would like something done about this problem, political pressure in the USA has not been strong enough yet to push the federal government into pursuing a consistent nationwide climate policy that would match the kind of broad-based approach taking hold in Canada or the European Union. Hence, Americans risk falling behind the rest of the world in terms of green technologies, energy-efficiency gains, urban planning, environmental protection, and public transport. While the ruling Republican Party has been captured by climate skeptics to the detriment of constructive policy-making, some of the party's elder statesmen have recently rang alarm bells about the need for a more balanced approach to global warming which has led them to push for adoption of a nationwide carbon tax. Led by former Treasury Secretaries James Baker and Henry Paulson, while also including economists George Schultz,

Gregory Mankiw, and Martin Feldstein (all of whom had played themselves leading roles in past Republican Administrations), this illustrious group responded to Trump's ascent to the presidency in February 2017 with a proposal for a $40/tCO$_2$ tax whose rates would rise continuously. The Baker plan would have replaced Obama's Clean Power Plan (which Trump has already committed to dismantle) as well as other climate regulations as the best "conservative climate solution" based on free-market principles. The tax would be collected at such entry points as the well, the mine, or the port where the various fossil fuels enter the domestic economy and in this fashion presumably also include imports. Thus collecting up to $300 billion in annual revenues, the proceeds of such a tax would be returned to consumers in the form of a "carbon dividend" amounting to a check of about $2000 for a family of four, a measure bound to anchor that tax's widespread acceptability.[21]

Another voice in support of a carbon tax has recently come from the International Monetary Fund whose research staff and leadership have made consistent arguments in favor of its widespread adoption since even before the Paris Agreement. That push is part of a broader IMF emphasis on the crucial role of fiscal policy in addressing the environmental challenges we face. This starts already with pointing to the huge subsidies (estimated to exceed $5 trillion per year) we globally still direct toward fossil fuels as we provide tax breaks for their exploration, keep prices down, and fail systematically to account for the negative externalities they cause as our primary sources of energy. Carbon taxes then provide in this context a necessary corrective, a powerful price instrument with which to encourage a switch to cleaner fuels and less energy use while at the same time raising large amounts of government revenues with which to pursue broader tax reform and/or fund other measures in support of greater climate resilience. In October 2015 Christine Lagarde, the head of the IMF, determined climate change to be a "macro-critical" issue, and she has argued ever since in favor of a strong carbon tax starting at $30 per ton as the most effective policy response to face this systemic challenge. One of the key arguments of the IMF is that such a carbon tax, while it should encompass the twenty leading countries responsible for 80% of all GHG emissions, need not be set at a uniform level. On the contrary, the IMF's own research has shown large differences between countries when it comes to the costs of internalizing their respective contributions to global warming and hence justifying very different carbon tax rates. For Saudi Arabia and Iran, for example, those costs (and hence ideally also carbon tax rates) would be very high not only because of their position as leading oil producers, but also in light of how artificially low their theocratic governments have kept domestic gasoline prices and electricity costs for decades. In India and China costs (and thus also rates for eventual carbon taxes) are kept high because of their excessive reliance on coal, a particularly dirty fossil fuel. At the other end of the spectrum you have Brazil with actually negative costs as provider of the world's largest carbon sink, the Amazon rainforest.[22]

The time for such a concerted effort on a global level is rapidly approaching. Right now only about a quarter of all GHG emissions are subjected to a price worldwide, and that carbon price averages less than $10/ton—less than a quarter of what it should be. Yet at the same time ninety countries have committed themselves in their NDCs to some form of carbon pricing, whether through a tax or via cap and trade. In 2020, when we assess progress concerning implementation of the commitments made in the initial NDCs and seek to adopt more ambitious policy goals, we may very well have an opportunity to make substantial progress in mobilizing leading carbon emitters into a coalition of the willing with regard to introducing a carbon tax on a large scale, perhaps pushed forward by the emergence of a debate on climate policy in the wake of the US election campaign. Several candidates, such as Governor Andrew Cuomo (D-NY) or even challengers to Trump in the Republican party (such as Ohio governor John Kasich), are waiting in the wings to do precisely that. At that point a carbon tax, if adopted by leading countries across the board, may even be integrated into existing cap-and-trade schemes by serving as the minimum-price floor which GHG emitters will have to pay for obtaining their emission permits. The question then is what would be a reasonable amount to charge suppliers of fossil fuels and other emitters? Luckily, the recent report of the Stiglitz-Stern Commission provides us with reasonably and convincingly argued guidance here.

NOTES

1. For details on the calculation of carbon budgets in IPCC (2014) and possible adjustments to bring those measures more up to date, see R. McSweeny and R. Pearce (2016), N. Evershed (2017), as well as Global Carbon Project (2017). J. Rogelj et al. (2016) and New Climate Institute (2016) discuss the implications of the Paris Agreement for targeted emission-reduction pathways and the climate budget.

2. Atmospheric CO_2 concentrations are measured daily by two independent monitoring programs, Scripps and the US' National Oceanic and Atmospheric Administration (NOAA), at Hawaii's Mauna Loa Observatory over 3400 meters over sea level. It should be noted that Trump's budget for fiscal year 2018 proposed a 16% cut for NOAA. The data collected from these two sources are discussed in the CO2.earth site. There you will also find data for yearly global warming trends discussed here. The global emissions data for 2016 are from R. Pidcock (2017).

3. I have found discussions of the social cost of carbon in S. Evans et al. (2017), Institute for Policy Integrity (2017) as well as W. Nordhaus (2016) particularly useful.

4. The Interagency Working Group on Social Cost of Carbon, convened in 2009 by Obama's Council of Economic Advisors and Office of Management and Budget, also included the Council on Environmental Quality, National Economic Council, Office of Energy and Climate Change, Office of Science and Technology Policy, Environmental Protection Agency, and the US Departments of Agriculture, Commerce, Energy, Transportation, and Treasury. The 2010 Technical Support Document of the IWG explains in great detail the methodology used in calculating its $SC\text{-}CO_2$ estimates.

5. See the Office of Management and Budget (2003) and the U.K. Treasury (2003) for such guidance.

6. The burning of fossil fuels, such as coal, emits fine particulates which harm human lungs. But Trump's EPA is arguing that emission reductions in areas where concentrations of such fine particulates are low to begin with, which is 12 micrograms of fine particles per cubic meter, do not yield any health benefits, because there is no risk below that threshold to begin with. See C. Mooney (2017) for more on this rule change and the other EPA initiatives we discussed to downplay the costs of climate change.

7. See CDP (2016) for an extensive discussion of internal-carbon pricing trends among corporations, by sector and region, from which I took those examples of specific firms. Only 30% of the firms surveyed disclose their internal carbon price.

8. For the commission's report see J. Stiglitz and N. Stern (2017).

9. For a meaningful discussion of the SCC, the MAC, and their interaction to yield the SPC, see R. Price et al. (2007).

10. The Coase Theorem, as formulated in R. Coase (1960), framed the internalization of the social costs as having the parties involved bargain to a mutually beneficial efficient outcome, as in his example of a rancher's cows invading the neighboring farmer's farmland to cause crop damage so that the need arises to decide whether to build a fence or accept the

crop damage. Who builds the fence, assuming it costs less than the crop damage, depends on the initial allocation of property rights which in the absence of transaction costs should not matter.

11. We should note here that Richard Sandor, a key financial innovator and close collaborator of Ronald Coase, set up his own carbon-market experiment. In 2003, he launched the Chicago Climate Exchange (CCX) as America's first voluntary, legally binding trading platform for emissions permits among over 400 participants with a combined aggregate baseline of 680 million metric tons of CO_2e and a goal of cutting emission levels for six GHGs by 6% in 2010.

12. The original idea of using Coase's Theorem in this way to create a market-based solution for having negative externalities paid for by their creators comes from D. W. Montgomery (1972). See in this context the meaningful discussions of transaction costs and incentives associated with emissions trading offered by R. Stavins (1995), E. Woerdman (2001), P. Heindl (2012) and J. Corea and J. Jaraite (2015).

13. For some specific case studies exemplifying this problematic aspect of carbon-offset projects in developing countries harming the local populations and/or their environmental surroundings there see Carbon Trade Watch (2009), the *Bulletin* of the World Rainforest Movement, and the powerfully argued book on climate change by N. Klein (2014), and M. Curtis (2017).

14. Besides the new ITMOs the Paris Agreement also reinforced other flexibility mechanisms, transforming Kyoto's JI and CDM provisions into stronger and more broadly applicable "Cooperative Approaches" and "Sustainable Development Mechanism" respectively, as discussed by S. Zwick (2016a). I would in the same vein also recommend S. Zwick (2016b) on the Paris Agreement's wide-ranging implications for climate-mitigation transfers between countries laid out in. Additional helpful commentary about the new market provisions contained in the Paris Agreement can be found in Center for Clean Air Policy (2016) and IETA (2016).

15. The carbon price information discussed here subsequently was collected from Web sites tracking the price trends of various cap-and-trade schemes, such as carbon-pulse.com, the "commodities" section of markets.businessinsider.com, or the c2es.org site of the Center for Climate and Energy Solutions.

16. Such emissions futures and options, while certainly useful to make speculative bets on carbon-price movements over the near future, also serve as useful tools to hedge exposure and mitigate environmental compliance risk, thereby making the cap-and-trade schemes work better.

17. Price floors in cap-and-trade schemes, as in those practiced in California, Quebec, or the RGGI, typically involve setting a reserve price at allowance auctions which automatically reduces the number of allowances being made available if the price is too low and unsold allowance are held in reserve for possible future release or eventual permanent withdrawal. Another way of implementing a price floor is to have a tax top up any actual auction price below the floor to the required minimum.

18. According to J. Camuzeaux and E. Medford (2016) and Carbon Disclosure Project (2017), of the more than 1250 listed corporations having adopted an internal carbon price many have chosen price levels significantly above current market prices in the 6€–$15 per ton range, all the way to Switzerland's pharmaceutical giant Novartis' $100/tCO$_2$e.

19. See G. Fouché (2008) and Swedish Institute (2016) for more details on Sweden's pace-setting initiatives to lower emissions across a wide range of activities and structures while turning those into a competitive edge over other nations in terms of energy efficiency and technological edge.

20. Good discussions of British Columbia's successful experiment with a carbon tax, which has led it to be recommended by the United Nations, World Bank, and OECD as a model for others to follow, can be found in *The Economist* (2014) and E. Porter (2016).

21. See J. Schwartz (2017) for more detail. Broader implications of a US carbon tax are well discussed in I. Perry (2015).

22. The IMF's increasingly urgent case for a globally applied carbon tax, set at a minimum of $30/tCO$_2$e with higher rates possible for individual member states in need of greater revenues or more intense mitigation efforts, has been made, among other sources, by C. Lagarde (2015) and International Monetary Fund (2017).

References

Camuzeaux, J., & Medford, E. (2016, December 12). How Companies Set Internal Prices on Carbon. *Climate 411* Blog. Environmental Defense Fund. http://blogs.edf.org/climate411/2016/12/12/how-companies-set-internal-prices-on-carbon/. Accessed 4 Oct 2017.

Carbon Disclosure Project. (2016). *Embedding a Carbon Price into Business Strategy.* https://b8f65cb373b1b7b15feb-c70d8ead6ced550b4d987d7c03f-cdd1d.ssl.cf3.rackcdn.com/cms/reports/documents/000/001/132/original/CDP_Carbon_Price_report_2016.pdf?1474899276. Accessed 25 Nov 2017.

Carbon Disclosure Project. (2017). *Commit to Putting a Price on Carbon.* https://www.cdp.net/en/campaigns/commit-to-action/price-on-carbon. Accessed 4 Oct 2017.

Carbon Trade Watch. (2009, December). *Fact Sheet 2: Carbon Offsets*. http://www.carbontradewatch.org/downloads/publications/factsheet02-offsets.pdf. Accessed 22 Sept 2017.

Center for Clean Air Policy. (2016). *CCAP Submission on Internationally Transferred Mitigation Outcomes*. New York: UNFCCC. https://unfccc.int/files/parties_observers/submissions_from_observers/application/pdf/696.pdf. Accessed 23 Sept 2017.

Coase, R. (1960). The Problem of Social Cost. *Journal of Law and Economics, 3*(1), 1–44.

Corea, J., & Jaraite, J. (2015). *Carbon Pricing: Transaction Costs of Emissions Trading vs. Carbon Taxes* (CERE Working Papers, 2015: 2). https://gupea.ub.gu.se/handle/2077/38073. Accessed 15 Sept 2017.

Curtis, M. (2017, October 3). Guest Blog: Are European Taxpayers Funding Land Grabs and Forest Destruction? *FERN Blog*. http://www.fern.org/node/6380. Accessed 4 Oct 2017.

Drupp, M., et al. (2015). *Discounting Distangled: An Expert Survey on the Determinants of the Long-Term Social Discount Rate* (Working Paper, No. 195). Leeds University's Centre for Climate Change Economics and Policy. http://piketty.pse.ens.fr/files/DruppFreeman2015.pdf.

Evans, S., Pidcock, R., & Yeo, S. (2017, February 14). Q & A: The Social Cost of Carbon. *Explainers* Blog, Carbon Brief. https://www.carbonbrief.org/qa-social-cost-carbon. Accessed 24 Nov 2017.

Evershed, N. (2017, January 19). Carbon Countdown Clock: How Much of the World's Carbon Budget Have We Spent? *The Guardian*. https://www.theguardian.com/environment/datablog/2017/jan/19/carbon-countdown-clock-how-much-of-the-worlds-carbon-budget-have-we-spent. Accessed 31 May 2017.

Fouché, G. (2008, April 29). Sweden's Carbon-Tax Solution to Climate Change Puts It Top of the Green List. *The Guardian*. https://www.theguardian.com/environment/2008/apr/29/climatechange.carbonemissions. Accessed 5 Oct 2017.

Global Carbon Project. (2017). *Global Carbon Budget 2016—Presentation*. http://www.globalcarbonproject.org/carbonbudget/16/presentation.htm. Accessed 31 May 2017.

Heindl, P. (2012). *Transaction Costs and Tradable Permits: Empirical Evidence from the EU Emissions Trading Scheme* (ZEW Discussion Paper, No. 12-021). http://www.zew.de/pub/zew-docs/dp/dp12021.pdf. Accessed 15 Sept 2017.

IETA. (2016). *A Vision for the Market Provisions of the Paris Agreement*. Geneva: International Emissions Trading Association. http://www.ieta.org/resources/Resources/Position_Papers/2016/IETA_Article_6_Implementation_Paper_May2016.pdf. Accessed 23 Sept 2017.

Institute for Policy Integrity. (2017, February). *Social Costs of Greenhouse Gases—Fact Sheet.* https://www.edf.org/sites/default/files/social_cost_of_greenhouse_gases_factsheet.pdf.

International Monetary Fund. (2017). *Climate, Environment, and the IMF.* Factsheet. https://www.imf.org/en/About/Factsheets/Climate-Environment-and-the-IMF.

IPCC. (2014). *Fifth Assessment Report: Climate Change 2014—Synthesis Report.* Cambridge, UK: Cambridge University Press. http://www.ipcc.ch/report/ar5/.

Klein, N. (2014). *This Changes Everything: Capitalism vs. The Climate.* New York: Simon & Schuster.

Lagarde, C. (2015, December). Ten Myths About Climate Change Policy. *Finance and Investment,* 64–67. http://www.imf.org/external/np/fad/environ/pdf/011215.pdf. Accessed 22 Sept 2017.

McKinsey. (2013, September). *Pathways to a Low-Carbon Economy: Version 2 of the Global Greenhouse Gas Abatement Cost Curve.* Report. http://www.mckinsey.com/business-functions/sustainability-and-resource-productivity/our-insights/pathways-to-a-low-carbon-economy.

McSweeny, R., & Pearce, R. (2016, May 19). Analysis: Only Five Years Left Before 1.5C Carbon Budget Is Blown. *CarbonBrief.* https://www.carbonbrief.org/analysis-only-five-years-left-before-one-point-five-c-budget-is-blown. Accessed 30 May 2017.

Montgomery, D. W. (1972). Markets in Licenses and Efficient Pollution Control Programs. *Journal of Economic Theory, 5*(3), 395–418.

Mooney, C. (2017, October 11). New EPA Document Reveals Sharply Lower Estimate of the Cost of Climate Change. *Washington Post.* https://www.washingtonpost.com/news/energy-environment/wp/2017/10/11/new-epa-document-reveals-sharply-lower-estimate-of-the-cost-of-climate-change/?utm_term=.eeb7ab8e5d02. Accessed 25 Nov 2017.

New Climate Institute. (2016a). *Climate Initiatives, National Contributions, and the Paris Agreement.* Berlin. https://newclimate.org/2016/05/23/climate-initiatives-national-contributions-and-the-paris-agreement/. Accessed 27 May 2017.

Nordhaus, W. (2016). Revisiting the Social Cost of Carbon. *Proceedings of the National Academy of Sciences of the United States, 114*(7), 1518–1523. https://doi.org/10.1073/pnas.1609244114.

Office of Management and Budget. (2003). *Circular A-4: Regulatory Analysis.* White House. https://www.transportation.gov/sites/dot.gov/files/docs/OMB%20Circular%20No.%20A-4.pdf.

Perry, I. (2015). *Implementing a US Carbon Tax: Challenges and Debates.* Washington, DC: International Monetary Fund. http://dx.doi.org/10.5089/9781138825369.071.

Pidcock, R. (2017, November 15). Analysis: What Global Emissions in 2016 Mean for Climate Change Goals. *Carbon Brief.* https://www.carbonbrief.org/what-global-co2-emissions-2016-mean-climate-change. Accessed 24 Nov 2017.

Porter, E. (2016, March 1). Does a Carbon Tax Work? Ask British Columbia. *New York Times.* https://www.nytimes.com/2016/03/02/business/does-a-carbon-tax-work-ask-british-columbia.html. Accessed 6 Oct 2017.

Price, R., Thornton, S., & Nelson, S. (2007). *The Social Cost of Carbon and the Shadow Price of Carbon: What They Are, and How to Use Them in Economic Appraisal in the UK.* MPRA Paper, No. 74976. https://www.gov.uk/government/uploads/system/uploads/attachment_data/file/243825/background.pdf. Accessed 10 May 2017.

Rogelj, J., den Elzen, M., et al. (2016). Paris Agreement Climate Proposals Need a Boost to Keep Warming Well Below 2 °C. *Nature, 534,* 631–639. https://doi.org/10.1038/nature18307.

Schwartz, J. (2017, February 7). 'A Conservative Climate Solution': Republican Group Calls for Carbon Tax. *New York Times.* https://www.nytimes.com/2017/02/07/science/a-conservative-climate-solution-republican-group-calls-for-carbon-tax.html. Accessed 6 Oct 2017.

Smith, S., & Braathen, N. (2015). *Monetary Carbon Values in Policy Appraisal: An Overview of Current Practice and Key Issues* (OECD Environment Working Papers, No. 92). http://dx.doi.org/10.1787/5jrs8st3ngvh-en.

Stavins, R. (1995). Transaction Costs and Tradable Permits. *Journal of Environmental Economics and Management, 29*(1), 133–148. http://scholar.harvard.edu/files/stavins/files/transaction_costs_jeem.pdf. Accessed 15 Sept 2017.

Stiglitz. J., & N. Stern (2017). *Report of the High-Level Commission on Carbon Prices.* Washinton, DC: World Bank Publishing. https://www.carbonpricing-leadership.org/report-of-the-highlevel-commission-on-carbon-prices/.

Swedish Institute. (2016). *Sweden Tackles Climate Change.* https://sweden.se/nature/sweden-tackles-climate-change/. Accessed 5 Oct 2017.

The Economist. (2014, July 31). British Columbia's Carbon Tax: The Evidence Mounts. https://www.economist.com/blogs/americasview/2014/07/british-columbias-carbon-tax. Accessed 6 Oct 2017.

Tol, R. (2009). The Economic Impact of Climate Change. *Journal of Economic Perspectives, 23*(2), 29–51.

U.K. Treasury. (2003). *The Green Book: Appraisal and Evaluation in Central Government.* https://www.gov.uk/government/uploads/system/uploads/attachment_data/file/220541/green_book_complete.pdf.

Woerdman, E. (2001). Emission Trading and Transaction Costs: Analyzing the Flaws in the Discussion. *Ecological Economics, 38*(2), 293–304. https://doi.org/10.1016/S0921-8009(01)00169-0. Accessed 16 Sept 2017.

World Bank. (2016). *State and Trends of Carbon Pricing, 2016.* Washington, DC. documents.worldbank.org.

Zwick, S. (2016a, February 1). *The Road from Paris: Green Lights, Speed Bumps, and the Future of Carbon Markets.* Ecosystem Marketplace. http://www.ecosystemmarketplace.com/articles/green-lights-and-speed-bumps-on-road-to-markets-under-paris-agreement/. Accessed 11 Nov 2016.

Zwick, S. (2016b, January 29). *Building on Paris, Countries Assemble the Carbon Markets of Tomorrow.* Ecosystem Marketplace. http://www.ecosystemmarketplace.com/articles/building-on-paris-countries-assemble-the-carbon-markets-of-tomorrow/. Accessed 11 Nov 2016.

CHAPTER 6

Climate Finance

The pricing of carbon addresses a huge market failure which threatens our planet. But it is also a vital piece in another puzzle, that of finance. The transition to a low-carbon economy, around which we will mobilize our fight against climate change in coming decades as the defining challenge of the twenty-first century, depends crucially on finance. Specifically, it depends on both the ability and the willingness of financial institutions and markets to play their part in this transition. There are many reasons why finance will be intimately involved. Climate change is not least an insurance problem, a question of protecting against major losses in the wake of events or trends triggered by global warming. And insurance is one of the central pillars of finance. Climate change is also associated with several unprecedented sources of risk capable in the aggregate of disrupting the entire system, and so we will have to adjust our recently launched systemic-risk measures and models accordingly. The emission-trading schemes being put in place in the wake of the Paris Agreement are in essence financial markets and will have to be organized as such. As we work on this transition to a low-carbon economy, we will have to wind down carbon-intensive activities (e.g., burning of fossil fuels) and de-activate structures associated with heavy emissions of greenhouse gases (e.g., coal-fired power plants). This disengagement, which may be carefully planned for or unfold in a disorderly fashion, will leave affected producers with large amounts of "stranded" (i.e., no longer freely usable) assets that will put downward pressure on those firms' valuations. How we will deal with these more or less inevitable losses and at the same time free up capital to help fund the massive scale of needed investments in low-carbon activities and structures will be one of the great challenges facing the entire system of global finance. Finally, the climate-change mitigation efforts of a majority of (low- to middle-income) countries, even just at the relatively modest level of their initial first-round NDCs, depend significantly on obtaining sufficient financial support from the rich countries. Climate change is at heart a distributional problem, affecting poorer regions and vulnerable groups more heavily while having been disproportionately caused by the richer nations and the well-to-do. This ethical consideration alone justifies fairly massive mobilization of financial resources being directed from richer countries to poorer countries. The Paris Agreement recognized all this by calling for "making finance flows consistent with a pathway towards low greenhouse gas emissions and climate-resilient development" [Article 2.1(c)] and setting up a $100 billion Green Climate Fund to transfer funds from developed to developing countries (see note 23 of Chapter 1).

Is finance going to be up to this challenge? Difficult to say! On the one hand, it has all the attributes to face the challenge of climate change. For one, financial institutions and markets are organized globally, and climate change is a planetary phenomenon. Finance is also a highly innovative system, mobilizing a lot of talented people to design knowledge-intense services and funding instruments wherever needs arise for such innovations. Financial markets are inherently flexible, responsive to signals, dynamic, typically well organized (while constantly evolving in response to lessons learned or problems addressed), and widely accessible. While there is a strong profit motive driving financial investors, most of them also have a manifest desire to weigh risks properly. Yet, on the other hand, financial markets and institutions are still marked by the horrors of the global crisis of 2007/2008 and its difficult aftermath. They not only suffered much damage, but ended up quite traumatized by what for many was a near-death experience. In addition, there has been a massive worldwide reregulation effort following this crisis which, while making the system undoubtedly safer and sounder thanks to higher bank capitalization levels, bigger liquidity cushions, lower leverage ratios, more resilient market structures, and better disclosure of information, has also put a large burden on those thus regulated. They are spending a lot more of their resources on compliance which adds to their post-crisis caution and makes them less aggressive venturing into new areas such as the challenges associated with climate change. Worst of all, banks and institutional investors may follow rules of thumb about risk-return trade-offs and capital allocation which fail to address the specific challenges posed by climate change and thus may prove irrelevant, if not altogether a hindrance, in helping with the transition to a low-carbon economy. For all these reasons, finance, as currently constituted, may not be ready at all for this huge and complex task. We will thus have to create, step by step, a climate-specific system of financial institutions, instruments, and markets—a *climate finance*.[1]

FUNDING A "GREEN" INFRASTRUCTURE BOOM?

Moving the world onto a lasting pathway of steadily falling GHG emissions is going to require a large amount of infrastructure investment. We will need new sources of energy supplies, smart and better integrated electricity grids, a major shift in the mix of power plants, lots of waste management facilities, a significant expansion of public transportation, better insulated buildings, a massive reforestation effort, natural area preservation on land and sea, a global GHG monitoring and measurement system, just to name a few priorities. This all concerns mitigation, pro-active steps of prevention. Wait until you get to adaptation! Coping with the damage caused by climate change is bound to be much more expensive than trying to prevent it through mitigation in the first place. We may have to build huge systems of sea walls, move entire cities inland, integrate hundreds of millions of climate migrants, figure out how to feed the planet's rapidly growing population while agricultural yields are falling, fight pandemics, and cope with extreme weather events becoming the "new normal" (e.g., reconstructing devastated areas). As this still far-from-exhaustive list of needed investments makes clear, we better adopt a flexibly broad notion of infrastructure to cope with the many facets of climate change. All of this amounts to a lot of added investment spending.

Yet ever since the Great Recession of 2007–2009, we have had a widespread shortfall of infrastructure investment spending, with the exception of China and a few other emerging-market economies. The share of infrastructure investment in GDP has declined in eleven of the G-20 countries over the last decade, including the European Union, Russia, the USA, and Mexico, and now stands globally at a historically low 3.8%.[2] This trend is part of a broader stagnation of global investment spending amidst significant excess capacities in a large number of industrial sectors and a strong investor preference for short-term, often purely financial engagements. We have now had a decade of extremely low, at times even negative, nominal interest rates across much of the maturity spectrum, and this indicates that the total demand of external funds for investment spending fails to match a large global savings pool and continuous liquidity injections by central banks. This has contributed to the slow-growth recovery from the traumatic 2007–2009 crisis. Right now would then be a very good time to boost investment spending significantly, for which the needed transition to a low-carbon economy is the perfect vehicle. Such a "green" investment boom is propitious in light of the very low interest rates even on long-term bonds and would boost economic growth worldwide.

Estimates for the investment volume needed to meet the objectives of the Paris Agreement vary, of course, because of the long time horizon involved and also the significant sources of uncertainty posed by climate change. For instance, the Global Commission on the Economy and Climate, headed by former Mexican President Felipe Calderón and climate economist Nicholas Stern, reported in October 2016 that the world needed to spend an estimated $90 trillion over the next fifteen years to replace ageing power plants, roads, buildings, sanitation, and other structures while also providing cleaner energy supplies and smart electricity grids. Such an effort would have to nearly double the current annual $3.4 trillion level of infrastructure investment to $6 trillion. The large upfront costs of such a huge worldwide effort would, however, eventually be more than compensated for by efficiency gains, fuel savings, health benefits, phasing out the annual $550 billion subsidies to fossil fuels, and avoidance of systemic damage from climate change. The Organization of Economic Cooperation and Development came to a very similar estimate of $95 trillion. But the OECD also emphasized a net boost to global GDP of 4.6% by 2050 from combining low-carbon infrastructure investments with green innovation and climate-friendly structural as well as fiscal reforms while limiting environmental damage from having pursued these transition actions decisively in a timely fashion. The Energy Transitions Commission, headed by Lord Adair Turner who had run Britain's Financial Services Authority until its abolition in 2013, proposed in April 2017 spending an estimated $15 trillion globally between now and 2030 on renewables and other low-carbon technologies as well as more energy-efficient equipment and structures while sharply reducing investments in fossil fuels. The ETC's four-pronged de-carbonization and energy-efficiency promotion strategy would require annual net investment increases of $600 billion (which is about 3% of total global investment spending per year).[3]

However much we will end up having to invest in a green infrastructure and energy efficiency to manage the transition to a low-carbon economy, all the reports agree that this huge undertaking needs to be launched in decisive fashion sooner rather than later. Yet there are many obstacles preventing us from doing precisely that. Protected by still-huge direct subsidies of half a trillion dollars per year and lots of political power, the fossil-fuel producers have a worldwide lock on the existing infrastructure which they are loath to cede to alternative suppliers of energy. One of the most important steps the world community must take by political agreement is to phase out these subsidies as rapidly as possible. But that would turn powerful states like Russia or Saudi Arabia into big losers, and so we would have to overcome their resistance as well as that of domestic oil, gas, and coal producers. At the same time, green investments into a decarbonized and more energy-efficient infrastructure are not sufficiently profitable in the absence of adequate carbon prices. Today, as already indicated in Chapter 5, we still exempt three-quarters of global CO_2 emissions from any social-cost pricing, and even the quarter of emissions we cover has on average a price that is less than a quarter of the estimated social cost it creates. We thus need to put into place an appropriate carbon price range such as the one suggested recently by the Stiglitz-Stern Commission (of $40–80 per ton by 2020, going up to $50–100 per ton a decade later). As long as we fail to do so, emission-cutting investments will not be undertaken to the extent needed, simply because they do not yield an adequately measured benefit. As we face the prospect of revisiting the NDCs of the Paris Agreement over the next few years to make their mitigation goals more ambitious, we have to find the political will to eliminate fossil-fuel subsidies, restrict the burning of fossil-fuel reserves, and put into a place a global carbon price. Such an international agreement, hard to do even under the best circumstances of political will and spirit of cooperation, is impossible without the USA returning to the fold.

Even we assume such a scenario where oil and coal are no longer subsidized while carbon is properly priced, and it is not obvious that we can muster sufficient funding for the needed hike in green investment spending. Banks are still reeling from systemic-crisis conditions they experienced in the late 2000s. Their recovery has been slow, amidst much reorganization and loss-taking. Governments all over the world also introduced extensive financial reregulation efforts from 2010 onwards which have subjected the banks to much heavier regulation of their capital, liquidity, leverage, and disclosure of information. For all these reasons, banks have become more hesitant lenders, even though central banks have at the same time flooded their domestic banking systems with zero-yield reserves by aggressively pursuing quantitative easing for nearly a decade now.[4] Commercial banks will in the end also find climate-related projects still subject to great uncertainty, and in the face of hard-to-measure yet considerable risks they will not want to lend as much as they should. Companies themselves may shy away from committing to risky, long-term projects, especially to the extent that they themselves prefer investments which typically pay off much more rapidly. It does not help that chief executives typically receive most of their remuneration in the form of stock options and bonuses both of which depend on quarterly earnings and hence incentivize managers toward excessive concern with what is happening to the corporate bottom line now. The same bias in favor of the short term also afflicts institutional investors such as mutual funds or hedge funds, even pension funds and insurance companies both of which should technically have a longer investment horizon.

Apart from their large upfront costs and their long-term nature, infrastructure investments also have a public-good aspect to the extent that they benefit society at large, both directly to users and indirectly by making the economy work better. Hence, they are traditionally responsibility of the government. But governments all over the world have struggled over the last decade getting their budget deficits under control as public debt has risen rapidly in relation to GDP. How then can we get them mobilized for the needed surge in green infrastructure investment in the face of self-imposed fiscal austerity? For one, it would be helpful for public officials to recognize that infrastructure investments would boost currently sluggish growth, have large multiplier effects, and ultimately pay for themselves through the efficiency gains and environmental improvements they bring to the economy. All these positives justify strong commitment to boost infrastructure spending, a type of spending also normally supported by the electorate. Governments need to go further in separating their budgets into one for operating expenses as well as income maintenance programs and another one for capital expenditures. The latter does not need to be subject to the same fiscal-austerity rules, especially if capital budget expenditures get amortized over the life of the (infrastructure) investments rather than expensed as they occur.

Much infrastructure investment, especially the kind of energy- and transportation-oriented projects we need to mobilize for the transition to a low-carbon economy, could be financed by public development banks—either national ones such as Brazil's National Economic and Social Development Bank (BNDES) or France's Caisse des Dépôts et Consignations, or multinational ones such as the African Development Bank, the European Investment Bank or, with a global reach, the World Bank. Sponsored by governments, those lenders have cheaper access to funds and can mobilize those resources without the same narrow profit motive as exists among private banks. They can thus fund large projects over the long haul, provided those promote national economic development. They can also partner with private sources of funds, encouraging those to engage more readily because of the public sector backing these partnerships imply. Public–private partnerships for infrastructure projects are surely going to play a significant role in the transition to a low-carbon economy, especially in areas where the projects concerned carry large potential benefits for the profit-seeking private sector (e.g., a new type of renewable energy source and alternative car technology). Finally, public development banks can also help local governments achieve better credit ratings and mobilize their own funds for urban renewal projects mitigating against climate change.[5]

We are now beginning to see major initiatives to launch new public development banks or expand existing ones to help fund the needed surge in transition-related investments. For example, already in 2010 a group of twenty-three national and sub-regional development banks got together to set up the International Development Finance Club (IDFC, idfc.org) to pool their expertise and resources in promoting, among other objectives, climate finance and green banking. In November 2014, the EU's Jean-Claude Juncker launched the European Infrastructure Investment Plan for which he set up the European Fund for Strategic Investments, co-financed by his European Commission and the European Investment Bank. EFSI has now committed by the end of 2017 about 80% of the plan's original €315 billion target. Since mid-2016, the OECD has pushed the idea of publicly financed "green investment banks" to help mobilize a much greater amount of private investment in low-carbon, climate-resilient infrastructure through the innovative use of transaction structures as well as risk-reduction and transaction-enabling techniques. In the meantime, China has dramatically expanded its domestic as well as global commitments to development finance, not least through its China Development Bank and many additional facilities it has recently put in place (e.g., Asia Infrastructure Investment Bank), which are also increasingly incorporating environmental objectives.[6]

At the heart of the ongoing global effort to provide funding facilities for climate finance is the creation of the Green Climate Fund (GCF), the centerpiece of the Paris Climate Agreement of 2015. Even though the GCF was already introduced in 2011 at the COP16 in Durban, the fund's central role emerged only four years later when the world community committed itself to this possibly large resource-transfer mechanism from rich nations to poorer nations which will also provide a coordinating umbrella for many national development banks and private institutions engaged in the worldwide construction of climate finance. The GCF is supposed to have raised $100 billion by 2020, an important goal not least in light of the fact that the more ambitious climate-policy objectives in the initial NDCs of many poorer countries depend on funding access which the GCF is supposed to mobilize for them before 2023. So far the GCF has raised $10.5 billion of which the Trump Administration will renege on two-thirds of America's initial $3 billion commitment. Its ability to raise public funds is also crucial inasmuch as the GCF is supposed to use those for leveraging more private finance (e.g., bank lending and pension funds) which otherwise would not commit to climate-finance projects in developing countries. It is not yet clear whether this leveraged private sector finance should be counted toward the $100 billion total or in addition. While being set up, there are still many questions about the precise functioning of the GCF. As is the case with the rest of green infrastructure investment finance, the GCF represents a worthy initiative in the creation of a worldwide institutional architecture for climate finance but is as of now nowhere near the scale required to address the problem effectively.[7]

GREEN BONDS

With an outstanding volume globally exceeding 100 trillion dollars, compared to an aggregate stock-market capitalization worldwide of 63 trillion dollars, corporate bonds will obviously have a large part to play in financing the low-carbon transition. Already a decade ago, in 2007, official bodies like the World Bank, European Investment Bank, and African Development Bank began pushing the idea of so-called *green bonds* to promote the financing of investments aimed at combating climate change. Initially, these kinds of environmentally linked bonds were mostly issued by either development banks, such as Germany's KfW, or utilities such as the French power company GDF Suez. In 2010, a nonprofit umbrella organization known as the Climate Bonds Initiative (www.climatebonds.net) was set up to give green bonds greater credibility, pushing especially third-party verification procedures. Investor confidence was an issue not least because green bonds have similar yields to normal corporate bonds, while carrying possibly greater risks. And then there is also the issue of "greenwashing," with companies pretending to do something positive for the environment so as to look good to investors increasingly favoring green objectives, without really doing anything substantial and lasting in that direction. The Green Bond Principles of 2014, which spelled out officially sanctioned features of what constitute different types of green bonds, strengthened the reputational quality of this useful funding instrument. And so its volume of new issues began to rise rapidly thereafter, growing from a yearly volume of just $35 billion in 2014 to over $93 billion in 2016 while expanding to a greater variety of issuers.

We have now (in late 2017) a new market in the making, one that might very well end up playing an immensely useful role in funding the transition to a low-carbon economy. As we approach $120 billion in new green-bond issues during 2017, we have also seen this year all kinds of new issuers taking advantage of such environmentally oriented financing—sovereign nations like France (issuing a €7 billion 22-year green bond at 1.75% in January 2017), cities like Mexico City (to pay for transit improvements and energy-efficient street lighting), and consumer-product companies such as Toyota (to finance leasing of electric cars) or Apple (to increase its already considerable share of renewable energy and use of greener materials). In the meantime, the fragmentation of practices and standards pertaining to green bonds still allows for much experimentation. Green bonds get issued not just for climate-change mitigation projects, but also to pursue other environmental objectives such as pollution control or water treatment. Some issues may have an ethical tinge inasmuch as they contain explicit prescriptions against using proceeds in socially harmful areas like tobacco or firearms. Britain's Lloyds Bank recently issued a so-called Environmental Social and Governance (ESG) bond, bundling loans for small- and medium-sized providers of renewable energy together with loans for healthcare providers in poor areas. The whole area of asset-backed securities, comprising at this point just a small fraction of all green-bond issues (only about 5%), could play an enormously useful role in the future to help promote lending to smaller firms or nonprofits developing climate-related products and services.

There is now a global bondholder activism taking root in parallel with the already more widespread and better-anchored shareholder activism. Investors want parallel voluntary standards, such as the aforementioned Green Bonds Principles or the Climate Bond Standards, to get better harmonized into a widely accepted industry-wide standard for green bonds which assures disclosure of measurable environmental impact and its certification by an independent third party so that the deceptive practice of greenwashing will be kept to a minimum. After all, the label "green" is self-applied, and there must be some objective, generally accepted mechanism to verify that such labeling is justified. Controversial issuers like Poland, continuing its push for coal, or France's leading utility company Engie, involved in a contentious hydro-power project in Brazil, have triggered a discussion whether green bonds should be judged by the issuer or by the projects they finance. Given the need to scale up this corner of the global bond market, we should allow the widest possible range of issuers, including high-carbon emitters such as oil companies, provided they use the proceeds from green bonds on emission-reducing or otherwise environmentally friendly projects subject to third-party verification. This is also the opinion shared by Moody's and Standard & Poor's. Of course, these rating agencies have a vested interest, since they both presumably see a major business opportunity in rating green bonds on a sliding scale as to the usefulness of their funding contribution and/or providing ESG assessment scores about the issuers. Such evaluations would set the stage for greater price differentiation among various "shades of green" bonds which up to now have moved closely in line with the global aggregate bond index. In the same vein, indices have sprung up recently for green bonds, notably those offered by S & P DJI and Barclays MSCI. In their wake, we have also seen green bonds' exchange-traded funds being introduced (e.g., by France's Lyxor and VanEck in the USA in March 2017), an innovation for which there may be potentially much investor demand. All these initiatives point to a rapid evolution toward a mass-market infrastructure.[8]

Where green bonds can surely make a very meaningful contribution is to help emerging-market and frontier economies meet their NDC commitments undertaken in the wake of the Paris Agreement.

It is encouraging in this context to see sovereign green-bond issues being launched by Bangladesh, Mexico, Morocco, or Nigeria. China has already taken a global leadership role since 2015, with one-third of all green bonds issued there during the last couple of years. Even though green bonds have been issued by companies (e.g., automobile maker Geely) and local governments, the major issuers in China so far have been banks trying to lessen their excessive reliance on shadow-banking sources and uses of funds. As is the case with much of China's financial system, Chinese green bonds follow their own regulatory guidelines that differ from international norms, in this case overseen by the People's Bank of China (for non-corporate issues) and the National Development and Reform Commission (for corporate issues). Those regulators impose less stringent application rules than, say, in Europe, let corporate borrowers use half of the proceeds for working capital or paying down loans, demand less information on how the proceeds have been used by the borrowers, and are more flexible in what they consider to be eligible projects (e.g., "clean" coal). Allowing foreign investors to trade Chinese bonds through Hong Kong by the end of 2017, in the so-called Bond Connect program, may oblige gradual harmonization with the emerging international norms pertaining to green bonds.[9] India too is about to discover the potential of green bonds. With bank lending in India paralyzed by large mountains of non-performing loans, its ambitious plans for renewable energy will surely come to depend heavily on green bonds instead. Even though sovereign governments are increasingly tapping the emerging green-bond market, this debt instrument is in its take-off stage still predominantly issued by companies and financial institutions. If you add to that also green securitizations, then the variety of issuers— as kept track of by Bloomberg New Energy Finance (bnef.com)—makes it already obvious today that green bonds have a potentially large role to play in post-Paris plans for a low-carbon transition as a global undertaking of the highest priority.

Changing Corporate Priorities

Capitalism relying on private enterprise and its profit motive will have to be part of the solution just as much as it can and has been part of the problem. The key here is to change the corporate governance and incentive structure in the right direction, toward long-term investment horizons centered on profitable, yet also socially useful productive investment.

Here, we have a long way to go. The current paradigm is pointing to the opposite—a short-term focus on the bottom line; managers obsessed with quarterly earnings as they are rewarded accordingly; hierarchical organization of the firm controlling its subsidiaries and affiliates so as to match the growing decentralization of production (in increasingly far-flung global supply chains) with the centralization of cash flows for control purposes; organization of the firm as a bundle of assets dominated by a highly financialized center as holding company; preference for external growth via mergers and acquisitions; too much focus on global tax minimization and not enough on the long-term reproduction of key assets (including the collective knowledge pool of the labor pool); ultimately a crowding out of long-term concerns crucial for growth such as renewal of the capital stock or research and development. This is neither the natural order of things nor does it have to be that way forever. The corporate paradigm is a matter of ideological choice, policy priorities, and the history of each country's triangular relation between industry, finance, and government. The varieties of capitalisms currently vying for global dominance, with the Europeans and the Chinese trying to pull even with the eroding Anglo-Saxon dominance of long-standing neoliberalism, also point to the ultimately political nature of the corporate paradigm and hint at the possibilities of change toward alternatives. We will argue later (in Chapter 8) how and why the challenge of climate change deserves a different type of capitalism, a more ecologically oriented capitalism, something we may wish to refer to as "eco-capitalism." For this to happen, we have to develop a new corporate paradigm of governance, organization, and priorities of objectives.

How did we get to this point of counterproductive corporate behavior? The global post-war boom of the 1950s and 1960s, the golden era of what the French Régulationists have characterized as the Fordist ("mass production") accumulation regime, had been grounded in the steady industrial expansion strategy of American, later also European and Japanese multinationals. The stagflation crisis, which took root worldwide after the first oil-price shock in 1973, enabled those firms to compensate for a longer-term decline of profitability and stagnant productivity by accumulating inflation-induced "paper" profits arising from the steady erosion of money's value when measuring costs at "historic" (i.e., earlier) levels and revenues at current dollar levels. When the Federal Reserve finally decided to slay the inflation dragon in consecutive rounds of monetary policy tightening from October 1979 onwards, those paper profits got squeezed out of corporate income statements. At that point, the US and British economies had gone through a decade of intensifying stagflation conditions, a new type of structural crisis combining rising unemployment and inflation which had put the post-war "retain and reinvest" strategy of corporate America under enormous stress. The destruction of their accounting profits left large numbers of corporations systematically undervalued in the stock market.

At that point, in the early 1980s, you also saw the emergence of a new investor class ready to pounce on those undervalued stocks. The Employee Retirement Income Security Act of 1974 had provided US pension funds with incentives to engage much more actively in the stock market. The Securities and Exchange Commission's decision in May 1975 to deregulate broker commissions encouraged larger-volume stock-market transactions and discount brokerages, which the spreading computerization of financial markets further facilitated. Mutual funds became more important in response to these changes, a trend that accelerated greatly after the emergence of money-market funds in 1975 and individualized 401(k) pension plans in 1978 in the wake of which many American households began to switch their savings out of low-yielding (because of ceiling-regulated) bank deposits and into funds. The Economic Recovery and Taxation Act of 1981, the heart of the supply-side revolution known as "Reaganomics," created a huge incentive for all that fund-channeled money to pour into the stock market by slashing marginal tax rates and capping the tax rate on capital gains at a historically low 20%. Enter Michael Milken and his investment bank Drexel Burnham Lambert promoting high-yield, low-rated bonds! Eager investors attracted by those promising yields, especially the newly deregulated savings and loans associations and mutual savings banks, snapped up these "junk" bonds in huge quantities. They thus became the perfect funding vehicle for a Drexel-orchestrated group of aggressive "corporate raiders" (comprising such legendary figures as Ivan Boesky, Sir James Goldsmith, Carl Icahn, or T-Boone Pickens) to exploit the widespread undervaluation of stocks by launching hostile takeover bids against some of America's most cherished corporate giants—Goodyear Tire, TWA, U.S. Steel, Gulf Oil, Revlon, and so forth. Companies under attack tried all kinds of other responses, most notably designing various "poison pills" which involved giving shareholders rights to purchase additional shares at a discount either before or after the takeover to make these bids more difficult to carry out. Key to the outcome of these battles was the competition between raiders and corporate managers for the support of large shareholders, giving institutional investors suddenly quite a bit of power. While typically a majority of shareholders would back top management, they often exacted a price for their support. Their emphasis on dividend payouts and/or more rapid appreciation of share prices obliged CEOs to pay closer attention to the corporate bottom line and rein in costs. Thus was born a new "downsize and distribute" paradigm, crystallized around corporate management's commitment to maximize shareholder values at the expense of other corporate objectives.[10]

This new paradigm was also given a solid ideological grounding, as part of a broader counterrevolution sparked by the near-simultaneous election victories of Margaret Thatcher and Ronald Reagan in 1979 and 1980. Milton Friedman, a leading defender of the free-market doctrine (and one of Reagan's key policy advisors), had already written much earlier that a corporation's sole "social responsibility" was to earn as much profit as possible "within the rules of the game" for its owners, the shareholders.[11] This argument was pushed further during the 1970s and early 1980s by so-called agency theory according to which the shareholders, as the principals, needed an active market for corporate control through takeovers to discipline well-entrenched corporate managers, the agents, whose companies performed poorly. Agency theorists thus justified the subsequent wave of hostile takeover bids by the raiders. Their argument also implied that the rate of return on equity be regarded as the ultimate measure of performance. Maximizing shareholder value was thus turned into a creed. And to overcome the supposed gap between principal and agent, the latter should be incentivized to see things from the point of view of the former. Agency theorists thus pushed for giving corporate managers stock options as their principal form of payment, a practice that took off during the stock-market boom of the 1980s and ended up raising executive pay scales very rapidly.[12]

Since then, the doctrine of shareholder value maximization has become the global standard for corporations to follow. The successful post-stagflation revival of the US economy, with a booming stock market at its center, made sure that other advanced capitalist nations would take notice and seek to copy this success formula. US institutional investors themselves, from pension funds to hedge funds, carried the idea across the globe as they diversified their equity portfolios geographically. During the late 1990s and early 2000s, their global reach extended to emerging-market economies whose burgeoning stock markets helped fund local champions ready to submit to the rule of shareholder value. Michel Aglietta (1998), a leading figure in the French Régulation School, has talked in this context about "capitalisme patrimonial."[13] In this phase of capitalism's evolution, middle-class families at large turn into financial investors via their pension plans and also as they favor moving their savings into mutual funds, thereby reinforcing the public's fascination with the stock market and broad ideological support for share prices as the yardstick of corporate success. At the top of this "rentier" class pyramid sit the corporate managers receiving amazingly generous executive pay packages, stuffed with stock options, from their compliant board of directors whose members they carefully picked themselves to assure unwavering support.

Those stock options typically get vested shortly, at which point they can be sold off. During the 2000s, a practice known as "back-dating" became popular which allowed managers to boost their returns by resetting their stock options to an earlier date when their share price was lower. As large amounts of new shares were issued for executive compensation packages, share buybacks would counteract such ownership dilution while also boosting share prices. Since those repurchases also proved a more flexible way than dividend payouts to return money to shareholders, it was not surprising to see their popularity explode (rising in the USA alone from $5 billion in 1980 to $349 billion in 2005 and $589 billion in 2016). Evidence abounds that annual executive compensation packages in the tens of millions of dollars for chief executive officers or chief financial officers have little to do with their actual performance, but are motivated instead by what CEOs or CFOs at other companies are getting paid—thus truly forming a class apart which has managed to capture a growing, by now often obscenely large, share of the nation's income or wealth. And we also know from convincing studies that many corporate leaders suffer from so-called management myopia seeking to boost quarterly stock-market returns rather than their firm's long-term growth potential.[14]

Shareholder value maximization has thus given rise to short-termism among corporate managers focusing on this quarter's current earnings rather than thinking about the long-term health of the company. This bias has encouraged de-linking of profit from investment, accumulation of cash hoards, large dividend payouts, manipulative stock-market tactics, expansion through mergers and acquisitions as opposed to capacity-expanding productive investment, the use of stock swaps in such fusions, tax evasion, outsourcing to lower-cost suppliers, and downsizing. As a matter of fact, the stock market often cheers announcements of mass layoffs as proof of management's cost-cutting (hence profit-enhancing) zest. We can run these practices for some time without too much damage to the firm's future growth prospects. But over time, the "distribute and downsize" mind-set will take its toll. Corporate managers will fail to appreciate sufficiently at their own peril that most growth-promoting activities, such as investing in new production capacity, conducting research and development, building up an experienced and motivated workforce, nourishing good relations with suppliers and customers, or gaining brand loyalty, happen over the long run and need to be planned for as such. When you think only about the "now" and "here," you are not going to give these activities enough attention. Even on a macroeconomic level these biases of short-termism play out badly as they translate into declining savings rates, inadequate levels of productive investment, slowing productivity growth, and worsening inequality. When it comes to climate change, short-termism is exactly the opposite of what is needed. And so we have to think about how to alter the mind-set of corporate managers and shareholders decisively.[15]

ENVIRONMENTAL, SOCIAL, AND GOVERNANCE CRITERIA

Milton Friedman, in his defense of shareholder value, admonished any notion of a company's social responsibility beyond the bottom line as bearing more costs than is worth pursuing, hence in effect a waste of effort. This bias has stuck for a long time. And its conclusion was reinforced by recurrent instances of divestment campaigns among activist investors, notably the widespread US campaign during the 1970s to divest from South African companies in protest against apartheid. Even though successful, this campaign left Wall Street with the impression that such "ethical" investment hurts returns by limiting the choices of investors. Such negative sentiment prevails even today to the extent that climate change evokes massive withdrawal from traditional energy and utility stocks as a result of which investment portfolios end up less diversified and deprived of lucrative stocks. But such prejudice is about to be proven precisely that, since it is becoming increasingly clear that the application of broader, socially responsible criteria actually boosts returns. Of course, what we are talking about here is no longer just "ethical" investment, narrowly and negatively defined as withdrawing from certain objectionable companies. Instead, we have to apply such environmental, social, and corporate governance (ESG) criteria broadly and positively as markers for possibly better-managed companies on the basis of which we can compose a profitable portfolio of equity shares that at the same time satisfies our desire for responsible investment choices. There are indications that this change in sentiment is taking root.

The so-called ESG criteria enable socially responsible investors to evaluate corporate behavior in terms of well-defined non-financial performance indicators and so compose a portfolio of shares in companies found to do well in these areas. Environmental criteria include GHG emissions, pollution, energy use and efficiency, waste, natural resource conservation, and exposure to as well as management of environmental risks (e.g., hazardous waste, oil spills, toxic emissions, regulatory compliance). Social criteria basically relate to the company's relations to various stakeholders, such as its employees, suppliers, customers, or the local communities it operates in. As regards a firm's workforce, for instance, relevant performance indicators would look at its absenteeism rate, the employees' maturity, training and qualifications, staff turnover, the level of inclusion in the firm's recruitment policies to indicate diversity, or employee benefit packages. Does the firm choose suppliers sharing its values? Does the firm pay sufficient attention to consumer protection? Are the firm's working conditions conducive to the health and safety of its employees? Does the firm invest in its local communities? Finally, the corporate governance criteria apply to how board members are chosen, the relationship between CEO and directors, the structure of management, the determination and levels of executive pay packages, which rights shareholders are given, how transparent and honest the firm's accounting practices are deemed to be, how it handles its litigation risks, and whether the firm abstains from political contributions to obtain favorable treatment. No matter how well defined the performance indicators in each of these areas, the ultimate judgment as to a firm's adherence to ESG factors is inevitably quite subjective. It thus depends on how deeply the investor gets to understand the firm she/he is evaluating, and this relation between the evaluator and the evaluated may in turn allow the investor to shape the firm's attitudes and efforts in each of the ESG areas. The whole purpose of this exercise is the notion that firms scoring well on the ESG criteria will in the end reap the competitive advantage of being better managed in terms of faster growth and higher profitability.

One of the important implications of the ESG criteria is the interconnectedness between environmental factors, social factors, and corporate governance mechanisms. A socially responsible firm cannot just focus on one of these factors without also engaging in the others. Otherwise, its approach is incomplete, perhaps even hypocritical, certainly not sufficiently responsible. More broadly speaking, we simply cannot disconnect the environment from the social-relations context within which firms operate nor separate either from how corporations are governed. These connections hold true for the micro-level of corporations as much as they also apply to the macro-level of nations. Environment, the social, and governance go hand in hand. In that sense, it is fair to say that the ESG criteria may well end up serving as a key building block for embedding transformed corporate behavior and structures into a progressive eco-capitalism capable of addressing climate change by reconnecting economics, society, and nature.

While social and governance issues had been discussed by Wall Street for quite some time (see, for instance, the remarkable impact after 1998 of the *Fortune 100 Best Companies to Work For* list), it was the prospect of climate change which pushed the ESG criteria to the forefront. In 2006, the UN-supported *Principles for Responsible Investment* (PRI) laid out six principles for companies to incorporate ESG criteria into their investment decisions and for investors to hold corporations accountable to follow these criteria. A decade later 1700 asset managers and other institutional investors combining over $60 trillion in assets under management, including such industry leaders as the CalPERS pension fund or the world's largest asset manager BlackRock, have signed up to these sustainable investment principles. While many of these signatories have yet to follow up with concerted action, the very presence of this large and global network has pushed ESG criteria onto the radar screen of the global investor community ever since. Their growing influence has been nourished further by such initiatives as the Network for Sustainable Financial Markets (2008), the Sustainable Stock Exchanges Initiative (2009), the Global Sustainable Investment Alliance (2012), or the Montréal Carbon Pledge (2014). But the most important propellant has been the gradually dawning realization, confirmed by more and more empirical evidence, that companies committed to following ESG criteria have stronger earning potential over the long haul than those that do not.[16] This realization has allowed environmental, social, and corporate governance issues become part of a company's "fiduciary duty" and thus legally defensible objectives in the pursuit of profit. That change is a vitally important prerequisite for the needed transformation of our system into eco-capitalism.

Between 2012 and 2016, the amount of assets under professional management incorporating ESG criteria quadrupled in the USA to $8.1 trillion, amounting to 22% of the total. A similar acceleration occurred in Europe as well.[17] Much of that spread has been facilitated by the introduction of a growing number of ESG-based indices, such as AXA Investment Managers' ESG scoring system for individual corporations, Sustainalytics' Global Compact 100 index tracking social and environmental sustainability of a selected group of companies, the FTSE$_4$Good Index Series measuring the performance of companies demonstrating strong ESG practices, or the various ESG-based indices promoted by MSCI and Standard & Poor's (in cooperation with RobecoSAM). These indices seek to turn behavioral characteristics of firms regarding their treatment of the environment, their workforce, or internal governance practices into measurable scores reflecting overall performance. After all, it stands to reason that a company's carbon emissions, labor issues along its supply chain, or boardroom selection procedures will be material to how it performs in the long run.

Besides ESG-based indices which measure how firms are organized and prioritize their key objectives, we also need indices assessing a portfolio's correct (re-)alignment with the low-carbon transition objective. Low-carbon indices, such as those provided by MSCI or Standard & Poor's, have taken advantage of recent advances in index-construction methodology by providing a range of portfolio-alignment options. Those allow investment funds to tilt their portfolios toward a lower carbon footprint at different speeds while earning at least a benchmark (market average) return and without having to face greater volatility. Rather than going for the all-or-nothing choice of withdrawing entirely from all fossil fuels, more gradual low-carbon indices identify worst offenders per sector and region to build portfolios without those. And they weigh more heavily toward already well-established emission-lowering products, such as companies pushing production of electric cars (e.g., Renault and Volvo), than more radical, yet insufficiently tested technologies (such as CCS). But here we will need a third group of transition-driving market indices, namely environmental-market indices looking at companies making green products fighting climate change, like solar panels or wind turbines. A good example would be the recently launched FTSE Environmental Markets Index Series comprising currently two indices aimed at environmental technology and business opportunities in such sectors, as renewables, energy efficiency, water technology, or waste and pollution control.

Yet at the same time, it is neither obvious how these ESG factors can be best measured reliably nor clear how they affect the bottom line directly. Only some firms report ESG-related data reliably, and the definitions they use often differ. Most only started recently reporting any ESG data which they typically update no more than just once a year. In the face of such incomplete and asynchronous data sets about extra-financial factors, there are a lot of opportunities to scrub such data for extra returns (so-called "alpha"). Funds also play around with so-called smart-beta strategies whereby they rebalance existing stock-market indices by tilting them toward ESG factors and so improve their risk-adjusted returns. As more of these alpha-extraction and ESG-tilted smart-beta strategies prove successful, they will provide impetus for yet more ESG data to be generated and indexed.

The ESG criteria are in the end only relevant to the extent that they actually incentivize corporations to change how they are structured and operate. The key is to pressure corporations into better behavior and longer time horizons. The Institutional Investors Group on Climate Change, comprising among its 140 members in eleven countries totaling assets worth €18 trillion some of Europe's largest pension funds and asset managers, fosters cooperation among large investors to push companies into better consideration of risks and opportunities arising from climate change. UK-based "Aiming for A" Coalition of institutional investors has successfully launched shareholder resolutions forcing high-carbon extractive and utility companies (e.g., Shell, BP, Glencore) to reveal more about how they intend to deal with the transition to a low-carbon world. There is growing consensus that large pension funds and other institutional investors need to return to investing directly rather than let their portfolios be handled by asset managers, not least in order to make their weight felt more effectively as owners while at the same time cutting out intermediation costs and the trade-at-all-cost mentality of those middlemen. Direct investors would probably more likely see themselves also as active investors, especially to the extent that they identify increasingly as "universal owners" owning the negative externalities of the firms they invest in and committing to maintaining a healthy planet for their members as part of their fiduciary duties.

That new generation of institutional investors also talks of "active ownership," while on the other side of that equation more firms want to turn into "public benefit" corporations whose charters include explicit commitments to environmental and social responsibility as well as a longer-term investment horizon.[18] So there is gradual progress toward transforming how companies and their investors organize their priorities.

Climate-Related Financial Disclosure

In his "Tragedy of the Horizon" speech of 29 September 2015 the chairman of the Financial Stability Board Mark Carney announced that he would set up a task force charged with formulating a voluntary standard for carbon emitters to report their current emissions and specify how they plan to get to net-zero emissions in the future. This initiative centered on the importance of information and its disclosure to motivate action. We need to know about GHG emissions, if we want to get them to net zero within a tightly prescribed carbon budget. Firms have to think about how climate change affects them and how they best can deal with this challenge, while investors will want to have such carbon-related disclosure to figure out valuations as well as capital allocation. Policymakers too need relevant data to assess the interaction between their policy measures and market responses. Beyond the usefulness of such carbon-tracking information, it is also important to push for a worldwide standard in the interest of comparability and reliability. And so the FSB initiative was also launched to provide such coherence at a time when there were suddenly a great variety of information-reporting initiatives being pushed—as, for example, the UN Principles for Responsible Investments (PRI), Britain's Carbon Disclosure Project, the efforts of the Carbon Disclosure Standards Board to offer firms a framework for reporting environmental information, and the US-based Sustainability Accounting Standards Board seeking to specify an accounting framework within which to report financially material sustainability issues.

A bit more than a year later, in December 2016, the so-called Task Force on Climate-related Financial Disclosures (TFCD), headed by Michael Bloomberg who had made climate change one of his passionate concerns while serving for twelve years as mayor of New York City, published its much-anticipated report with recommendations for an international reporting framework. The main idea behind this effort was to provide a standardized framework for climate-related financial disclosures by affected organizations that could help interested parties—shareholders, investors, lenders, insurers, regulators, etc.—assess how these organizations evaluate risks as well as opportunities linked to climate change and intend to cope with either. Attempting to classify climate-related risks, the TFCD identified policy and legal risks, technology risks, market risk, and reputation risk as transition risks while distinguishing between acute and chronic physical risks. Among climate-related opportunities, the TFCD provided detailed sets of examples with regard to resource efficiency, energy sources, products and services, markets, and what it termed "resilience." Both catalogs of risks and opportunities have identifiable and specific ways of impacting a firm's income or balance sheet which the TFCD spelled out so that corporations can figure out how best to take account of these climate-related effects on their bottom line.[19]

The TFCD focused its recommendations for disclosure on four separate, yet interrelated areas. With regard to governance, a corporation should be expected to discuss management's role in managing and assessing climate-related risks and opportunities as well as its board's oversight thereof. With respect to strategy, a corporation will need to identify those risks and opportunities over the short, medium, and long term, then evaluate their respective consequences for the organization's businesses, strategy, and financial planning. Here, the task force also pushed for assessing the implications of different scenarios, including pursuit of a 2°C scenario. The third set of recommended disclosures addressed risk management, specifically how the organization would identify, assess, and manage climate-related risks and integrate that process into its overall risk management. Finally, apart from disclosing its Scope 1, Scope 2, and, where appropriate, Scope 3 GHG emissions, the organization will also describe the metrics it uses for those climate-related risks and opportunities in line with its strategy and risk management and, in addition, define the targets against which it measures its performance pertaining to those risks and opportunities. The TFCD provided fairly detailed guidance for what needed to be disclosed by all sectors in these four areas of governance, strategy, risk management, as well as metrics and targets. It then added supplemental guidance for non-financial sectors heavily impacted by climate change, such as energy or transportation, as well as for various financial institutions, notably banks, insurance companies, asset managers, and asset owners (i.e., pension funds and other institutional investors).

One crucial aspect of the task force's recommendations was to have everyone think through the implications of pursuing the Paris Agreement's ≤ 2 °C objective, and for that purpose it suggested here in great detail how to conduct scenario analysis.[20] This method of figuring out a range of possible outcomes, which will play out over the medium to long term under conditions of great uncertainty, applies obviously well to climate change for which several scenarios are imaginable. Of course, the choice of scenarios is crucial here. Besides being obliged to think through the 2° scenario, firms may also project forward business as usual, analyze the pathway implied by the NDCs already put in place, and perhaps add a fourth scenario assuming gradually tightening national mitigation commitments over the coming decade. Each of these different scenarios will carry its impactful risks and opportunities weighing on the bottom line of producers and their investors. These scenarios can be analyzed by means of data and quantitative models or in qualitative terms highlighting trends and relationships for which there is not yet a lot of reliable data available.

The TFCD report foresees widespread adoption of its recommended climate-related disclosure guidelines within four years of its publication, that is by 2021. This dissemination push operates on two parallel levels. For one, its recommendations are meant to provide a global standard how affected organizations—be they producers or financial institutions tied to them—ought to report the impact of climate change. The idea is to align those TFCD-recommended reporting guidelines with what various select disclosure frameworks pushed by different non-governmental organizations working on climate change have already tried to do. This means working with relevant NGOs, including the Climate Disclosure Standards Board (CDSB), the Global Reporting Initiative (GRI), the Institutional Investors Group on Climate Change (IIGCC), the International Integrated Reporting Council (IIRC), the Sustainability Accounting Standards Board (SASB), and the World Resources Institute (WRI), to get a harmonized approach that has been widely endorsed and offers state-of-the-art reporting. Moreover, whatever emerges from this harmonization effort in favor of a globally applicable standard will then have to be incorporated into national law for eventual mandatory adoption. This presumes that national governments will be ready to impose climate-related financial disclosures as part of their overall reporting requirements and accounting rules. Once again it is France that has shown its leadership in that area of policy-making with passage in August 2015 of the aforementioned "Law on Energy Transition for Green Growth" (see Chapter 3). That law provides a detailed set of performance targets for France's transition to a low-carbon economy in terms of changing its energy mix, improving energy efficiency, and retrofitting its buildings. Its now-famous Article 173 defines a mandatory reporting framework for listed corporations, banks, institutional investors (e.g., mutual funds and pension funds), and asset managers which serves as a model for other countries making climate-related financial disclosures mandatory in line with TFCD recommendations.[21]

The provisions of Article 173, going hand in hand with the TFCD guidelines, put finance right at the center of our global mobilization against climate change. Banks will have to have a climate component in their regularly required stress tests. Large asset managers and institutional investors, worth more than 500 million euros in assets, have to assess their portfolio's exposure to both physical and transition risks posed by climate change. They also have to account for actions taken to meet (national and/or their own) low-carbon goals and targets. All asset managers and institutional investors, even the smaller ones, must report on their incorporation of ESG criteria. Publicly traded companies have to disclose in their annual reports the environmental impact of their activities, climate-related risks they face, and the measures taken to manage those risks. While its reporting requirements are surely ambitious, the law gives the parties affected a lot of flexibility in figuring out how best to meet them. It adopts a so-called comply or explain approach according to which investors can decide what methodology they prefer as long as they justify its use. Experimentation is thus explicitly encouraged, a good thing when considering that we are really just at the beginning of climate-related financial disclosure with a long learning curve to go. In the end, we expect investors to adopt best-practice solutions, which the law itself promotes by requiring two-year reviews of reporting practices. While the flexibility built into Article 173 extends to what details investors may provide across the board, the law does specify extensive reporting of GHG emissions (past, present, future; direct, indirect), of policy risks linked to climate targets, of changes in availability and prices of natural resources, and of climate change's impact on assets (including exposure to extreme-weather events). The combination of TFCD and France's Article 173 provides a framework for climate-related financial disclosures going forward.

CLIMATE CHANGE AS AN INSURANCE PROBLEM

In his much-quoted "Tragedy of the Horizon" speech of September 2015 Mark Carney laid out in detail four unique risks associated with climate change.

- To begin with, the very passage of time wasted carries its own risk. The longer we wait to act against climate change, the worse it will get—either in terms of the damage caused by not having addressed this issue in timely fashion or in terms of how much more radical our responses will have to be later when compared to acting now.
- Physical risks stem from the destabilization of the climate itself. We have already mounting evidence of the growing frequency, far greater intensity, and frightful damage potential of catastrophic weather events such as stronger storms or worse droughts. We can also literally see sea levels rising (in places like the tip of Manhattan or Venice) or deserts spreading (as in the Western USA or Central Asia) both of which are bound to affect large numbers of people in lasting and life-transforming fashion. We are expecting 700 million climate-related migrants by 2050. Of course, we also have to foresee intermediate climate-change manifestations damaging our habitat physically such as floods or wildfires. Finally, we will surely face permanent losses in biodiversity as well as the threat of pandemics as new contagious diseases emerge and spread. We have in recent years already witnessed the onset of this worrisome trend with the arrival of the Avian Flu, the Ebola virus, or the Zika virus.
- Liability risks arise when the damage and suffering caused by climate change prompts parties affected to seek compensation from those they hold responsible for having caused these adverse conditions.[22] We can expect such claims aimed at emitters to emerge, perhaps even spread dramatically, in the future as climate change takes root across the globe. As of now, it is not at all clear from a legal standpoint, under what kind of law such cases will be processed successfully. To the extent that this is an intergenerational issue, how will future generations be able to seek restitution from us? To the extent that this is a geographic and distributional issue, how will poorer regions get compensated by more highly industrialized regions having emitted proportionately more in support of their consumption norms and imposing those emissions on the rest

of the world? While we have yet to figure out the legal context for future damage claims (e.g., binding emission targets), liability risk is here to stay and likely to be significant.

- Transition risks take root, as the name implies, in the transition to a low-carbon economy when changes in public policy, technology, investor preferences, or physical events force the repricing of assets. For example, oil, gas, or coal reserves can no longer be burned and so become "stranded" assets once the world decides it needs to get out of fossil fuels. To the extent that these now-unusable reserves become subject to what Carney termed "jump-to-distress" pricing, they affect the valuations of the energy companies. The same may happen to utilities when they have to close down high-emission power plants before those have been fully amortized. As a matter of fact, we can imagine a large variety of carbon-intense assets to become devalued and so burden their owners with financial losses.[23] Whether these write-offs follow a slow and reasonably predictable path or such repricing occurs instead in chaotic fashion will make a big difference in how such adjustments affect financial stability. Better to do them early and slowly rather than later and in a panic! Another source of transition risks is connected to technological change, both in terms of what works and what does not, as well as its potentially transformative impact when it gets implemented on a large scale such as can be expected from carbon capture and storage (CCS) technology carrying considerable take-off risk.

In light of these unique risks, it is fair to say that climate change must also be understood as a unique insurance problem, especially in terms of scenarios that currently still constitute "fat-tail" events (e.g., a hurricane hitting New York City in late October, as happened with Sandy in 2012; four major hurricanes in a row devastating the Caribbean and Texas within a month in 2017) possibly becoming the norm in the not-so-distant future. Insurance companies will have to prepare for this eventuality by constantly upgrading their forecasting models and scenario analysis, adapting their business models to changing circumstances in terms of coverage, claims, and premiums, and implementing a forward-looking capital regime that leaves them sufficiently well capitalized to withstand significantly higher losses in the wake of global warming. The great danger is that the process of climate change outpaces the insurers' capacity to respond which points to the importance of *re-insurance* whereby coverage risks are passed on to reinsurers. We must also consider public backstops for insurance which nationalize certain costly risks beyond the reach of individual insurance coverage (e.g., as is already taking place with flood insurance in low-lying areas facing rising sea levels). Of course, as has become evident in the US-sponsored National Flood Insurance Program (NFIP), if the government subsidizes flood insurance too much, it ends up encouraging over-building in flood-prone areas as an unintended consequence. Insurance premiums for buildings in low-lying areas have to be high enough to limit this counterproductive incentive.

As insurers face a steadily growing burden protecting their clients from climate-related risks, they can be expected to rely more heavily on so-called *catastrophe bonds* especially if and when it will have become more difficult to purchase traditional reinsurance. Such cat bonds emerged first in the mid-1990s in the wake of the enormously costly Hurricane Andrew hitting South Florida (August 1992) followed by the Northridge earthquake triggering extensive damage across Southern California (January 1994). They took off after Hurricane Katrina destroyed New Orleans in August 2005. This unique type of bond helps insurers cover the costs of catastrophic events which may trigger billions in payout claims from policy holders. Issued by insurers via special-purpose entities they would set up for that reason and distributed to investors through investment banks hired to help with the issue, cat bonds transfer the insurance risk posed by catastrophic events to investors holding those bonds. Usually having fairly short maturities (up to three years) and rated below investment-grade as a high-risk instrument (typically double-B or B), cat bonds carry high yields with spreads ranging from 3 to 20% above the LIBOR interbank rate to attract investors. If there is no trigger event before the cat bond matures, like a major hurricane causing sufficient damage beyond a threshold in a specified region, then the investors gain—basically pocketing the coupons paid in interest by the issuer. If, on the other hand, there is a trigger event like the aforementioned hurricane defining the underlying "catastrophe" as such, then the investors face a loss of principal which gets transferred to the insurer to be used to cover claims arising from this catastrophic event. The key here is how to price the risks of extremely rare and unprecedented events correctly, in the absence of precedence and pattern, so that these risks can be spread out over a large number of investors willing to take the gamble from which they expect to gain.[24]

While traditional insurance policies or cat bonds protect against high-risk, low-probability events, such as floods or hurricanes, businesses also need to transfer weather-related risks associated with low-risk, high-probability scenarios. Take, for instance, an unusually warm winter leaving electricity utilities with excess supplies of oil or natural gas as people heat their homes less, or an unusually cold summer discouraging people from traveling and hence causing fewer reservations at hotels or airlines. Early on, already in 1999, the Chicago Mercantile Exchange introduced for that reason *weather futures* (and options on these futures) reflecting monthly and seasonal average temperatures calculated by means of a specific index for fifteen American and five European cities. There is an index for the winter measuring daily temperature deviations from a benchmark level in terms of a heating degree day (HDD) value, and one for the summer in terms of a monthly cooling degree day (CDD) value. The index value added up over a month then determines how much the contract pays out to the buyer of such protection. The major users of such weather futures are energy companies, agricultural producers, restaurants, and businesses involved in tourism or travel.

Climate change promises to induce a much wider application of such weather-related futures and options, going even beyond standardized exchange-traded contracts to include privately negotiated, individualized contracts made between two parties "over the counter." Think in this context of climate-related swaps tied to events and trends on which both sides make their bets. Sellers of such swaps offer protection in the form of payouts when well-defined triggers materialize while receiving regular premium payments from the buyers of such protection in the meantime. Such arrangements can be customized more flexibly to particular situations, may mobilize much larger amounts of capital for the assumption of climate-related risks, and encourage a more responsive pricing dynamic in a situation likely to be in flux for a long time to come. Whether futures or swaps, weather derivatives will cover many more areas across the globe and also a greater variety of weather-related phenomena beyond daily temperature fluctuations deviating from the monthly average, such as wind speeds, precipitation, or hours of sunshine. Wind- or sunshine-based weather derivatives are obviously of great usefulness to suppliers of renewable energy who may also be obliged by banks to obtain such weather risk management products as a condition attached to their loans. Traditional energy firms use these products for the opposite reason, namely as a protection against falling sales when renewable-energy-supply alternatives enjoy periods of above-average sunshine or wind speeds. All things considered, it will be crucially important to mobilize, beyond traditional insurers and government backstops, large amounts of financial capital to assume weather-related risks in exchange for possible gains.

The Discount Rate, Capital Budgeting and Other Foibles of Financial Economics

One of the key issues facing climate finance is that several key notions of mainstream financial economics fail to apply. Take, for instance, the idea that the stock market is efficient in the sense that it processes all available information accurately and so typically sets correct prices.[25] Yet we have right now a fairly massive case of mispricing, concerning the share prices of the oil majors such as BP or Exxon Mobil. It is widespread knowledge that fossil fuels will have to be divested from, leaving oil and gas companies with trillions in unusable reserves. And yet share prices for these energy giants remain high despite all those "stranded assets" about to appear on their books. Barely any adjustment has been made! One reason, noted first by Daniel Kahneman and Amos Tversky (1979), may be loss aversion prompting investors to hang on to shares even though they are bound to decline in value. More generally, these two psychologists have systematically explored a variety of cognitive biases in the face of uncertainty all of which raise serious doubts about the innate rationality of investors. Based on their empirical findings, which have formed the basis of what after Kahneman's Nobel Prize in Economics in 2002 has emerged as *behavioral finance*, we are led to conclude that irrational behavior predominates. People assess probabilities often incorrectly, filter out information they do not agree with, overestimate their own thinking capacity while underestimating the advice of experts, respond more strongly to the most recent information, and shape their attitude about new phenomena by framing those in terms of what they are already familiar with.[26] All of these biases, and others possibly as well, may prevent investors from taking the steps needed for an orderly transition to a low-carbon economy. One of the most powerful biases is herd behavior which makes us move in tandem with others. This may be advantageous when wishing to accelerate change (as with adopting ESG criteria, for example). But it is burdensome when accentuating pro-cyclical forces, especially panics which we may well face when finally ready to acknowledge stranded assets.

Another mainstream concept put in doubt when it comes to climate finance relates to the question how to evaluate investment projects we undertake to fight climate change, such as installing renewable energy sources, retrofitting buildings, developing clean cars, or carbon capture and storage technology. Normally, when comparing current cost outlays with future revenues, we must consider the time value of money to take into account the difference in timing between cash out- and inflows.

In other words, we have to discount estimated future revenues to present value to make the two cash flows comparable in light of the fact that the time value of money renders future cash flows less valuable than those occurring now. Using what has become known as the capital asset pricing model (CAPM), or capital budgeting, corporations calculate the net present value of an investment project as the sum of discounted future revenue flows, net of investment costs. Investors use the same discounted-cash-flow method to determine the "fair" (or fundamental) value of a security like a share or a bond both of which carry their own series of cash inflows over time in need of discounting (e.g., interest for bonds, dividends for shares, and possible capital gains at the end of the holding period).[27]

Unfortunately, we cannot apply this capital asset pricing model easily to climate-change mitigation projects. We already have a problem here of lacking historic precedence. We do not have a lot of experience when it comes to green investments, and so it will be difficult to get reliable estimates of future returns or costs based on past behavior. The problem of estimating future returns in the absence of past experience with projects of a similar kind gets compounded by the long time horizons involved in climate change. Typical CAPM calculations involve no more than ten years. But projects associated with climate-change mitigation and adaptation are most likely to stretch over much longer horizons, up to fifty years for, say, power plants. Even when stretched over just thirty years, still a very long horizon under normal conditions but at best average for the issue of climate change, the majority of what determines the net present value of such a project consists of valuing cash flows arising many years from now. A third problem projecting future cash flow estimates in climate-mitigation projects has to do with the fact that a lot of their future benefits take the form of reducing social costs, meaning GHG emissions, for which we do not have a proper price yet and whose eventual price is going to be by definition somewhat arbitrarily picked as a matter of political compromise.

As if the evaluation of climate-mitigation investment projects was not already fraught with enough complications, there is also the problem of risk. As Mark Carney has pointed out so eloquently in his "Tragedy of the Horizon" speech, climate change involves special (physical, liability, and transition) risks many of which we have never faced before. The very nature of these risks—global, cumulative, dynamically interactive, and engulfing both the micro-level of individual actors or projects and the macro-level context of their embeddedness in a broader environmental setting subject to continuous change—will require systemic-risk models of much greater sophistication than the ones we have developed over the last decade for the financial system in response to the systemic crisis hitting the global banking system in 2007/08. We are just beginning to contemplate this challenge of systemic climate-risk modeling, adding scenario analysis and climate stress tests into the mix for better damage and loss-potential estimates.[28]

Yet another major issue flying into the face of mainstream financial economics concerns the choice of the appropriate discount rate for climate-mitigation projects. Remember, we need to discount future cash flows to present value, and this requires us to choose a discount rate. That rate, at which we discount future cash flows to present value, represents from the investor's point of view his cost of capital. We can apply the cost-of-capital notion to the discount rate for two reasons. Either the investor has borrowed money used for the investment at a certain interest rate and then would want to have that investment yield a return in excess of that "hurdle" rate. Or the investor could think of the discount rate as an opportunity cost of returns that could have been earned elsewhere which the project under consideration should at least match to justify being prioritized. In the end, the appropriate discount rate is a measure of how much we value the future today. If the discount rate is too high, we devalue the future too much as a result of which we are less inclined to do much about it. The problem is especially pronounced with regard to climate change because of its long time horizons. When cash flows get stretched over very long periods of time, then even fairly small alterations in the discount rate have a big impact on how much those far-away cash flows would be worth to us today. Let us assume that we would like to avoid 100$ in climate-change damage thirty years from now. If we applied a standard discount rate, say 3%, then we would be willing to spend 41$ today to achieve that goal of avoiding $100 in damage three decades later. If we were to raise the discount rate to 5%, we would end up only willing to spend 23$ today. But with a discount rate set at just 1%, we would be willing to spend a significantly higher amount of 74$ today.

The choice of the discount rate is thus a matter of concern in the fight against climate change. At this point, the typical discount rate used in the USA averages between 3 and 4%. In the standard CAPM formula, risks are incorporated into the investment evaluation by raising the (risk-adjusted) rate at which we discount future cash flows. Given the high-risk nature of climate change, one could easily defend a risk-adjusted discount rate of 5% if we followed the mainstream CAPM logic, perhaps even as high as 7% (as recently suggested by Trump's EPA). Such risk adjustment would have especially dramatic effects for projects combating climate change which are typically of a long-term nature. When applied to projects whose cash flows stretch out over a long period, higher discount rates have a disproportionately large impact of reducing the present value of future returns. Many low-carbon transition projects would thus simply never be undertaken, if we allowed the standard capital asset pricing model to apply.

This then leads us to the question of how to justify especially low discount rates, kept here at 1% or even less, for projects designed to address climate change. The reasoning has to start with radical uncertainty. If we do not do anything about climate change, we stand a tangible chance of making the planet in large parts uninhabitable for our children and grandchildren as average temperatures rise more than 5°C and sea levels more than two meters from current levels. Even a ten-percent probability for such a catastrophic scenario makes the "business as usual" preference of the climate sceptics morally indefensible, given that we could have done something to prevent inflicting such an unmitigated disaster on innocent future generations. That prospect of avoidable disaster, no matter how small its chance as long as it is tangible, obliges us to act, and so the Precautionary Principle applies here (even in its strong version). It is precisely the certainty of this uncertainty, that anything is possible including planetary catastrophe, which makes the usual risk calculus, by which we try to turn an intangible reality, that of radical (i.e., inherently unknowable) uncertainty about the future, into something measurable, namely risk, inapplicable. Even the smallest chance of the greatest risk, namely a planetary catastrophe threatening half of the human species, obliges us to assume that it is indeed possible and act accordingly. Once there, we have effectively moved beyond the usual uncertainty <-> risk connections and risk-return trade-offs. That is a moral obligation we owe future generations, hence beyond the daily calculations of finance and dictates of a profit-driven market economy.

The very uncertainty associated with climate change justifies super-low discount rates also from a standpoint of insurance. Even today (mid-July 2017) we are willing to accept a yield of 1.3% on a 30-year German bond, and that is after a tripling of that rate (from 0.35%) in just one year prompted by more optimistic assessments of the euro-zone's performance capacity. So we are in effect willing to accept a 1.3% return stretched over three decades as the price for having as close to total certainty as possible, assuming that the German state, with its responsible fiscal stance, political stability, quality of moderate leadership in the international arena, and underlying strength of its economy, represents that "best of all possible worlds" projected forward into the foreseeable future. Making sure that we can survive reasonably well in the face of a 1.5–2°C temperature hike instead of exposing ourselves to the dangers of unbridled global warming ought to be worth at least as much certainty as that, hence justifying a similarly low rate as compensation for relative safety.

Ultimately, we have to regard the question of applying a low discount rate to low-carbon transition initiatives as an ethical issue. Future generations are born into a world we leave them, either a miserable one because of our inaction or a still tolerable one because of our timely actions at a sufficiently large scale. Future generations have no say in that matter; only we do. It is a moral imperative to leave as good a world behind as possible. In that sense, we need to recognize the containment of climate change (and subsequent avoidance of disaster for our children and grandchildren) as a global public good from which no one should be excluded. It is this social-benefit aspect of the effort that justifies applying a non-market (i.e., low) discount rate as a matter of public policy to make sure we give enough value now to future prevention. How to resolve this contradiction between a high private discount rate set in the marketplace and a low social discount rate set by policy remains to be seen. But without a solution to this problem, we will not be able to mobilize finance in the fight against climate change.

NOTES

1. The Paris Agreement explicitly introduced the concept of "climate finance" and set up a twenty-member Standing Committee on Finance to oversee its launch, incorporating in the process a number of existing and new funds under what is termed the Financial Mechanism. These include the Global Environmental Facility (GEF), the Green Climate Fund (GCF), and a so-called Adaptation Fund yet to be set up. A good overview of this official UN-sponsored part of "climate finance" can be found on the Web site of the United Nations Framework Convention on Climate Change http://unfccc.int/focus/climate_finance/items/7001.php#intro.

2. All the data in that paragraph on the global infrastructure investment shortfall are from McKinsey Global Institute (2016).

3. For more on the reports cited here, see Global Commission on the Economy and Climate (2016), O. Milman (2016), Organization of Economic Cooperation and Development (2017), and Energy Transitions Commission (2017).

4. Quantitative easing, an unorthodox monetary-policy strategy deployed on a significant scale by the Federal Reserve after November 2008 and subsequently also adopted elsewhere, consists of massive securities purchases by the central bank which thereby pumps cash reserves into the banking system to drive down interest rates (and also exchange rates) so as to incite more borrowing, higher asset prices, and revived economic activity. For more on QE, see J. Benford et al. (2009), J. Stein (2012), and *The Economist* (2015).

5. The crucial and productive role possibly played by national and multinational development banks in the transition to a low-carbon economy has been highlighted in the important report on climate finance provided by the Canfin-Grandjean Commission (2015) under the auspices of France's Hollande Administration in the run-up to the COP21 Conference at the end of 2015.

6. See S. Gordon (2017) for more details on the implementation of the Juncker Plan and functioning of the EFSI. The global push for green investment banks is outlined in Organization of Economic Cooperation and Development (2016). A good discussion of new financial instruments public funds could use to mobilize a multiple of private funds for green investments can be found in J. Brown and M. Jacobs (2011). China's emerging global leadership role in development finance has been well documented in K. Gallaher et al. (2016).

7. For more on the Green Climate Fund, see its Web site www.greenclimate. fund as well as the United Nations Framework Convention on Climate Change (2014).

8. See the very informative piece by K. Allen (2017) on the wider meaning of these controversies pertaining to how to judge green bonds. And C. Floods (2017) provides useful information on efforts to improve the institutional setting for the green-bond market.

9. See L. Hornby (2017) for more details on the rapidly evolving Chinese market for green bonds.

10. For an interesting discussion of the corporate paradigm shift from the post-war "retain and reinvest" strategy to the "downsize and distribute" posture embodied by the prioritizing of shareholder value maximization, see W. Lazonick and M. O'Sullivan (2000).

11. The precise quote by M. Friedman (1962, pp. 133–4), much cited by his many admirers all over the world, was "There is one and only one social responsibility of business–to use it resources and engage in activities

designed to increase its profits so long as it stays within the rules of the game, which is to say, engages in open and free competition without deception or fraud." M. Friedman (1970) restated the argument of shareholder value's centrality in terms of agency theory, ending the article with the same quote above.

12. Among key contributions of agency theory, see S. Ross (1973), M. Jensen and W. Meckling (1976), as well as E. Fama and M. Jensen (1983).

13. In this context, it is also worth mentioning the analysis (in English) by M. Aglietta (2000) and R. Boyer (2000) of shareholder value maximization at the center of a new finance-led, post-Fordist growth regime taking root globally in the 1990s.

14. A. Edmans, V. Fang and K. Lewellen (2017) found in a large-scale study that corporate managers are much more likely to slash investment spending, lower research and development budgets, push for positive earnings guidance, and actively seek upward revisions of analyst forecasts in the same quarter that their stock options vest than during other times, all to boost their returns when they can sell their shares at thereby increased share prices.

15. J. Authers (2015) suggests that, notwithstanding the deeply entrenched dominance of shareholder capitalism and its grounding in legal doctrine in the United States, there are growing signs that this paradigm is being challenged on a variety of levels and from several directions.

16. See Bloomberg (2017) as well as Deutsche Asset Management (2017) for how pervasive sustainable finance, with ESG at its core, has become to the modus operandi of the business-and-market information giant. See also S. Grene (2012) on the market's dawning realization of a distinction between SRI and ESG and the frentic work of providing more ESG data to market participants at the time. For empirical studies showing a positive correlation between solid ESG practices and operational or stock price performance, see the Sustainability and Finance Symposium co-sponsored by pension fund CalPERS and University of California-Davis in June 2013, G. L. Clark et al. (2014), J. Kynge (2014), or S. Murray (2015).

17. See EUROSIF (2016) and USSIF (2016).

18. The ways specific types of funds integrate ESG criteria differently are discussed in F. Reynolds (2015), M. Scott (2015), or S. Grene (2016). See J. Authers (2015b) for more details on these different initiatives in the direction of institutional investors' increasing collaborative push for corporate responsibility in the face of climate change.

19. Mark Carney (2015) characterized climate change as a "tragedy of the horizon" inasmuch as the long-term nature of this challenge extends far beyond the normal planning horizons of policy-makers and investors. The Task Force's report can be found in TFCD (2016).

20. For an interesting discussion how France's largest bank BNP Paribas uses scenario analysis, an important innovation in financial-disclosure methodology pushed by the TFCD, to look at the impact of "de-carbonizing" its clients' investment portfolios see A. B. H. Soulami (2017).

21. Useful discussions concerning the innovative details of the path-breaking Article 173 disclosure requirements can be found in S. Rust (2016), Forum pour l'Investissement Responsable (2016), and Principles for Responsible Investment (2016).

22. D. Jergler (2015) discusses why climate change litigation is so much more common in the USA than elsewhere. For a global review of climate change litigation trends and issues, see United Nations Environment Programme (2017).

23. C. Clouse (2016) reports on the extent of the so-called carbon bubble comprising potentially "stranded" assets across several crucial industrial sectors which have yet to be re-priced downward, and on how the prospect of large losses when that bubble bursts is beginning to trouble investors, even those otherwise skeptical of climate change.

24. M. Lewis (2007), author of such best selling books about Wall Street as *Liars' Poker* or *The Big Short*, reports on the two key thinkers behind the modeling and pricing of extreme-risk events such as hurricanes and how their contributions have helped launch the cat bond market.

25. This so-called efficient-market hypothesis at the core of mainstream financial economics, highlighting the stock market's informational efficacy, was put forth by E. Fama (1965, 1970).

26. There are possibly many more well-established cognitive biases shaping investment behavior to the detriment of addressing the climate-change challenge, such as status-quo preference, trend-chasing bias in favor of projecting past data forward, or familiarity bias. See in this context also H. K. Baker and V. Ricciardi (2014) as well as H. K. Baker et al. (2016).

27. When financial markets are (informationally) efficient, the actual market price of such a security is assumed to equal precisely that "fair" value comprising the sum of discounted cash flows. The Capital Asset Pricing Model (CAPM) underpinning this argument, deservedly considered one of the central pillars of modern finance, has been much discussed, as in those original contributions of W. Sharpe (1964) or F. Black et al. (1972).

28. Official recognition of climate change as a systemic-risk challenge, especially when associated with a belated and hence panicky transition to a low-carbon economy, can be found in European Systemic Risk Board (2016). For a broader discussion of climate change as a systemic risk, see M. Aglietta and E. Espagne (2016). The idea of a climate stress test as part of systemic-risk management of climate change has been well formulated by S. Battiston et alii (2017).

REFERENCES

Aglietta, M. (1998). La globalisation financière. In CEPII, *L'économie mondiale 2000* (pp. 52–67). Paris: La Decouvérte.

Aglietta, M. (2000). Shareholder Value and Corporate Governance: Some Tricky Questions. *Economy and Society, 29*(1), 146–159. https://doi.org/10.1080/030851400360596.

Aglietta, M., & Espagne, E. (2016, April). *Climate and Finance Systemic Risks, More Than an Analogy? The Climate Fragility Hypothesis* (CEPII Working Paper, No. 2016–10). http://cerdi.org/uploads/sfCmsNews/html/2958/201705_Espagne_Climate.pdf.

Allen, K. (2017, May 25). Sellers of Green Bond Face a Buyer's Test of Their Credentials. *Financial Times.*

Authers, J. (2015a, October 23). Vote of No Confidence in Shareholder Capitalism. *Financial Times.*

Authers, J. (2015b, December 15). Climate Talks Mark Turning Point for Investors. *Financial Times.*

Baker, K., & Ricciardi, V. (2014, February–March). How Biases Affect Investor Behavior. *European Financial Review.* www.europeanfinancialreview.com/?p=512.

Baker, K., Filbeck, G., & Ricciardi, V. (2016, December–January). How Behavioral Biases Affect Finance Professionals. *European Financial Review.* http://www.europeanfinancialreview.com/?p=12492.

Battiston, S., et al. (2017). A Climate Stress-Test of the Financial System. *Nature Climate Change, 7*, 283–288. https://www.nature.com/articles/nclimate3255?WT.feed_name=subjects_business.

Benford, J., et al. (2009). Quantitative Easing. *Bank of England Quarterly Bulletin*, 90–100. http://www.bankofengland.co.uk/publications/Documents/quarterlybulletin/qb090201.pdf.

Black, F., Jensen, M., & Sholes, M. (1972). The Capital Asset Pricing Model: Some Empirical Tests. In M. Jensen (Ed.), *Studies in the Theory of Capital Markets* (pp. 79–121). New York: Praeger.

Bloomberg LP. (2017, May 17). *2016 Impact Report.* https://data.bloomberg-lp.com/company/sites/28/2017/05/17_0516_Impact-Book_Final.pdf. Accessed 8 July 2017.

Boyer, R. (2000). Is a Finance-Led Growth Regime a Viable Alternative to Fordism? A Preliminary Analysis. *Economy and Society, 29*(1), 111–145. http://dx.doi.org/10.1080/030851400360587.

Brown, J., & Jacobs, M. (2011). Leveraging Private Investment: The Role of Public Sector Climate Finance. *ODI Background Note.* Overseas Development Institute. https://www.odi.org/sites/odi.org.uk/files/odi-assets/publications-opinion-files/7082.pdf.

Canfin-Grandjean Commission. (2015). *Mobilizing Climate Finance: A Roadmap to Finance a Low-Carbon Economy.* http://www.cdcclimat.com/IMG/pdf/exsum-report_canfin-grandjean_eng.pdf.

Carney, M. (2015, September 29). *Breaking the Tragedy of the Horizon—Climate Change and Financial Stability.* Speech at Lloyd's of London. http://www.bankofengland.co.uk/publications/Pages/speeches/2015/844.aspx. Accessed 2 November 2015.

Clark, G. L., Feiner, A., & Viehs, M. (2014). *From the Stockholder to the Stakeholder: How Sustainability Can Drive Financial Outperformance.* Oxford: Smith School of Enterprise and Environment, Oxford University. https://doi.org/10.2139/ssrn.2508281.

Clouse, C. (2016, October 27). The Carbon Bubble: Why Investors Can No Longer Ignore Climate Risks. *The Guardian.* https://www.theguardian.com/sustainable-business/2016/oct/27/investment-advice-retirement-portfolio-tips-climate-change-financial-risk.

Deutsche Asset Management. (2017). *Sustainable Finance Report, Issue #2.* https://www.db.com/newsroom_news/2017/cr/deutsche-asset-management-report-the-gathering-forces-of-esg-investing-en-11553.htm. Accessed 14 July 2017.

The Economist. (2015, March 9). *What Is Quantitative Easing?* https://www.economist.com/blogs/economist-explains/2015/03/economist-explains-5. Accessed 23 Mar 2017.

Edmans, A., Fang, V. W., & Lewellen, K. (2017). Equity Vesting and Investment. *Review of Financial Studies, 30*(7), 2229–2271. https://doi.org/10.1093/rfs/hhx018.

Energy Transitions Commission. (2017). *Better Energy, Greater Prosperity.* http://energytransitions.org/sites/default/files/BetterEnergy_Executive%20Summary_DIGITAL.PDF.

European Systemic Risk Board. (2016, February). *Too Late, Too Sudden: Transition to a Low-Carbon Economy and Systemic Risk* (Reports of the Advisory Scientific Committee, No. 6). https://www.esrb.europa.eu/pub/pdf/asc/Reports_ASC_6_1602.pdf. Accessed 17 Dec 2017.

EUROSIF. (2016). *European SRI Study 2016* (7th ed.). eurosif.org/sri-study-2016/ . #SRIStudy2016 #ESG. Accessed 14 July 2017.

Fama, E. (1965). The Behavior of Stock Market Prices. *Journal of Business, 38*(1), 35–104. https://doi.org/10.1086/294743.

Fama, E. (1970). Efficient Capital Markets: A Review of Theory and Empirical Work. *Journal of Finance, 25*(2), 383–417. https://doi.org/10.2307/2325486.

Fama, E., & Jensen, M. (1983). Separation of Ownership and Control. *Journal of Law and Economics, 26*(2), 301–325.

Floods, C. (2017, May 8). Green Bonds Need Global Standards. *Financial Times.*

Forum pour l'Investissement Responsable. (2016). *Article 173-VI: Understanding the French Regulation on Investor Climate Reporting* (FIR Handbook No. 1). http://www.frenchsif.org/isr-esg/wp-content/uploads/Understanding_article173-French_SIF_Handbook.pdf. Accessed 21 June 2017.

Friedman, M. (1962). *Capitalism and Freedom*. Chicago: University of Chicago Press.

Friedman, M. (1970, September 13). The Social Responsibility of Business Is to Increase Its Profits. *New York Times Magazine*. http://www.colorado.edu/studentgroups/libertarians/issues/friedman-soc-resp-business.html. Accessed 30 June 2017.

Gallagher, K. et al. (2016). *Fueling Growth and Financing Risk: The Benefits and Risks of China's Development Finance in the Global Energy Sector* (Working Paper 5-16). Boston University Global Economic Governance Initiative.

Global Commission on the Economy and Climate. (2016). *The Sustainable Infrastructure Imperative*. https://newclimateeconomy.report/2016/.

Gordon, S. (2017, March 29). Juncker's European Investment Plan: The Rhetoric Versus the Reality. *Financial Times*.

Grene, S. (2012, September 7). Early Days But Cautious Optimism on ESG. *Financial Times*.

Grene, S. (2016, January 24). Quants Are New Ethical Investors. *Financial Times*.

Hornby, L. (2017, May 4). China Leads World on Green Bonds But Benefits Are Hazy. *Financial Times*.

Jensen, M., & Meckling, W. (1976). Theory of the Firm: Managerial Behavior, Agency Costs, and Ownership Structure. *Journal of Financial Economics, 3*(4), 305–360.

Jergler, D. (2015, March 12). U.S. Dominates Climate Change Litigation. *Climate Control*. http://www.insurancejournal.com/news/national/2015/03/12/360370.htm. Accessed 17 June 2017.

Kahneman, D., & Tversky, A. (1979). Prospect Theory: An Analysis of Decision Under Risk. *Econometrica, 47*(2), 263–292. https://doi.org/10.2307/1914185.

Kynge, J. (2014, April 24). Sustainable Investors Outstrip Emerging Market Benchmarks. *Financial Times*.

Lazonick, W., & O'Sullivan, M. (2000). Maximizing Shareholder Value: A New Paradigm for Corporate Governance. *Economy and Society, 29*(1), 13–35.

Lewis, M. (2007, August 26). In Nature's Casino. *New York Times Magazine*. http://www.nytimes.com/2007/08/26/magazine/26neworleans-t.html?pagewanted=print. Accessed 18 June 2017.

McKinsey Global Institute. (2016, June). *Bridging Global Infrastructure Gaps*. Report. https://www.mckinsey.com/industries/capital-projects-and-infrastructure/our-insights/bridging-global-infrastructure-gaps.

Milman, O. (2016, October 6). World Needs $90tn Infrastructure Overhaul to Avoid Climate Disaster, Study Finds. *The Guardian*. https://www.theguardian.com/environment/2016/oct/06/climate-change-infrastructure-coalplants-green-investment. Accessed 17 Jan 2017.

Murray. S. (2015, June 2). Sustainability Measurement: Index Looks to Connect Investors. *Financial Times*.

Organization of Economic Cooperation and Development. (2016). *Green Investment Banks: Scaling Up Private Investment in Low-Carbon, Climate-Resilient Infrastructure.* http://www.oecd.org/env/cc/green-investment-banks-9789264245129-en.htm.

Organization of Economic Cooperation and Development. (2017). *Investing in Climate, Investing in Growth.* http://www.oecd.org/environment/cc/g20-climate/.

Principles for Responsible Investment. (2016). *French Energy Transition Law: Global Investor Briefing.*

Reynolds, F. (2015, June 7). Hedge Funds Warm to Responsible Investment Principles. *Financial Times.*

Ross, S. (1973). The Economic Theory of Agency: The Principal's Problem. *American Economic Review, 63*(1), 134–139.

Rust, S. (2016, February 1). France Aims High with First-Ever Investor Climate-Reporting Law. *Investment and Pensions Europe.* https://www.ipe.com/countries/france/france-aims-high-with-first-ever-investor-climate-reporting-law/10011722.fullarticle. Accessed 21 June 2017.

Scott, M. (2015, December 15). Investors Seek Ethical Benchmarks. *Financial Times.*

Sharpe, W. (1964). Capital Asset Prices: A Theory of Market Equilibrium Under Conditions of Risk. *Journal of Finance, 19*(3), 425–442.

Soulami, A. B. H. (2017, May 17). Using Scenario Analysis to Mitigate Climate Risks. *Environmental Finance.* https://cib.bnpparibas.com/sustain/a-four-step-process-for-modelling-climate-risk_a-3-946.html. Accessed 21 June 2017.

Stein, J. (2012, October 11). Evaluating Large-Scale Asset Purchases. *Speech.* https://www.federalreserve.gov/newsevents/speech/stein20121011a.htm.

TFCD. (2016, December 14). *Recommendations of the Task Force on Climate-Related Financial Disclosures.* Basel (CH): Financial Stability Board. https://www.fsb-tcfd.org/wp-content/uploads/2016/12/16_1221_TCFD_Report_Letter.pdf. Accessed 18 Feb 2017.

United Nations Environment Programme. (2017). *The Status of Climate Change Litigation—A Global Review.* http://columbiaclimatelaw.com/files/2017/05/Burger-Gundlach-2017-05-UN-Envt-CC-Litigation.pdf. Accessed 17 June 2017.

United Nations Framework Convention on Climate Change. (2014). *Green Climate Fund.* http://unfccc.int/bodies/green_climate_fund_board/body/6974.php.

USSIF. (2016). *US Sustainable, Responsible, and Impact Investing Trends.* Washington, DC: US SIF Foundation.

CHAPTER 7

Carbon Money

Despite the undeniable progress we have made in mobilizing the world community of nation-states to take the challenge of climate change seriously, we are still only at the very beginning of what will surely be a decades-long struggle. This is above all the case with finance, whose institutions and markets have barely begun to think through how they might best contribute to facilitating the needed transition to a low-carbon economy. Their effort so far is highly uneven, depends greatly on an effective carbon-price mechanism not yet in place, requires information disclosure for which we are just beginning to design some rules, and has to have sizeable resource transfers from rich to poor countries which the former are not yet quite ready for. But what would really push climate finance forward is if we linked it to money creation. For over eight decades, we have had a monetary regime in which most of our money gets created in acts of credit extension within the banking system ("credit-money"), thus anchoring finance around an elastic currency to fuel its rapid expansion along the dual channels of leverage and asset inflation. We can trace a growing variety of such credit-money forms over the last half century and then see how each of those boosted growth of that segment of finance tied to its creation and circulation. A properly designed type of money tied to the reduction of greenhouse gas (GHG) emissions could then be expected to boost climate finance to the scale needed. Such carbon money would also help incentivize cutting GHG emissions to the extent that such a reduction in social costs gets turned into a stream of revenues accruing to the actors initiating such reductions.

© The Author(s) 2018   209
R. Guttmann, *Eco-Capitalism*,
https://doi.org/10.1007/978-3-319-92357-4_7

We can build a whole new business model around that. And the carbon money advanced would at the same time help transform our economy into a more efficient, cleaner, more equitable system centered on the notion of sustainable development. Carbon money may thus very well have a crucial role to play in any credible plan for creating a carbon-neutral economy by mid-century.

WHY CARBON MONEY?

Money is at the center of our capitalist economic system. We live after all in a cash-flow economy in which all economic activities—exchange, production, finance—are organized as spatially and temporally interdependent monetary circuits woven together into a vast web of transactions. Money gets transferred between economic actors, structuring the social space. And money gets invested forward, thereby structuring time. All investments require money being spent now in order to make more money later. Above all, money is also a social institution through which we express and measure value, thus subjecting goods and services we produce as "commodities" for sale to a process of valuation (giving it a price) and validation (selling it for profit). The social validation is one of the crucial aspects of our economic system, since for anyone to earn income they have to sell something that someone else wants sufficiently to pay the required price for. In this way, money subjects everyone to a monetary constraint of having to sell before being able to buy, making it the central fulcrum of sanction and incentive.[1]

When it comes to climate change and our struggle to deal with this issue, we will do well if we managed to commodify carbon dioxide and then monetize efforts to reduce this pollutant. The emission of greenhouse gases, such as carbon dioxide, is a systemic threat through its multifaceted impact on our climate which in turn transforms our environment largely for the worse. We have treated this threat so far as a negative externality, in other words as a social cost which has up to now not entered into the calculus of our production- or consumption-related costs and benefits. Such a social cost needs to be internalized and so made part of our market transactions. Our internalization efforts so far have followed the classic route in terms of providing a market-based approach a la Ronald Coase in the form of "cap-and-trade" schemes for carbon markets or an emissions tax a la Arthur Pigou. Neither of these efforts has been anywhere near where they should be. Existing carbon markets everywhere suffer from chronic excess supply, hence low prices. And carbon taxes have proven difficult to put into effect at sufficiently high levels to make a difference. These policy failures are not only due to lack of political will, but also relative to the enormity of the social cost to be internalized. We are talking here about a social cost with the capacity to destroy much of our habitat as we know it. This kind of negative externality, huge to the point of posing a systemic threat, needs perhaps more radical solutions than the adjustments at the margins proposed by either Ronald Coase or Arthur Pigou. And it is precisely here where the idea of carbon money starts making sense.

The imposition of carbon money should propel our global battle against climate change to a whole new level, much closer to where it will need to be if we are to meet the Paris Agreement's objective of keeping the cumulative temperature increases to $\leq 2\ °C$. For one, there would be much greater impetus to have an appropriate carbon price everywhere and make that price matter in facilitating the reorganization of our system toward a carbon-neutral economy prioritizing sustainable development. To the extent that such an ubiquitous carbon price becomes a strategic variable guiding production and consumption decisions, it will provide us with adequate signals as to the relative social costs and benefits of high-carbon versus low-carbon alternatives. Polluters would end up having to pay a high enough and steadily rising price for their GHG emissions. At the same time, efforts aimed at lowering such emissions would be rewarded, turning the reduction of these huge social costs into a source of "private" revenue yielding the pollution reducers a sum of carbon money that was initially created to finance those efforts. Such endogenous creation of carbon money would extend to paying fossil-fuel producers and other dirty sources of GHG emissions for liquidating their "stranded assets" which we will have to get rid of in the pursuit of our Paris Agreement objectives. And as we anchor carbon money—its creation, circulation, and valuation—at the center of our movement to a zero (net)-carbon economy, climate finance will grow with it to the point of possibly giving us a new, more progressive financial system in sync with the new eco-capitalism we are trying to create here. We know from the analyses by French Régulationists of capitalism's evolution (see the aforementioned works by Michel Aglietta, Robert Boyer, or Alain Lipietz) that any new accumulation regime, such as the ecologically oriented capitalism we are hoping for in lieu of our contemporary finance-led capitalism, has up until now always involved a new monetary regime comprising different money forms, their management of the state's monetary authority ("monetary policy"), financial regulation, and international monetary reform. And so will it have to be also in the case of carbon money and eco-capitalism.[2] We are already beginning to see first signs of such a new money form tied to climate mitigation as put forth in the Kyoto Protocol and Paris Agreement, supplanted during the last couple of years by some more ambitious proposals discussed in the middle part of this chapter before concluding with my own proposal.

INTERNATIONAL EMISSION REDUCTION CERTIFICATES: A PRELIMINARY FORM OF CARBON CURRENCY

As already mentioned in Chapter 5, the global body dealing with climate change—the United Nations Framework Convention on Climate Change (UNFCCC)—has long pushed market-based mechanisms for GHG emission reductions. Its Kyoto Protocol of 1997 provided the framework guiding international mitigation efforts from 2008 to 2012 (the "first commitment period" aiming at a 5% reduction of GHG emissions from 1990 levels). This has been extended for an eight-year "second commitment period" from 2013 to 2020 during which GHG emissions would be reduced by at least 18% from 1990 levels. As part of the Kyoto framework, richer countries could meet their legally binding emission-reduction targets not least also by sponsoring verifiable GHG reduction projects elsewhere, typically where such efforts would cost less ton for ton. Up to now, there have been two such emission-reduction transfer mechanisms. The Joint Implementation (JI) mechanism provides transfers between signatories (so-called Annex 1 countries). And the Clean Development Mechanism (CDM) fosters cooperation with developing countries that are not part of the Kyoto Protocol.

Such emission-reduction transfer mechanisms make a certain amount of sense. It does not matter where GHG emissions are cut, since the stock of those heat-trapping gases accumulates across the entire atmosphere. Yet, the geographic distribution of mitigation costs is not equal at all. Costs tend to be lower in less developed countries where such mitigation projects may at the same time also yield stronger benefits in terms of promoting sustainable development. Take, for example, the provision of clean-burning cookstoves in developing countries cutting down on carbon emissions as well as other health hazards arising from exposure to the smoke of open fires while at the same time reducing deforestation and freeing time previously dedicated to the arduous task of collecting wood. Moreover, these transfer mechanisms allow richer countries to fund emission-reduction efforts in poorer countries where such initiatives could not have been launched on their own. More marginal countries are thereby drawn into the global climate-change mitigation effort.[3]

Kyoto's two cross-border "flexibility mechanisms" discussed here introduced standardized emissions offset instruments, so-called Emission Reduction Units (ERU) for JI-sponsored mitigation projects and Certified Emission Reduction (CER) units for CDM-guided projects, which in effect have served as an early type of carbon currency. Each ERU or CER is worth a ton of carbon dioxide (or CO_2 equivalent) whose emission the thereby sponsored project has helped to prevent. Those ERUs or CERs get earned as saleable credits from successfully launched emission-reduction projects, subject to a verification procedure, and can then be counted toward a country meeting its Kyoto targets. For this to work, either instrument has to be fully convertible, one to one, with the Assigned Amount Units (AAUs) of which each Annex 1 country of the Kyoto Protocol gets a predetermined amount calculated on the basis of its 1990 levels of GHG emissions. Let us assume that two countries A and B were initially assigned 100 AAUs each. Now, country A hosts a JI project for country B which reduces CO_2 emissions by an officially verified 10 tons. Country A has to convert 10 of its AAUs into 10 ERUs which it then transfers to country B, leaving it with 90 AAUs, while country B now has the equivalent of 110 AAUs to its name. When it comes to CDM projects, Annex 1 countries committed to emission-reduction targets can achieve these goals also by sponsoring emission-reducing projects in developing countries which earn them CERs on top of their AAUs. The developing countries hosting CDM projects are under no obligations to reduce emissions, but see it to their advantage to make a voluntary contribution in this fashion.

During Kyoto's first commitment period from 2008 to 2012, there were 582 JI projects launched, yielding 658 million ERUs of which 80% were issued in 2012. But this sudden rush into JI projects had much to do with just two countries, Russia and Ukraine, seeking to turn their initial oversupply of AAUs under the Kyoto Protocol into what came to be called "hot air laundering" by exporting their AAU surpluses which at that point had become barely tradeable inside those countries. So Russia and Ukraine made up over 88% of all the ERUs issued. And their dumping, which those two countries achieved by bypassing the official centralized verification procedure in favor of domestic certification of often dubious projects, depressed ERU prices all the way down to 0.2 euros by 2013. That price was so low that it rendered the few additional, truly emission-reducing JI projects unprofitable even though they were eminently worthwhile from a climate-mitigation point of view. This dismal outcome of the JI experiment prompted needed reforms at the beginning of the second commitment period, subjecting all future JI projects to the official independent verification procedure and making the issue of new ERUs contingent on verified emission reductions.[4]

The CDM, used to promote climate-change mitigation activities in developing countries that are not bound to Kyoto's emission-reduction targets, had a somewhat more promising launch, thanks not least to its more credible governance structure. By the end of 2012, there were 6392 CDM projects, and the total amount of CERs issued at the end of the first commitment period totaled 1.481 billion. The mechanism has continued to grow quite rapidly during the first half of the second commitment period, with CERs at the end of 2017 amounting to 3.742 billion according to data provided by the UNFCCC. Even though still project based, the CDM initiatives were more broadly conceived as helping developing countries to achieve sustainable development goals beyond just climate change. They have had the added benefit of getting the private sector involved in developing countries to participate in climate-change efforts and thereby gain valuable experience. The issue of CERs also helps finance the so-called Adaptation Fund which is intended to support climate adaptation projects in developing countries that are parties to the Kyoto Protocol, with the share of proceeds allocated to the AF amounting to about 2% of the CERs issued. By late 2015, the AF had obtained circa $331 million for projects in 54 countries. A major problem with the CDM has been the preponderance of its use by certain large emerging-market economies whose rapid growth in recent decades has made them fairly well-to-do middle-income countries. China alone has been responsible for over half of the CERs as host party. India, South Korea, and Brazil absorb together yet another third of the total, leaving the rest of the world with less than a quarter of all issues. This means that other developing countries in Latin America, Africa, the Middle East, and Southern Asia, many of which are highly exposed to the consequences of climate change, have been deprived of adequate access to the financial resources and transfer of know-how mobilized by the CDM.[5]

Kyoto's two transfer mechanisms have both been flawed from their very inception. The Joint Implementation mechanism became captured by transition economies, in particular the two largest ones (Russia, Ukraine), for whom the 1990 baseline for GHG emission levels coincided with the turn away from a centrally planned command economy. The subsequent transition, a process characterized by depressed activity and becoming less dependent on heavy industry, led to lower emissions so that the initial allocation of AAUs for these countries was based on overestimating their projected GHG emission levels. This over-allocation added to the excess supplies of emission permits promoted by the ETS of the European Union, causing the Kyoto Protocol to end up with a 11 gigatonne "hot air" loophole of bogus credits whose laundering undermined its effectiveness.[6] The difficulty of properly monitoring JI-sponsored projects and assuring the additionality of mitigation initiatives under either JI and CDM also had to be addressed as the world sought to move to a new climate-mitigation regime under the Paris Agreement of 2015. This agreement had to provide for an entirely new mechanism, besides avoiding the carryover of Kyoto's bogus "hot air" credits. To begin with, now that the whole world was ready to sign on to a climate-mitigation treaty, the logic of the CDM helping non-signatory developing countries no longer applied. In addition, the Paris Agreement did not specify legally binding emission-reduction targets, but was built instead around voluntary climate-policy plans. This change rendered the whole idea of credits such as ERU or CER difficult to put into practice. Finally, the Paris Agreement also wanted to avoid the administrative weaknesses of the Kyoto mechanisms to promote a more effectively verifiable process capable of promoting actual emission reductions.

Given these constraints, it was not surprising that the Paris Agreement of 2015 came up with an entirely new mechanism for transferring emission reductions between countries, supposed to be in place by 2021. In its Article 6, the agreement introduced so-called Internationally Transferred Mitigation Outcomes (ITMOs) whereby countries can trade emission reductions among each other to meet their respective targets as specified in their NDCs.[7] These ITMOs are no longer necessarily project bound. Instead, they may involve financial and technical resource transfers in support of a country's specific climate-policy objectives. A key aspect here is Article 6.1's explicit requirement to tie ITMOs to "sustainable development and environmental integrity" which raises the legitimate question what these broader contextual requirements may mean as the world commits to mitigate climate change collectively by cooperating with each other—a question we shall address in greater detail in Chapter 8 concluding this book. Article 6.1 also specifies the pursuit of "higher ambition" as a prerequisite for such ITMOs. In other words, this new mechanism is not conceived as an offset against the "buying" country's foregoing its own emission reductions at home, as was the case with JI and CDM. Instead, ITMO-related mitigation measures must come in addition to (and go beyond) the already committed emission cuts. We should note in this context that most countries tried to spell out this "higher ambition." Seventy-eight percent of all initial NDCs committed to in the wake of the Paris Agreement specified both unconditional emission-reduction targets and "conditional" ones which were often twice or thrice as ambitious in terms of emission-reduction targets than the former, assuming sufficient financial and technical resource transfers from richer to poorer countries.[8]

It is clear when looking at its Article 6 provisions that the Paris Agreement projects a much more far-reaching international cooperation effort with regard to climate-change mitigation which it sees develop along three different tracks.

- The so-called cooperative approaches under Articles 6.2 and 6.3 foresee international carbon markets. This might mean launching national or regional cap-and-trade schemes and linking them together, as California and Quebec have done (with Ontario joining them a bit later) or China and South Korea are thinking of doing. Yet, it is not at all clear how easy it will be to create such supranational carbon markets via linkages. The individual market infrastructures of existing schemes are highly specific and difficult to harmonize, possibly necessitating a more globally standardized approach in the future. But such standardization, while helping smaller countries and their markets avoid getting gobbled up by bigger ones, ignores the need for differentiating carbon prices depending on the region. And, related to that, how do you keep account of the fact that various countries have different climate-change ambitions? Such differences may require proportional conversion ratios, with the more ambitious country discounting the value of the less ambitious issuer's emission allowances relative to its own (e.g., one Swiss allowance trading for two ETS EU allowances). This approach has been spurred on by New Zealand in particular and also received the backing of the G-7 industrialized nations.

- A second mechanism specified in Articles 6.4 and 6.7 provides for a so-called Sustainable Development Mechanism (SDM) which at times has also been referred to as the "Emission Mitigation Mechanism" (EMM). Whatever its name, we are talking here about a centralized platform for transferring emission-reduction certificates much like the CDM of the Kyoto Protocol discussed earlier in this section. The only difference is that the Paris Agreement, and hence also its SDM, no longer distinguishes between developed and developing countries as was the case with the CDM. Even after preliminary discussions in Marrakesh (Morocco) during COP22 in November 2016 and in Bonn (Germany) during COP23 a year later, it is still not clear how ITMOs and SDM will relate to each other.

- Article 6.8 also introduces the possibility of "integrated, holistic, and balanced non-market approaches." This alternative may prove especially attractive to leftist leaders skeptical of the market mechanism, such as Bolivia's Evo Morales or Venezuela's Nicolas Maduro. But its more profound long-term importance may lie in the scope of experimentation for multilateral cooperation this provision implicitly encourages by how broadly it defines its scope, including finance, technology transfer, and capacity-building while specifying both public and private participation. The latter includes a range of non-state actors. It is quite possible that this non-market provision might end up fostering a lot of innovative climate-mitigation approaches in areas that do not easily lend themselves to the logic of the marketplace, such as urban planning or state-sponsored acceleration into renewable energy.

The Article 6 mechanisms have yet to be worked out in detail, and so it is by no means clear how much of a role they are going to play in helping countries meet their NDC objectives or pursuing subsequently more ambitious goals over the coming decade.[9] They can either be badly set up and lack coherence, thus ending up as a distraction which at best makes only a marginal contribution to addressing the problem of climate change. Or they are set up well and then start playing an increasingly useful role in fostering a collective effort at progressively "higher ambition." Much depends here on the effectiveness of the monitoring, reporting, and verification procedure to be set up for the administration of the NDCs, the ITMOs, and the SDM. A first step in that direction came already under Kyoto which created centralized bodies to that effect, the Joint Implementation Supervisory Committee and the CDM Executive Board. While the JISC has only recently started its evaluation work in earnest, the CDM EB has already accumulated more experience in assessing the eligibility of qualifying projects, verifying emission reductions (or enhancement of removals by identifiable carbon sinks) with the help of independent, properly accredited evaluators known as Designated National Authority (DNA), and recording qualifying emission-reduction projects in so-called registries while also keeping track of all Kyoto units (i.e., AAUs, ERUs, CERs) in an International Transaction Log. It can thus serve as a model for the new Paris Agreement recording and accounting system, some key differences between the two climate-policy regimes notwithstanding. For one, in the Paris regime we will no longer have to rely on targeting new projects which would not have been undertaken otherwise, the so-called additionality requirement which has turned out difficult to prove. Instead, the three transfer mechanisms under Paris must have an "overall mitigation impact." This provision of having a proven net-mitigation impact, rather than being simply an offset as under Kyoto, means that ITMOs can only be launched once seller countries can prove to have met their NDC targets. This "supplementarity" requirement should be easier to administer than Kyoto's "additionality" requirement, but depends greatly on establishing credible baselines against which progress toward meeting NDC targets can be measured. A crucial question will be the extent to which the net-mitigation impact applies only to the NDCs at large or can be disaggregated to targets set for specific sectors which, if the case, would allow ITMOs to kick in earlier and be issued in larger volumes. The new verification superstructure will also need good accounting rules to comply with the Paris Agreement's emphasis on avoiding "double counting."[10]

CARBON OFFSETS

While much of the Kyoto Protocol will be soon supplanted by the Paris Agreement provisions and mechanisms, there is at least one initiative that may survive the former's life cycle—the private trading of carbon credits used as offsets to enhance one's contribution to the global mitigation effort. Kyoto's aforementioned flexibility mechanisms provide for tradeable emission-reduction certificates which, besides allowing developed countries to meet their Kyoto commitments, can also be bought by private companies for use as offsets toward their own emission-reduction targets they have committed to under mandatory or voluntary cap-and-trade schemes. This option has proven especially popular among European companies needing to meet their emission caps under the EU ETS. They bought approximately 1 million CERs and ERUs between 2008 and 2012 as offsets, which depending on the sector concerned amounted up to 20% of their verified emissions.[11] When firms buy those offsets, they are financing projects reducing GHG emissions somewhere else, such as reforestation, improving energy efficiency of buildings, promoting less polluting transportation, upgrading power plants, or fostering the spread of renewables.

The firms using such offsets have supposedly taken this step only after exhausting their own mitigation efforts. At some point, it will simply have become cheaper to fund GHG emission reductions elsewhere via purchases of CERs or ERUs rather than continuing to pursue increasingly expensive reductions on their own. While many firms may operate responsibly in such cost-balancing fashion, others may simply purchase offsets to make themselves look better to their public ("greenwashing") and/or to keep their own GHG emissions at a high level without feeling guilty. Even if the firms concerned do not consciously use carbon offsets to greenwash or as an excuse to overindulge, the very presence of this option is bound to diminish the urgency many firms may feel to push their own efforts at reducing their carbon footprint to the maximum. Carbon offsets let them off the hook by allowing them to pay someone else to reduce emissions. Still, as we have seen in Chapter 3 with the example of Microsoft, the world's leading firms are now increasingly prioritizing carbon neutrality to combine improved competitiveness with an effort to re-brand themselves as a socially responsible firm. The use of carbon credits allows them in this context to claim having moved closer to carbon neutrality. The biggest corporate users of such carbon offsets are either highly polluting, such as airline companies, or engage in activities of low-carbon intensity for whom offsets are a cheap alternative to help reduce emissions, such as banks and other financial institutions. As emissions-trading schemes are bound to spread in the wake of the Paris Agreement, voluntary carbon offsets are likely to grow as a by-product to which they may actually be directly tied as already the case with the

EU ETS or California's cap-and-trade scheme. It should be noted that heavy polluters in the US or China have at times already bought carbon offsets as a hedge against future mandatory cap-and-trade regulations which would sharply increase the price of such offsets, and such a speculative motive has also spurred carbon-offset purchases by hedge funds and trading desks of investment banks.

Carbon credits used as offsets fetch highly variable prices, depending more on location and nature of project than on the intermediation mechanism transferring those credits. In 2016, for example, those prices ranged from \$0.5 per tCO_2e to \$50 per tCO_2e. While the price maintained a fairly stable \$10–15 to 10–12€ per ton on average between 2008 and 2012, it declined sharply in the following years all the way to an average of \$3 or 2.5€ per ton in 2016. Such low prices may indeed encourage demand, notably also of the speculative kind mentioned above. While the price decline matches that of the world's carbon markets in recent years, they are indeed at a historical low right now and bound to increase in the not-so-distant future as the world grapples with the challenge of imposing a sufficiently high carbon price across the globe in order to motivate innovation and accelerate movement toward a low-carbon economy.[12]

Projects eligible for generating carbon offsets cover a wide range of mitigation activities. Those can be grouped together under the broad categories of renewable energy (solar, wind, hydro, biofuel), methane collection and combustion (involving farm animals, landfills, industrial waste), energy efficiency (e.g., cogeneration plants, buildings), LULUCF (land use, land-use change, and forestry) projects providing natural carbon sinks (forests, soil management) and also including REDD+ (Reduced Emissions from Deforestation and Forest Degradation) initiatives, or other industrial pollutants such as hydrofluorocarbons (HFCs) or perfluorocarbons (PFCs). Ideally, all these projects would meet the "additionality" requirement of not having been undertaken otherwise, but this is difficult to prove conclusively. Another important consideration concerns their transparency, in terms of both access to reliably accurate information about the projects and subsequent details of the transaction being recorded. Finally, there must be a trustworthy, preferably independent third-party verification of actual emission reductions achieved by the project in question. We have several different certification regimes in place already, such as CDM Gold Standard or the Voluntary Carbon Standard, which audit the projects and also assess those in terms of their social benefits to their local communities. But such proliferation risks confusion and makes it difficult to assure uniform quality standards. Surely, some projects thus funded are of poor quality, possibly hurting the reputation of the entire carbon-credit enterprise.

An important extension of the voluntary carbon markets concerns individuals, small businesses, and other non-corporate users. Major emitters of greenhouse gases, such as utilities or airlines, may offer their customers carbon offsets in compensation for their high-carbon consumption, such as when taking a flight. A flight from New York to Paris may then be neutralized by $20 in carbon offsets, just as an average US family may pay $30–40 per year to offset a year's worth of gas and electricity use. Climate-activist NGOs, such as Carbon Retirement, may purchase and then retire carbon credits from an emissions-trading scheme like the EU ETS. There are a growing number of online carbon-offset platforms, such as Element Markets, Terra Pass, *Native*Energy or Carbonfund.org, where individuals and small business can calculate their own carbon footprint by means of easily accessible carbon-tracking calculators and then buy whatever amounts of carbon credits they desire at the quoted unit price. These micro-transactions are often tied to similarly small-scale mitigation efforts, like switching to LED lightbulbs, acquiring cookstoves, installing water filters, enhancing methane capture at a landfill, or planting trees.[13]

Even though such online carbon-offset schemes may be ethically questionable inasmuch as they make business owners or consumers feel less guilty while letting them continue polluting, they may still become a fixture going forward. This should become even more so the case to the extent that we engage over the coming decade in a more serious pursuit of keeping within a finite carbon budget. We may at that point see fit to break down the total global emissions amount into progressively smaller parcels so that individual actors, such as businesses or consumers, become increasingly aware of their own actual carbon footprint relative to where they should be in compliance with the overall carbon budget. Such micro-management of the carbon budget presumes, of course, increasingly precise direct-measurement instruments such as portable spectrometers which can track GHG emissions at the source and also provide accurate estimates for all kinds of indirect Scope 3 emissions (e.g., employee commuting, business travel, third-party distribution and logistics, production of purchased goods, emissions from use of sold products). In such a pollution-focused world, where carbon and other greenhouse gas emissions are measured regularly and the pursuit of specific targets in compliance with an overall carbon budget is widespread, we may wish to reward individual efforts to cut emissions below set target levels and/or penalize those that willfully ignore their role to play in the overall mitigation effort.

DIGITAL CARBON TOKENS

During 2017, while I was writing this book, one of history's great speculative bubbles took the unit price of the cryptocurrency Bitcoin from its yearly low of $786 on January 11 to an end-of-year closing price of $13,122 (after reaching a yearly high of $19,843 on December 11, just three weeks earlier). This explosive rise, accompanied by the kind of market volatility loved so dearly by speculators, came about despite recurrent central bank attempts to ban this new type of digital money, bankers' denunciations of the phenomenon as a "fraud," and internal splits within Bitcoin's governance structure. Even though such a bubble is bound to burst, a bust perhaps already in the works amidst the hair-raising dive during the last three weeks of 2017, it will leave something more durable in its wake as asset bubbles are prone to do even after they have collapsed—in this case the mainstreaming of privately issued cryptocurrencies circulating within their own payments system beyond government control to fuel an online e-commerce platform.[14]

One reason why Bitcoin became such a spectacular focus of worldwide speculative fever has been its grounding in a new digital network technology which promises to transform how we organize the Internet as a vehicle of economic and social activity—the so-called *blockchain* which records Bitcoin transactions accessible to all for verification. Blockchain's distributed ledger technology (DLT) revolutionizes how we record and distribute information on the Internet in decentralized networks which do not need trusted third parties for verification and can be organized without centralized control. Every node in such a network automatically processes the information provided on its own, for everyone else to confirm via a shared ledger. The work of securing digital relationships—authentication of identities, authorization of access, recordkeeping of transactions, and protecting the infrastructure from cyberattacks—is thus no longer done by a centrally located gatekeeper, but instead automatically "supplied by the elegant, simple, yet robust network architecture of blockchain technology itself."[15]

By creating transparent and incorruptible databases without the need for intermediation, these blockchains lend themselves to the organization of decentralized peer-to-peer (P2P) networks. The latter organize their online activities through so-called smart contracts built into DLTs which are the computerized transaction protocols guiding participants' actions. Smart contracts are visible to all users of a specific blockchain, while also allowing for a remarkable degree of customization in response to an individual user's specific needs. This innovation is bound to be a transformative technology, as evidenced by the headlong rush of commercial banks into figuring out how best to use DLTs across their entire spectrum of activities and product offerings, including extensive automation of their back-office operations. In other words, blockchain-based P2P networks containing elaborate smart-contract protocols will disrupt existing chains of exchange and so transform a very large number of activities organized online. Let us illustrate with a few ideas already in the pipeline or on the drawing board of DLT developers. For instance, our standard centralized top-down electricity generation and distribution system under the control of public utilities may soon give way to decentralized bottom-up P2P networks where someone with a solar panel may sell excess power to his neighbor in exchange for payment in Bitcoins or another crypto-currency. DLTs may be usefully applied to global supply chains (tracing interactions, organizing intra-chain payments), tracking of donor funds (not least for the carbon offsets discussed in the previous section), introduction of personal identification systems, use of mobile cash transfers, bestowing of land titles (e.g., in urban slums, earned as a sort of reward) in developing countries to regularize the existence of the otherwise marginalized poor, and an entirely open-ended plethora of other application possibilities. In the end, strategic intermediaries like brokers, bankers, lawyers, or accountants may become obsolete and/or transformed on the basis of deepening their informational-service offerings.

So far the most explosive use of DLTs has been in the area of cryptocurrencies taking the form of *initial coin offerings*. Such ICOs are a new form of crowdfunding where e-commerce entrepreneurs or software programmers launching a new project seek contributions from investors whom they reward in return with digital tokens that can be used within the network associated with that start-up. Most of those ICOs have used a virtual currency called Ether as the funding currency which circulates within the global Ethereum network. In contrast to its cousin Bitcoin, Ethereum's software is designed to facilitate the development of apps for complex financial transactions no longer needing third-party intermediation, the creation of online marketplaces, and the introduction of programmable transaction protocols known as smart contracts of which the tokens created by the ICOs are just one expression. Each digital coin is slightly different from the next, but they can all be stored, traded, and transferred within the Ethereum platform. Investors have gobbled up these coins, to the point of pouring $3.3 billion into ICOs during 2017 alone, in a speculative craze driven by hope that the future trading value of this or that coin will shoot up shortly for massive capital gains—a bubble clearly tied to the Bitcoin mania of late 2017. While the ICO bubble will surely burst one day as many of the projects thus funded do not come to fruition and investors rack up losses, it too will leave something useful behind—a new organizational form of online entities best described as "crypto cooperatives" grouping digital entrepreneurs, programmers, and investors into a coin-specific network which they all seek to expand working together.[16]

It is not all that far-fetched to imagine how DLT may open up a whole new world in our global struggle against climate change. In November 2017, Canada's Trudeau Administration announced a $1.6 million seed-fund investment into a blockchain project, facilitated by the International Emissions Trading Association (IETA) and Ottawa-based Climate CHECK, to help Pacific Alliance nations (Chile, Colombia, Mexico, and Peru) combat climate change by aiding them with emissions data collection, cost reductions, and easier access to climate finance for funding of emission-reduction projects. This kind of project points to the use of DLTs as a foundational technology for a far deeper societal mitigation-effort infrastructure as already embodied by the knowledge network Climate Ledger Initiative (CLI, climateledger.org) cosponsored by the Massachusetts Institute of Technology, the UNFCCC, and the World Bank. Obviously, we can imagine using blockchains for much better monitoring, reporting, and verification (MRV) of emissions, a *sine qua non* condition for our global climate-mitigation effort. And we surely will want to use DLTs to address the whole question of carbon pricing, in terms of both how to set such "prices" accurately and also how to differentiate them according to specification of location and other programmed conditions. Decentralized distribution of renewable energy in new networks may be another fruitful area of DLT application in climate policy, as may be promotion of environmentally friendly transportation or sustainable land use.

One concrete example of using blockchains and digital tokens in the fight against climate change is Veridium Labs, an environmental FinTech start-up located in Hong Kong which was set up in July 2017 as a collaborative effort by environmental incubator EnVision Corp and blockchain software developer ConsenSys. Veridium's overall goal is to create an online marketplace for "natural-capital" assets through the issue of tokens and a blockchain-based pricing mechanism so that this hitherto illiquid and largely unpriced asset class can be mobilized more effectively for sustainable supply-chain solutions as well as other sustainable development projects. This online natural-capital market known as Veridium Ecosystem would engage leading Fortune 500 companies, such as Microsoft, Intuit, Zurich, Allianz, Marks & Spencer, SAP, and Gazprom, as well as socially responsible financial institutions (notably pension funds and investment funds) to buy environmental offset credits fueling the demand side of its natural-capital marketplace. Its supply side would be composed of various types of environmental asset tokens (e.g., REDD+ credits, water rights, biodiversity credits), meaning the issue and distribution of digital tokens to those acquiring such underlying conservation credits which can then be exchanged for either cash or other cryptocurrencies. As a first project in this direction, Veridium introduced in October 2017 the issue of 15 million TGR (Triple Gold REDD+) tokens, backed 100% by so-called REDD+ forest carbon-offset credits linked to the Rimba Raya Biodiversity Reserve in Borneo (Indonesia), which buyers can purchase with Bitcoin, Ether, or fiat currencies. Veridium issues a second token, known as VRD and serving as a membership reward token, which provides access to the Veridium Ecosystem. One hundred VRDs are worth one TGR. While still in its infancy, we can already see here in this example the contours of an online marketplace for natural-capital assets, whose monetization via tokens as in the case of Veridium's TGRs supports the needed pricing mechanism and liquidity pipeline to unlock the huge potential of this hitherto marginalized asset class for the systematic "greening" of our economy.[17]

OTHER CARBON MONEY IDEAS

Veridium provides a glimpse into a future when we will have come to rely on new kinds of online marketplaces, multiparty cooperation arrangements, and cryptocurrencies to mobilize climate-mitigation efforts in the direction of sustainable development. It has taken only five years to move us from swaps of officially regulated emission-reduction projects via privately organized carbon credits ("offsets") to creating markets for "natural-capital" assets fueled by digital tokens. We see here the beginning of a very dynamic period during which climate change becomes an ever-growing vector reshaping various facets of our economy. And this is a good thing! We need to move beyond the ultimately unproductive, even dangerous, but also likely scenario of continuing to do very little and then panicking ten, fifteen years from now amidst growing signs of impending climate-change disaster into a regulation-driven, tax-heavy "top-down" approach obliging key sectors to undertake deep emission cuts rapidly. Much better is a more gradual approach, assuming that it can be scaled up progressively, of intensifying climate-mitigation efforts where we push the capitalist market logic into a different direction of ever-deeper ecological framing. This is best advanced by monetizing such mitigation efforts. We have in recent years seen a variety of proposals to that effect.

Given how difficult and ineffective it has been to set up cap-and-trade schemes, it may behoove us to take a look at possible alternatives. One such idea is Richard Douthwaite's "cap and share" proposal aimed at reducing transportation-related emissions. Douthwaite, an environmental economist based in Ireland and co-founder of the Foundation for the Economics of Sustainability (FEASTA), suggested in 2006 a new policy for rolling back the use of fossil fuels in the interest of climate stabilization.[18] His proposal starts from the idea that GHG emissions need to be capped and that any such cap would have to be enforced strictly at progressively lower levels to stabilize the climate. A second idea played out here is that the earth's atmosphere is a public good, a common resource, to which everybody should be entitled to have equal access. From these ideas follows that, once an annual GHG emissions cap has been agreed to, all citizens would be allocated the same number of tradeable emissions permits. Everyone could then sell off their permits through the banking system to coal, oil, and gas companies to cover whatever amount of carbon dioxide these energy producers would be authorized to emit from burning fossil fuels. Government-appointed inspectors match permits with CO_2 emissions by those fossil-fuel companies and then cancel them, thereby enforcing the cap which would be imposed at progressively lower levels from year to year. In this way, citizens would get some compensation from rising energy prices as we phase out fossil fuels in the interest of climate stabilization.

A second set of proposals concerns schemes which offer rewards in the form of coupons or currency for reductions in GHG emissions, so-called emission-reduction currency systems. These ERCS are quite different from the UNFCCC's official emission-reduction transfers helping countries to meet their caps or the voluntary carbon offsets allowing purchasers to compensate for their own pollution inasmuch as they serve as incentive to change behavior and do not confer any right to pollute. There are several such schemes in a trial stage, each one far from being anything but experimental. While some of these, such as RecycleBank or EarthAid, offer redeemable reward points, I am more interested in schemes pushing carbon-related currencies. One such idea is Next Nature Lab's ECO coin which can be earned and spent on sustainability-promoting activities. Another is local community-based currencies tied to emission reductions, such as the Green Power Certificates circulating in Tokyo's suburb Edogawa, the Maia Maia ERCS in Western Australia issuing a local currency called Boya to reward communal efforts such as planting trees and reducing power bills, or Ireland's Liquidity Network sponsored by the same foundation (FEASTA) Richard Douthwaite belonged to.

We have already witnessed early examples of how initial coin offerings and blockchain technology might be used for emission reductions, as evidenced by Veridium Ecosystem. There is also Swiss nonprofit company Poseidon's blockchain platform Stellar.org which would enter fully verified carbon credits and link those to tokens used to fund emission-reducing micro-projects, thereby creating a new kind of market mechanism bringing carbon pricing to all kinds of everyday transactions as the value of these tokens would fluctuate in proportion to Poseidon's carbon-credit sales revenue. Another example of a new blockchain-based digital carbon currency is Swiss-based Blue Foundation's Bluenote, introduced in November 2017. The Bluenote token gets issued to finance emission-reducing investments in retrofitting commercial buildings and other energy-efficiency projects helping cities. What is most interesting about this initiative is its introduction of an open and transparent blockchain platform where any number of urban environmental projects can upload their raw performance data and obtain a community-verified calculation of their emission reductions and other social impacts.[19]

MICHEL AGLIETTA'S CARBON CERTIFICATES

As with many aspects of climate policy, it behooves us to take a closer look at carbon-money ideas in France where academics, research institutes, and think tanks have invested a relatively large amount of time and effort into debating post-Paris proposals for accelerating the global effort against climate change. Of special interest here is the work of one of France's greatest economists alive, Michel Aglietta, the founder of the French "Régulation" School (see Chapter 2) and also a specialist on the question of money. Still very active in terms of publications and participating in key policy debates even though formally retired, Aglietta began to get interested in the question of climate change in the late 2000s and soon found himself pondering the question how to incentivize global mitigation efforts by linking projects aimed at reducing emissions to a new type of money with which to monetize such investments. As important as his specific ideas in that regard is the broader context within which he places his proposal for carbon money. After all, Aglietta is a man thinking about the capitalist system as a whole and in terms of its long-term evolution, thus embedding the question of climate change in what it means to have an economy prioritizing sustainable development as opposed to our current system's domination by private gains and short-term orientation.[20] We will return to that shift in the next chapter, concluding the book, when we specify the outlines of this new kind of "eco"-capitalism.

Within his broader argument in favor of a more sustainability-oriented type of capitalism as a new accumulation regime yet to be constructed, Aglietta (2015a, b) stresses the crucial role finance will have to play in the needed transition to a low-carbon economy. But the current configuration of what goes nowadays for finance is entirely inadequate to deal with this challenge. None of its main tenets—the efficient-market hypothesis, shareholder value maximization, and "mark-to-market" accounting—is equipped to address the unique characteristics of climate-mitigation investments. This is as true for the need to retire a huge stock of high-carbon assets before the end of their life cycle as it is for the imperative of funding uniquely risky low-carbon replacements on a massive scale over a very long time horizon. Even though such a huge shift out of "brown" assets and into "green" assets would boost growth amidst post-crisis stagnation while at the same time also preventing certain disaster from climate change, the financial system we have in place today cannot make this happen. Just the discount rate typically applied to productive investments, especially when adjusted for risk, is far too high to carry out those kinds of long-term investment projects implied here. The dictates of shareholder value maximization keep the time horizon of investments too short. Another problem is typically large up-front costs while returns only arrive quite a bit later. Supposedly, "efficient" stock and bond markets are actually inefficient when it comes to offsetting private costs against social benefits, a crucial issue when considering climate-mitigation efforts. And in this context it is decidedly unhelpful not to have a reliably high or omnipresent carbon price with which to work out such offsets. Ongoing efforts centered on intrinsically imbalanced emissions-trading "markets" (cap-and-trade schemes) or unevenly applied carbon taxes will certainly not go far enough to resolve the challenge of carbon pricing effectively. Something else is needed, and Aglietta sets out to present the broad outlines of such an alternative with which to drive our global effort against GHG emissions forward.[21]

Crucial to his argument is that we need to incentivize socially beneficial investments which capture the externalities so that their social returns can be incorporated just as much as the private returns accruing to investors as their profit. One such category of socially beneficial investments is those aimed at lowering GHG emissions. Such re-orientation toward inclusion of externalities requires obviously a corporate governance quite different from the currently dominant paradigm of shareholder value maximization, and recent efforts in the direction of environmental, social, and governance (ESG) criteria (discussed in Chapter 5) are an important first step toward that goal. In addition, there would have to be a new system of social accounting to complement current corporate accounting rules. With regard to climate-mitigation projects, we would also need to have a viable price for carbon (and other greenhouse gases) to make their social returns sufficiently tangible for adequate calculation. Aglietta proposes here an original approach that does not depend on an intrinsically inadequate carbon-market price set in cap-and-trade schemes or a politically unfeasible carbon tax. Instead, he suggests setting a shadow price as a notional value, made to rise over time by political agreement and applied to climate-mitigating investment projects going forward for estimation of their respective social returns. The idea here is that reducing GHG emissions benefits society and should therefore be considered as something of social value worth rewarding. We would then be able to estimate the social returns of emission-reducing investments in terms of how much GHG abatement they provide, subject to verification by an independent evaluation authority that would have to be set up for that purpose on a global scale. Aglietta refers to such GHG-abatement projects as "carbon assets" whose value is defined by the amount in tons of avoided CO_2-equivalent emissions multiplied by the prevailing shadow price of carbon per ton, checked and certified by the evaluation authority set up to assess and verify carbon assets.

Such climate-mitigating GHG-abatement projects are inherently so risky that they are difficult, if not altogether unlikely, to be launched on their own. They typically involve very long time horizons, unproven new technologies, large up-front costs, and a great deal of uncertainty as to their outcome. In addition, they will often replace existing high-carbon infrastructure at some loss, either because the latter still would have some years to live and/or because the latter may provide the same output (e.g., electricity) more cheaply when calculated in terms of the narrow (private-cost) market price. And these switch costs would have to be included in the GHG-abatement investment which make the transition to a low-carbon economy so much more expensive and difficult to finance. For all of these reasons, it will be essential to offer government guarantees without which those carbon assets would simply not be created sufficiently or at all. As part of their NDCs in compliance with the Paris Agreement, national governments would commit over consecutive five-year periods to guaranteeing a specified amount of carbon assets, which Aglietta would like to be administered by an international supervision body harmonizing this process for all participating states by specifying the carbon-asset allocation rules among the states and fixing time horizons, sectors, and technologies for validation of project types.

Once each country has specified, in cooperation with the international supervision body, its list of GHG-abatement projects validated as carbon assets, then its government can actualize its guarantees of such investment projects. Aglietta proposes here two alternative intermediation channels through which to affect such guarantees. One involves creation of a publicly funded Green Fund issuing top-rated bonds sold to institutional investors (e.g., pension funds, mutual funds, sovereign-wealth funds) all over the world and thereby inclined to tap the huge pool of savings piling up globally. We can think of those GF securities, in Aglietta's terminology, as "climate bonds on carbon assets" inasmuch as they would be backed by the projects they would help to finance when the Green Fund lends funds thus raised to specialized financial institutions financing the various approved GHG-abatement projects creating carbon assets. GF could also lend money to that country's development bank(s) or invest directly in project bonds. You would thus end up with a publicly backed and highly diversified carbon-asset portfolio and so turn GHG-abatement projects into an attractive asset class for risk-averse institutional investors.

Aglietta's other proposal for state guarantees of carbon assets is more radical, involving the central bank in a direct monetization scheme. While several economists (such as Ferron and Morel 2014 or Campiglio et al. 2017) have already pushed for a "green" quantitative easing involving dedicated central bank purchases of green bonds, Aglietta's proposal goes much further than that. He would have the central bank directly monetize GHG emission reductions by issuing so-called *Carbon Certificates* (on its liability side) which serve as reserves or collateral for the financial institutions providing the funds for the validated projects. The latter's risk from these investments is thus socialized. Unlike quantitative easing involving central bank purchases of existing securities in secondary markets, we are talking here about central banks monetizing credit which finances new real investments in the form of carbon assets. The central bank can book the value of these carbon assets thus created on its asset side, providing so an appropriate counterweight to the carbon certificates on its liability side. In other words, green project developers would receive government-guaranteed carbon certificates up to the amount of emission reductions, with the central bank refinancing loans to those developers up to the amount of the carbon certificates.

Such monetization is not inflationary to the extent that the projects concerned have their predetermined value guaranteed by the state and the (carbon shadow-) price involved here follows a path fixed in advance. The only question of concern is failure to create the kind of carbon asset validated in advance because of less-than-expected emission reductions in which case there is asset devaluation triggering equivalent destruction of carbon certificates and a sharing of losses between the firm launching that project and the lender funding it. It may also happen that the government ends up not following the upward-sloping path for the social value of carbon originally agreed to (perhaps because it proves politically too costly to do so amidst resistance in the electorate against thereby increased costs). If, for example, the carbon price is lower in ten years than the predetermined level along the path set in advance, then the monetized GHG-abatement projects may well end up unprofitable and so risk defaults by the green project developers on their loans. Such failures to repay trigger the government guarantees and so turn the (devalued) carbon assets into a central bank claim against the government, causing either higher public debt or, in case of monetization, possibly also inflation. Aglietta's plan thus obliges governments to choose between doing the right thing in the face of political headwinds, namely to raise the carbon price adequately over time, or face negative macroeconomic consequences.[22]

Jean-Charles Hourcade's (2015) proposal, based on essentially the same intermediation and monetization architecture as Aglietta's, refines the carbon-asset monetization process further. Hourcade starts his argument by pointing to the very real "danger line" inherent to many carbon-mitigation projects thanks to their typically significant up-front costs which impose large net cash outflows during the early stages of the project's life cycle that your typically short-term-oriented investors will be hesitant to accept. Having a sufficiently high carbon price may help counter this problem to some degree by allowing bigger returns to kick in faster, but not enough to be a sufficient remedy on its own. We thus depend on additional financial incentives to unleash the needed wave of such low-carbon projects. As with Aglietta's plan, Hourcade foresees a notional and gradually rising carbon price used to calculate the social value of low-carbon investment projects in terms of the $CO_2(e)$ emissions they cut, that which he terms VCRA—the value of carbon remediation assets (CRAs). This may be somewhere between $55 and $140 per ton of CO_2 by 2030 for the 450 ppm scenario envisaged by the Paris Agreement's $\leq 2\ °C$ goal to be realistically achievable, not too far off the aforementioned $50–100 per ton target range of the Stern-Stiglitz Commission. Once you have emission-reduction investments assessed and their anticipated VCRA creditably estimated by an international MRV agency, then we can connect these eligible projects to a carbon-based monetary facility set up to help launch them. At the heart of this facility is the allocation (up to a predetermined amount) of government-backed carbon certificates (CCs) to the low-carbon projects and valued at the VCRA. And these CCs can be used as partial repayment of the loan launching that project, once the underlying emission reductions tied to the project in question have been verified by the MRV agency. Banks can convert the CCs received as partial repayment into CRAs carried on the books of the central bank. This asset swap basically creates a government-guaranteed asset, the CRAs, which allows the central bank to offer new credit lines refundable with certified reductions of CO_2 emissions—a true form of carbon money inclined to make green climate-mitigation projects much easier to finance and less risky to carry.

FULLY MONETIZING CARBON

The Aglietta-Hourcade proposal for a carbon-based monetary facility using government guarantees and central bank monetization of debts funding carbon assets is a major advance toward proper grounding of climate finance so that it can take off and play its important part in the transition to a low-carbon economy. I am fully in favor of its adoption by a "coalition of the willing" at the earliest possible time, hopefully by 2021 when the Paris Accord will replace the Kyoto Protocol to start a new climate-policy cycle. But I do have some reservations about their proposal as well, and those may ultimately warrant proposing an even more far-reaching carbon-money alternative the broad outlines of which I will present here in conclusion of this chapter.

One issue I have with the Aglietta-Hourcade proposal is with their insistence of having a steadily rising shadow price for carbon. True, such price increases make carbon assets more valuable over time, and this gain compensates for the disproportionate discounting of returns accruing far into the future—a possibly important consideration given the long-term nature of most emission-mitigation investments making up our carbon assets. But this advantage becomes much less important, if not altogether negligible, if we applied zero (or very low) discount rates to this kind of socially beneficial investment as has been suggested by Nicholas Stern and other economists specializing in matters of climate policy. Instead, it may well be that the prospect of rising carbon prices further into the future may encourage potential green investors to launch their projects later when the prospects of gains are larger in the wake of projected price increases. Such a disincentive against action now would be calamitously counter-productive. Another problem concerns the switching from "brown" high-carbon assets to "green" low-carbon assets. Aglietta briefly mentions this issue at the beginning as one inclined to increase the costs of the transition to a low-carbon economy greatly and, with it, the funding needs for a greening of the economy's productive apparatus. But then the issue disappears from their discussion which in the end only concerns new investments in emission-reducing projects without taking account of the structures these are meant to replace. Would not the retirement of polluting high-carbon assets also warrant issue of carbon certificates inasmuch as such a step reduces emissions as well? That question, one of utterly strategic importance, warrants an answer. Finally, I am not sure you would want to saddle central banks with carbon assets whose valuation depends directly on real-economy trends and fluctuations, as proposed by Michel Aglietta and Jean-Charles Hourcade. Central banks are unique institutions of macroeconomic policy-making whose intervention space at the center of whatever monetary regime currently prevails requires a certain autonomy and distance from the production dynamics of a capitalist economy in forward motion. Their engagement with the distribution of carbon certificates and/or valuation of carbon assets risks greatly complicating, if not altogether undermining, their ability to manage their other objectives such as financial stability or balancing of monetary policy.

We may for these three reasons wish to rethink the institutional architecture and modus operandi of carbon money. I am starting from the (very likely) proposition that some time before 2030 the world community of nations will realize that climate change represents a profound systemic threat to the survivability of our species, as currently organized, and that our efforts to contain this challenge have been woefully inadequate up to now. We will, in other words, become sufficiently concerned about impending disaster to push ourselves collectively into a higher gear of preventive action, knowing at that point that we have one more chance to forestall the worst before it is truly too late. It is at that point of readiness for collective action, crystallized around a widely shared sense of urgency as to the timeline involved before having to reach (net zero) carbon neutrality, that we will look at a dramatic policy choice between tightly enforced regulatory restrictions to pursue strict antipollution limits and a deeper structure of powerful incentives guiding market-regulated behavior toward a more sustainable type of capitalism. If it is the latter we wish to emphasize, there is no stronger incentive structure than putting money at the center of a rapid transition to the kind of low-carbon economy we need. But at this point the new regime of carbon money will have to go quite a bit further than even the comparatively ambitious carbon-certificates proposal of Michel Aglietta and Jean-Charles Hourcade. It would have to cover emissions in order to make emitters pay for their pollution, provide an answer to the needed liquidation of stranded high-carbon assets, help finance both mitigation and adaption investments, encourage the spread of a business model for environmental services turning the ongoing reduction of social (GHG emissions) costs into a profitable revenue stream, and perhaps even reward individual actors for keeping their daily activities within their carbon budgets.

The first decisive element of our carbon-centered monetary regime is the need to capture the social cost of carbon fully by having the emitters pay for their pollution. Right now we are putting pretty much all of our chips on cap-and-trade schemes which, if we include the recently launched market for emissions permits in China, together cover about 20% of all carbon dioxide emissions. The Paris Agreement foresees a substantial extension of that scale and even a linking up of national carbon markets to internationalize this effort. Over a decade of experience with this type of climate policy has shown us clearly that cap-and-trade schemes are complicated to set up and even more difficult to make work effectively. Even if they "work" more or less well, they fail to reduce carbon emissions significantly. Their entire thrust is toward letting polluters continue emitting greenhouse gases, provided they pay a (fairly small) price for this allowance. One can already conclude that cap-and-trade schemes are probably not worth their effort, and the task of abandoning them in favor of a stronger carbon-pricing mechanism will most likely be at the center of the anticipated acceleration of carbon-mitigation efforts I made reference to earlier. We will then probably go more aggressively for carbon taxes, which will have to be set high enough and made to rise progressively (hopefully according to a predetermined formula for better transparency and predictability) in line with the projected social cost of carbon linked to climate-related damage estimates. This is above all a political decision. The example of Sweden's 137€ per ton of CO_2 tax shows convincingly that economies can indeed maintain high levels of growth and employment while significantly cutting GHG emissions across the board, and such effective decoupling makes us optimistic about the potential effectiveness of carbon taxes.[23] Much depends on how this emissions tax is set and collected in different parts of the world, and we can envisage integrating the carbon tax into existing carbon markets where identified emitters have to acquire emissions permits at the tax-based price per ton of CO_2e to the extent that they want to continue to emit those greenhouse gases. Of course, also important will be what the carbon tax revenues will be used for. A portion of those revenues will have to go to protecting low-income households from higher energy and electricity prices. Yet, another portion may be directed toward promotion of low-carbon investments by funding promising mitigation and (later also) adaptation projects, directing funds especially toward those projects replacing high-carbon assets and processes. Since carbon taxes will provide a large source of globally collected revenues, they can also usefully finance capital, technology, and resource transfers from rich to poor countries at the center of the global struggle against global warming. Be that it as it may, carbon taxes can replace or help reduce other, less productive (from the point of view of either equity or efficiency) taxes, such as job-restricting payroll taxes or regressive consumption taxes, and so serve as the fulcrum for broader tax reform.

A second point worth considering is to make a distinction between the social cost of carbon (SCC) in terms of the damage caused by its emission to the environment and the avoidance of this damage beyond a tolerable limit as a public good. After all, the Paris Agreement (in Paragraph 109) already "[R]ecognizes the social, economic and environmental value of voluntary mitigation actions and their co-benefits to (....) health and sustainable development." For Brazilian environmental activist Alfredo Sirkis (2016), this clause points to the need for what he terms "positive pricing of carbon reduction" and the need to consider such emission reductions as "units of convertible financial value." This is exactly the point picked up by Michel Aglietta and Jean-Charles Hourcade in their proposal of creating such units in the form of money-like "carbon certificates. " But there is another way to go about it. For the UNFCCC, such "positive pricing" applies to "mitigation action" in which case it might make more sense to apply the carbon-reducing monetization process to the costs of mitigation investments and so help finance those projects directly.

If we adopted this approach, we would look at the marginal abatement cost (MAC) curve which shows the additional costs of emission reductions we incur from mitigation projects up to the point where we achieve the Paris goal of stabilization at an atmospheric GHG concentration level of, say, $450\,\mathrm{ppm/tCO_2e}$. While today different companies and governments all set their own specific, essentially arbitrary and widely divergent shadow prices of carbon to guide their decision-making, it is indispensable to harmonize this process toward working with a more uniform MAC. Aglietta's proposal for establishing an international supervisory body for mitigation project evaluation and approval, which for argument's sake we shall call IMPI (for International Mitigation

Projects Initiator), could be applied here very usefully. If set on a country-by-country level, the MAC ultimately depends not just on that country's climate-policy goals, but also on assumptions about the mitigation efforts of other countries and about technological progress projected over a decade or more, say to 2030. We could use the methodology developed by McKinsey to construct country-specific greenhouse gas abatement cost curves which IMPI would then use to figure out at what level to set the MAC for our targeted stabilization level. At first, it probably would do so on a country-by-country level, but then aim over time at a globally consistent level, assuming most or all countries participate in this effort under the auspices of the UNFCCC. Whatever the level chosen, we need to keep in mind that the MAC curve slopes downward: the lower our GHG concentration stabilization target (on the x-axis), the greater must be our emission-reduction effort involving increasingly costly mitigation projects (on the y-axis).[24] This downward slope, in contrast to the upward slope of the aforementioned SCC curve, also has important implications for improved technology lowering mitigation costs over time and for encouraging climate-stabilization action now rather than later.

The MAC curve will play a crucial role in the project evaluation work of the international supervisory body referred to here as IMPI. Think of IMPI as the UNFCCC's global action arm, an international organization comprising scientists, business executives, government officials responsible for sustainable development, and non-governmental organizations representing "civil society" and other stakeholders, charged with central coordination of the worldwide mitigation effort under the Paris Agreement or its eventual successor(s). IMPI's work would be to identify the range of available mitigation projects per sector and country, along the lines of McKinsey's greenhouse gas abatement cost curves which rank mitigation-investment projects of different sizes across different sectors from least costly/most beneficial to most costly up to the point of sufficient abatement to stabilize atmospheric GHG concentrations at the desired level. These cutoff points determine the MAC level chosen by IMPI, beyond which abatement costs will not be fully covered by the government support extended to these projects. The basic idea here is to encourage as much emission-reducing investment as needed in as short a period of time as possible and consider such activity a public good warranting government engagement. While the public sector can undertake a portion of that activity directly itself, be it through infrastructure investments or setting up public development banks funding strategically important climate-mitigation projects and technologies, the kind of support envisaged here is provided more indirectly through guarantees and liquidity support facilities. IMPI-approved projects will obtain bank funding more easily and at lower interest rates, if so supported by government. Once IMPI has determined the MAC level and the selection of qualifying projects, it will determine their respective carbon-certificate allocations and distribute those to its member states where those CCs will be injected into monetary circulation via the domestic banking system. The number of CCs per project should take into account a matrix of determinants comprising anticipated cost-benefit balance, the timing of either (in particular the relative weight of up-front costs), and its technology diffusion potential. Valued at the set MAC level, these CCs represent a certain amount of (carbon) money with which to express the estimated public-benefit contribution of a particular qualifying mitigation-investment project. We should also note that the CCs come with a guarantee, like a loan guarantee, backed by a public-insurance fund set up by IMPI's member states so that the CCs can circulate as if they were "real" money.

The IMPI-issued CCs enter a country's circulation through its central bank on whose balance sheet they appear as liabilities equivalent to bank reserves whom they top on the asset side of banks or other specialized climate-mitigation lenders (akin to mortgage banks for real estate or finance companies lending to consumers). These lenders then fund the IMPI-approved emission-reducing mitigation projects, which may also include carbon sinks such as reforestation, soil management, or turning landfills into a source of energy. The creditors make normal loans and then on top of those add the low- or no-interest CCs for use by their borrowers. We can think of the CCs essentially as akin to the IMF's Special Drawing Rights inasmuch as they are a limited-circulation money form earmarked only for one specific purpose, but otherwise convertible into a specified range of currencies.[25] The green project investors receiving these CCs as part of their bank financing can convert and spend them as money in pursuit of their investment project even though these certificates remain on the books of central bank and commercial bank lenders as a claim. To the extent that the thereby funded mitigation investment does indeed eventually yield social benefits in the form of GHG emission reductions, those CC claims get canceled having achieved their objective. In other words, the green project investors do not have to repay that portion of the loan comprising the CCs to the extent that they have achieved emission reductions the public-good value of which these certificates are meant to express.

What is the point of all of this? We are trying here to internalize into the cost-benefit calculations underpinning needed mitigation investment a positive externality arising from the systematic reduction of a social cost, that is, greenhouse gas emission reductions to which we attach a value rooted in the aforementioned MAC. In other words, pollution abatement is a social benefit for which its producers, the green project investors, get rewarded in the form of reduced debt servicing charges. The CCs represent carbon money which gets destroyed in a process of ex post social validation when it has achieved its aim, the reduction of emissions. Modern credit-money, that is, private bank money issued in the wake of credit extension (i.e., when banks make loans), implies an anticipation of the money-issuing lender, say the bank, that the borrower's project thus funded will succeed so that the thereby generated income can be shared (as profit of the borrower and interest of the lender) and the debt repaid. It is when the loan has been repaid, meaning that the initial ex ante private validation by the money-issuing lender has been confirmed ex post, and that the money gets destroyed. State backing of such money creation in the banking system, through the elastic provision of reserves ex nihilo by the central bank via its lending to banks or open-market purchases of securities, amounts to an ex ante social (pseudo-) validation through which risks (and possibly also eventual losses) are socialized in support of the private bets made by the commercial bank lenders and their borrowing clients.[26] The validation is always "pseudo," because the state's central bank only presumes future success of the private bets it underwrites, but cannot assure it in advance. Nevertheless, its presumption makes eventual success more likely. It is only ex post facto that the bets made in funding a project with the issue of new money can be effectively validated, at which point the money thereby created gets destroyed (as the bank loan gets repaid). Our proposed carbon-certificate circuit expresses exactly the same logic, except that here we are not following tangible income creation accruing to private sources as profit and interest, but instead the creation of a public good to which we have attached tangible value. The IMPI-issued CCs, which get transferred to national central banks and from there passed on to commercial banks with the aim of being lent to green project investors, represent an ex ante validation of future emission reductions which such carbon money is supposed to help accomplish. To the extent that they do, having thus been validated ex post, the CCs get destroyed—not by repayment of the private loan, but by affording society the social benefit this carbon money has helped to create.

One issue yet to address in presenting this proposal for emission-reducing carbon money is what central banks shall carry on their asset side to match the carbon certificates which they inject into the reserve base of banks. Aglietta had proposed real-economy carbon assets, and I had objected to that suggestion as unsuitable for central banks to carry. Instead, central banks should carry project (or "green revenue") bonds created by securitizing those CC-carrying bank loans funding the IMPI-approved mitigation projects. Using conservative "covered-bond" underwriting principles, such securitizations would turn lower-rated green project loans into top-rated green-revenue bonds which should divert a significant portion of the huge global "savings" pool currently engaged in all kinds of speculative endeavors (e.g., currency trading, carry trade, high-frequency stock trading) into more productive, planet-saving investments.[27] The central banks would obtain those tranches of the green-revenue bonds comprising the CC part of the underlying loans, thus having those CCs fully monetized in line with their backing by government guarantee. As the aimed-for emission reductions kick in and the CCs get destroyed, central banks would simply write down and then write off their green-revenue-bond tranches on the asset side of their balance sheets at the same time. That same covered-bond procedure of securitizing green project loans would have the commercial bank lenders simultaneously write down the respective CC portions of their loans (on their asset side) and their green-revenue bonds (on their liabilities side) as their borrowers' emission reductions materialize.

One of the advantages of this alternative carbon money system is its incentive for early mitigation action. While IMPI may not lower MAC levels the next time it revisits the question (as it is supposed to do regularly, let us say in biannual intervals) perhaps due to lack of progress, the overall commitment is to lower MACs over time as we move forward toward the targeted atmospheric greenhouse gas concentration level of $450 \, \text{ppm}/\text{tCO}_2\text{e}$ on a sustainable pathway of emission reductions. A lower MAC pathway is justified by addressing the abatement challenge effectively, reducing thereby the social cost so as to warrant gradually lowering the subsidies implicitly offered through the CCs, and also in the wake of technical progress in mitigation-related technologies reducing abatement costs over time across many sectors and activities. While we may have to revisit this question if and when using carbon money to finance future adaptation projects to address the damage caused by climate change, as of now dealing solely with mitigation we presume a gradually declining MAC. Of course, the prospect of lower MACs should motivate green project investors to act now rather than later when the subsidies provided by the CCs will be worthless. This is a crucial point, making my proposal possibly more effective than that of Aglietta and Hourcade. Another difference in favor of my version is its explicit use to deal with stranded assets. Since the removal of fossil-fuel-based structures (e.g., coal-fired power plants, cars with internal combustion engines fueled by gasoline or diesel) or practices (e.g., methane-emitting agribusiness or waste disposal) also lowers pollution, those initiatives too deserve transfers of CCs reflecting the estimated reduction of emissions. Such one-time payoffs may compensate polluters for the losses suffered from early retirement of not-yet-fully amortized "brown" assets and thus take care of the huge problem of how to deal with stranded assets which has motivated so much agitation by climate deniers up to now.

Left to consider is what kind of carbon money system to put in place once we have stabilized GHG emissions so as to keep economic activity at that level of (net zero) carbon neutrality. Here, we have to presume that all economic actors have assigned carbon budgets, much like they have financial budgets. Fixing and allocating these carbon budgets will be a major political and governance challenge, but let us argue that this issue can be addressed based on past patterns and considerations of equity just as we also have to assume that we can measure emissions on a micro-level of individual actors in compliance with their assigned budgets. We are clearly nowhere near there yet. But we are beginning to see socio-technological solutions to this problem emerge, from blockchains and the Internet of things to sensors, artificial intelligence, and algorithms for processing Big Data. Just as individual actors can spend below their financial-budget limit and so save to build wealth, we can imagine rewarding actors for keeping below their carbon-budget limits. Those rewards may conceivably come in the form of digital tokens, let us call them EcoCoins, which get paid out in compensation for avoided emissions below the allotted ceiling over time. Once again, as with the government-guaranteed and fully monetized carbon certificates discussed earlier, we are internalizing here a social benefit in pecuniary fashion as a private reward. This may extend to emission reductions in the wake of the IMPI-approved green investment projects in excess of the initially allotted carbon certificates which would trigger supplementary earnings of EcoCoins by the amount of the excess. Issue and distribution of those EcoCoins may be done by the national affiliates of IMPI, possibly backed by the aforementioned carbon certificates. The precise modalities of how these EcoCoins operate in terms of their issue, circulation, valuation, and convertibility can be addressed once we have a better sense more generally of what to do with the emerging phenomenon of cryptocurrencies and how best to regulate them. Be that as it may, it could be an attractive idea to have the EcoCoins serve as the monetary foundation for a block-chain-based online commons comprising sustainable development activities which may include not only environmental services but also other socially beneficial human development advancement. Let us just say that EcoCoins, once earned through lowered GHG emissions, can then be spent within this "social and solidarity economy" network or exchanged for generally accepted money forms at a lower rate of convertibility (to act as a disincentive against leaving the SSE network). An added advantage of such an online digital token form of carbon money for everyone is that we can thereby address the complex web of indirect Scope 3 emissions by consumers and employees over which emitters typically have not that much, if any control. EcoCoins would allow us to integrate this otherwise difficult-to-reach part of the climate-mitigation effort. Finally, and as a crucial part of the SSE network fueled by such online carbon money, we would create a new business model providing socially beneficially services which turn the reduction of a social cost, namely GHG emissions, into a stream of revenues for the service provider, be they a "public-benefit" company, cooperative, association, foundation, or social enterprise.

Obviously, there is still quite a bit of innovation and "trial-and-error" experimentation to get done before we will have put into place such a regime of carbon money making us pay for emissions, funding our mitigation efforts toward a zero-carbon economy, and rewarding us continuously for keeping within our carbon budget. But you can see the first sprouts of such a system emerge already, as I have tried to indicate in this chapter. And the pace of change is bound to accelerate in years to come, as the worldwide mobilization against the threat of climate change intensifies. Carbon money is bound to have an important transnational dimension as it addresses a problem the world community of nations has to address together. Its implementation will therefore inevitably get bound up with ongoing changes in the international monetary system as the world moves away from the long-standing dollar standard. How this process unfolds has to do with the larger question what kind of capitalism we will wish to aim for in this struggle for the preservation of our environment and that depends on our meaning of sustainable development.

NOTES

1. Mainstream economists, following the so-called Monetarist tradition (see M. Friedman 1956), treat money as a good, albeit one with unique supply and demand functions, to define it as an exogenous stock variable which exists separately from all the other goods and services. J. M. Keynes (1936) presented an alternative view of money as an asset of which investors hold a certain portion in order to meet their liquidity preference. By contrast, I look at money from a more interdisciplinary (historical, sociological, social, and psychological, beyond just economic) perspective as a social institution, much in the tradition of K. Marx (1867/1992) or G. Simmel (1900/1978).
2. I have done much work myself on the notion of "monetary regimes" in R. Guttmann (1994, 1997).
3. The JI and CDM procedures were defined in Article 12 of the Kyoto Protocol of 1997 and, together with "International Emission Trading" in the form of cap-and-trade initiatives (see Chapter 5), are referred to as the "Flexibility Mechanisms." All these are complex institutional arrangements, which are all well discussed in the Kyoto Protocol Web site of the UNFCCC (http://unfccc.int/kyoto_protocol/items/2830.php).
4. A good summary of the JI experiment, in particular the reasons behind Russia's and Ukraine's dumping of ERUs and the nefarious consequences of these actions during the 2012/2013 collapse of ERU prices, can be found in Carbon Market Watch (2013).
5. The CDM data are from the mechanism's own Web site cdm.unfccc. int, specifically the link http://cdm.unfccc.int/Statistics/Public/ CDMinsights/index.html. There is also a good deal of useful information about the CDM in the Guardian Environment Network (2011).
6. For the transition economies' bogus "hot air" credits and their laundering to the European Union which absorbed about 550 million of those, see Carbon Market Watch (2015).
7. The relevant provisions to that effect are specified as follows in Article 6.2: "Parties shall, where engaging on a voluntary basis in cooperative approaches that involve the use of internationally transferred mitigation

outcomes towards nationally determined contributions, promote sustainable development and ensure environmental integrity and transparency, including in governance, and shall apply robust accounting to ensure, inter alia, the avoidance of double counting, consistent with guidance adopted by the Conference of the Parties serving as the meeting of the Parties to the Paris Agreement."

8. For more on the implications of this "conditionality" as pertains to the NDCs, see New Climate Institute (2016), J. Strand (2017).

9. For updates on the ongoing post-Paris discussions within the COP framework about the three cooperation mechanisms for collectively elaborated emissions reductions, see R. Bhandary (2017), V. Rattani (2017).

10. IETA (2016), C. Hood and C. Soo (2017) both offer more details on the Article 6's recording, transparency, and accounting challenges.

11. The use by European firms of CERs and ERUs as offsets counting toward their emissions caps under the European Union's Emissions Trading System is discussed in great detail by M. Sato et al. (2016).

12. K. Hamrick and M. Gallant (2017) offer a comprehensive overview of the voluntary carbon offsets, including an interesting discussion of their price variability.

13. For more on these carbon-offset providers and how they operate, see B. Palmer (2016) or K. Spors (2017).

14. After first discussing electronic money more generally and in its different manifestations in R. Guttmann (2003), I presented an early analysis of the Bitcoin phenomenon as a breakthrough digital-currency technology in R. Guttmann (2016, pp. 213–7).

15. This quote is from N. Bauerle (2017) on the coindesk.com site which provides a trove of useful information on the latest developments concerning cryptocurrencies and blockchains. Another useful quote from Climate Ledger Initiative (see I. Callaghan 2017) defines blockchain as follows: "*A blockchain is essentially a public, cryptographically-protected, distributed ledger spread across a network of thousands of computers. This 'database' contains records of every transaction that ever takes place on it and it is constantly reconciling itself. In this way, it is virtually impossible to corrupt transactions that take place because, if anyone tried to change the record of a transaction, the entire system would be out of balance and immediately identify the inconsistency.*" See also *The Economist* (2015) as well as M. Iansiti and K. Lakhani (2017).

16. For a concise explanation of initial coin offerings, see N. Popper (2017) and *The Economist* (2017).

17. The Veridium Ecosystem experiment (see veridium.io) is well discussed in Fortune (2017) as well as A. Gati (2017).

18. It should be noted that R. Douthwaite (1999) also discussed how to democratize money, especially adopting different forms of money for different uses and taking its creation away from the government-banking duopoly, so that it can play a useful role in fostering sustainability and promoting more balanced economic development.

19. The Poseidon carbon-market project is discussed in M. Cuff (2017), while D. Pimentel (2017) provides more information about the Bluenote project.

20. His most comprehensive discussion of climate policy in the broader context of sustainable development and inclusive growth can be found in M. Aglietta (2015a, b).

21. A very similar proposal for monetizing government-guaranteed climate-mitigation projects to facilitate their launch has also been presented by France's leading environmental economist Jean-Charles Hourcade (2015).

22. For a more detailed account of this aspect of Aglietta's plan obliging governments to "keep skin in the game" in favor or steadily rising carbon prices, see Jean Pisany-Ferry (2015). The Aglietta Plan is also well presented, especially in terms of interconnection graphs and balance-sheet illustrations of the monetized carbon-certificate financing mechanism, in M. Aglietta et al. (2015).

23. As reported by L. Sala (2017), Sweden's progressively rising carbon tax has promoted a reduction in carbon emissions by 25% between 1990 and 2015 while the country's GDP grew by 69% over the same period.

24. An interesting discussion laying out the differences between social cost of carbon, shadow price of carbon, and marginal abatement cost can be found in R. Price et al. (2007). As to the methodology behind McKinsey's greenhouse gas abatement cost curves based on estimates of costs and significance of different mitigation investments in power, manufacturing, buildings, transportation, forestry, and agriculture/waste, see P.-A. Enkvist et al. (2007).

25. The International Monetary Fund (IMF) issues a finite number of so-called Special Drawing Rights and allocates them to its member states according to a distribution formula. Those SDRs, which count as a country's official reserves and can be used to settle official debts between countries, are a basket comprising fixed percentages of US dollars, euros, British pounds, Japanese yen, and Chinese renminbi against which they can be exchanged and so turned into actual money.

26. I owe this whole idea of ex ante (private and social) validations by (commercial and central) banks underpinning the issue of modern credit-money, further discussed in more detail in R. Guttmann (1994), to the brilliant work of S. De Brunhoff (1978).

27. In contrast to the asset-backed securities typically found in the USA, covered bonds common in Europe have the loans backing them remain on the balance sheet of the issuing bank. This means that investors in a covered bond retain their access to the "cover pool" of loans. So if the issuing bank should ever go bankrupt, the bondholders would still receive their regular interest-rate payments and get their principal back when the bond matures.

REFERENCES

Aglietta, M. (2015a, January). *The Quality of Growth: Accounting for Sustainability.* (AFD Research Papers, Agence Française de Developpement, No. 2015-01). https://www.afd.fr/sites/afd/files/2017-09/01-papiers-recherche-Quality-of-Growth_Aglietta.pdf.

Aglietta, M. (2015b). What Is the Quality of Growth? Sustainability and Inclusiveness. In L. Haddad, H. Hiroshi, & N. Meisel (Eds.), *Growth Is Dead, Long Live Growth: The Quality of Growth and Why It Matters* (pp. 19–56). Tokyo: JICA Research Institute. https://www.jica.go.jp/jica-ri/publication/booksandreports/jrft3q0000002a2s-att/Growth_is_Dead_Long_Live_Growth.pdf.

Aglietta, M., Espagne, E., & Perrissin Fabert, B. (2015, February). *A Proposal to Finance Low Carbon Investment in Europe.* La Note d'Analyse, No. 24, France Strategie. http://www.strategie.gouv.fr/sites/strategie.gouv.fr/files/atoms/files/bat_notes_danalyse_n24_-_anglais_le_12_mars_17_h_45.pdf.

Bhandary, R. (2017, August 25). Markets and Non-market Approaches for International Cooperation in the Paris Agreement: Open Questions in the International Negotiations. *Climate Policy Lab.* https://www.climatepolicylab.org/news/2017/8/25/markets-and-non-market-approaches-for-international-cooperation-in-the-paris-agreement-open-questions-in-the-international-negotiations. Accessed 25 Dec 2017.

Bauerle, N. (2017). What Is Blockchain Technology? *CoinDesk Blockchain 101.* https://www.coindesk.com/information/what-is-blockchain-technology/. Accessed 31 Dec 2017.

Callaghan, I. (2017, October 28). Blockchain to Unblock Climate Finance? *NDCI Global.* http://ndci.global/blockchain-unblock-climate-finance/. Accessed 31 Dec 2017.

Campiglio, E., Godin, A., Kemp-Benedikt, E., & Matikainen, S. (2017). The Tightening Links Between Financial Systems and the Low-Carbon Transition. In M. Sawyer & P. Arestis (Eds.), *Economic Policies Since the Global Financial Crisis* (pp. 313–356). London: Palgrave Macmillan. https://doi.org/10.1007/978-3-319-60459-6_8.

Carbon Market Watch. (2013, March 4). Joint Implementation: CDM's Little Brother Grew Up to Be Big and Nasty. *Newsletter #2*. https://carbonmarket-watch.org/2013/03/04/joint-implementation-cdms-little-brother-grew-up-to-be-big-and-nasty/. Accessed 19 Dec 2017.

Carbon Market Watch. (2015). *Avoiding Hot Air in the 2015 Paris Agreement* (Working Paper, No. 8). https://carbonmarketwatch.org/wp-content/uploads/2015/08/Hot-air-in-the-2015-Paris-agreement_clean_FINAL.pdf. Accessed 13 Dec 2017.

Cuff, M. (2017). New Blockchain-Based 'Carbon Currency' Aims to Make Carbon Pricing Mainstream. *BusinessGreen*. https://www.businessgreen.com/bg/news/3017564/new-blockchain-based-carbon-currency-aims-to-make-carbon-pricing-mainstream.

De Brunhoff, S. (1978). *The State, Capital and Economic Policy*. London: Pluto Press.

Douthwaite, R. (1999). *The Ecology of Money*. Cambridge, UK: Green Books. http://www.feasta.org/documents/moneyecology/contents.htm.

Enkvist, P.-A., Nauclér, T., & Rosander, J. (2007, February). A Cost Curve for Greenhouse Gas Reduction. *McKinsey Quarterly*. (https://www.mckinsey.com/business-functions/sustainability-and-resource-productivity/our-insights/a-cost-curve-for-greenhouse-gas-reduction. Accessed 11 Apr 2017.

Ferron, C., & Morel, R. (2014). *Smart Unconventional Monetary Policies (SUMO): Giving Impetus to Green Investment* (Climate Report, No. 46). http://www.cdcclimat.com/IMG/pdf/14-07_climate_report_no46__smart_unconventional_monetary_policies-2.pdf. Accessed 24 Mar 2016.

Fortune. (2017, August 29). *This Hong Kong Tech Startup Wants to Build a Natural Capital Marketplace*. http://fortune.com/2017/08/29/hong-kong-tech-environmental-credit/0.

Friedman, M. (1956). *Studies in the Quantity Theory of Money*. Chicago: University of Chicago Press.

Gati, A. (2017, August 31). Environmental Fintech Veridium Labs Announces Crowdsale. *Bankless Times*. https://www.banklesstimes.com/2017/08/31/environmental-fintech-veridium-labs-announces-crowdsale/. Accessed 2 Jan 2018.

Guardian Environment Network. (2011, July 26). *What is the Clean Development Mechanism?* https://www.theguardian.com/environment/2011/jul/26/clean-development-mechanism.

Guttmann, R. (1994). *How Credit-Money Shapes the Economy: The United States in a Global System*. Armonk, NY: M. E. Sharpe.

Guttmann, R. (1997). *Reforming Money and Finance: Towards a New Monetary Regime*. Armonk, NY: M. E. Sharpe.

Guttmann, R. (2003). *Cybercash: The Coming Reign of Electronic Money*. New York: Palgrave Macmillan.

Guttmann, R. (2016). *Finance-Led Capitalism: Shadow Banking, Re-regulation, and the Future of Global Markets.* New York: Palgrave Macmillan.

Hamrick, K., & Gallant, M. (2017). *Unlocking Potential: State of the Voluntary Carbon Markets 2017.* Ecosystem Marketplace. https://www.cbd.int/financial/2017docs/carbonmarket2017.pdf.

Hood, C., & Soo, C. (2017). *Accounting for Mitigation Targets in Nationally Determined Contributions Under the Paris Agreement.* OECD/IEA Climate Change Expert Group Papers. http://dx.doi.org/10.1787/63937a2b-en. Accessed 26 Dec 2017.

Hourcade. J.-C. (2015). Harnessing the Animal Spirits of Finance for a Low-Carbon Transition. In S. Barrett. C. Carraro, & J. De Melo (Eds.), *Towards a Workable and Effective Climate Regime* (pp. 497–514). Washington, DC: VOXeBook, Center for Economic and Policy Research. http://voxeu.org/sites/default/files/file/hourcade.pdf.

Iansiti, M., & Lakhani, K. (2017, January–February). The Truth About Blockchain. *Harvard Business Review.* https://hbr.org/2017/01/the-truth-about-blockchain. Accessed 21 July 2017.

IETA. (2016). *A Vision for the Market Provisions of the Paris Agreement.* Geneva: International Emissions Trading Association. http://www.ieta.org/resources/Resources/Position_Papers/2016/IETA_Article_6_Implementation_Paper_May2016.pdf. Accessed 23 Sept 2017.

Keynes, J. M. (1936). *A General Theory of Employment, Interest and Money.* London: Macmillan.

Marx, K. (1867/1992). *Capital* (Vol. 1). London: Penguin Classics.

New Climate Institute. (2016). *Conditionality of Intended Nationally Determined Contributions.* https://www.transparency-partnership.net/documents-tools/conditionality-intended-nationally-determined-contributions-in-dcs. Accessed 23 Dec 2017.

Palmer, B. (2016, April 28). Should You Buy Carbon Offsets? National Resources Defense Council. *Our Stories.* https://www.nrdc.org/stories/should-you-buy-carbon-offsets. Accessed 30 Dec 2017.

Pimentel, D. (2017). Blue Foundation Creates New Blockchain Token for Zero-Emission Projects. *BlockTribune.* http://blocktribune.com/blue-foundation-creates-new-blockchain-token-zero-emission-projects/. Accessed 5 Jan 2018.

Pisany-Ferry, J. (2015, December 4). Finance Can Save the World from Climate Change. *European CEO.* https://www.europeanceo.com/business-and-management/finance-can-save-the-world-from-climate-change/. Accessed 19 Sept 2017.

Popper, N. (2017, October 27). An Explanation of Initial Coin Offerings. *New York Times.* https://www.nytimes.com/2017/10/27/technology/what-is-an-initial-coin-offering.html. Accessed 31 Dec 2017.

Price, R., Thornton, S., & Nelson, S. (2007). *The Social Cost of Carbon and the Shadow Price of Carbon: What They Are, and How to Use Them in Economic*

Appraisal in the UK. MPRA Paper, No. 74976. https://mpra.ub.uni-muenchen.de/74976/1/MPRA_paper_74976.pdf. Accessed 10 May 2017.

Rattani, V. (2017, November 8). COP 23: Are Countries Ready for the New Market Mechanism? *Down To Earth.* http://www.downtoearth.org.in/news/cop-23-are-countries-ready-for-new-market-mechanism–59061. Accessed 25 Dec 2017.

Sala, L. (2017, December 5). The Swedish Carbon Tax: How to Tackle Climate Change in an Efficient Way. *Traileone.* http://www.traileoni.it/?p=4915.

Sato, M., Ciszewska, M., & Laing, T. (2016). *Demand for Offsetting and Insetting in the EU Emissions Trading Scheme* (Working Paper, No. 237). Grantham Research Institute on Climate Change and the Environment. http://www.lse.ac.uk/GranthamInstitute/wp-content/uploads/2016/06/Working-Paper-237-Sato-et-al.pdf. Accessed 28 Dec 2017.

Simmel, G. (1900/1978). *The Philosophy of Money.* Abingdon, UK: Routledge.

Sirkis, A. (2016). Preface: The Challenge of Moving the Trillions. In A. Sirkis et al. (Eds.), *Moving the Trillions: A Debate on Positive Pricing of Mitigation Actions* (pp. 10–20). Paris: CIRED. http://www2.centre-cired.fr/IMG/pdf/moving_in_the_trillions.pdf.

Spors, K. (2017, November 1). The Lowdown on Buying Carbon Offsets. *Green Business.* https://smallbiztrends.com/2010/09/lowdown-on-carbon-offsets.html. Accessed 30 Dec 2017.

Strand, J. (2017). *Unconditional and Conditional NDCs Under the Paris Agreement: Interpretations and Their Relations to Policy Instruments* (CREE Working Paper, No. 09/2017). https://www.cree.uio.no/publications/CREE_working_papers/pdf_2017/strand_carbon_pricing_cree_wp09_2017.pdf. Accessed 23 Dec 2017.

The Economist. (2015, October 31). *The Trust Machine: The Promise of the Blockchain.* https://www.economist.com/news/leaders/21677198-technology-behind-bitcoin-could-transform-how-economy-works-trust-machine. Accessed 11 Nov 2015.

The Economist. (2017, November 9). *The Meaning in the Madness of Initial Coin Offerings.* https://www.economist.com/news/leaders/21731161-there-ico-bubble-it-holds-out-promise-something-important-meaning. Accessed 15 Nov 2017.

CHAPTER 8

Sustainable Development and Eco-Capitalism

We are, of course, far away from full-fledged carbon money even though we can clearly see first sprouts of such a system emerge in the realm of academic journals or entrepreneurial initiatives. And this state of affairs is true more generally. Our global struggle against climate change still has a long way to go. But we are also beginning to make a concerted effort to launch this struggle in earnest, propelled forward by implementation of the provisions of the Paris Climate Agreement of December 2015. As I have tried to argue throughout the book, climate change poses a long-term systemic threat to human habitats and economic activity which we now recognize to be an urgent matter of policy-making as crystallized in the various climate-mitigation plans countries have put forth in compliance with the Paris Agreement. We know what we have to do, phasing out fossil fuels in favor of renewable sources of energy, boosting energy efficiency dramatically, providing alternative less-polluting means of transportation, overhauling strategic industrial processes, transforming agriculture, improving how we build structures or design cities, and managing waste so that we might turn it into a source of energy. And we understand that none of these structural changes will happen fast enough and go sufficiently far unless we impose an appropriate carbon price, adapt our financial system adequately to the exigencies of climate change ("climate finance"), and monetize incentives and projects aimed at reducing greenhouse gas emissions ("carbon money"). We have seen first steps in that direction, as discussed earlier in the context of environmental, social, governance (ESG) criteria moving corporate governance away

© The Author(s) 2018
R. Guttmann, *Eco-Capitalism*,
https://doi.org/10.1007/978-3-319-92357-4_8

from shareholder value maximization, green bonds offering long-term funding of climate-mitigation projects, or digital tokens issued in connection with environmental-protection projects.

We have made remarkable progress over the last few years to provide a more concerted and serious effort aimed at putting the world economy onto a long-term path of falling GHG emissions. But given where we will have to be in a couple of decades from now if we are to meet the long-term objectives laid out in Paris, these efforts so far do not yet amount to much. Fossil fuels are still dominant nearly everywhere and enjoy powerful political influence in many countries. Renewables are making headway, but not rapidly and far enough. We are not providing sufficient research and development effort to widen the range of affordable renewables (e.g., biofuels, geothermal) or improve energy efficiency dramatically. Our efforts to mobilize resource transfers on a sufficiently large scale from rich countries to the poorer countries most exposed to climate change have at best been halting so far. Financial institutions and markets, together with their regulators, have been too busy recovering from a huge global crisis a decade ago and implementing a new post-crisis regulatory framework (comprising Dodd–Frank, Basel III, etc.) to make much progress toward implementing the needed elements of climate finance expeditiously. Nor is it helpful that the USA, having elected a climate denier to the White House, is rolling back federal policy commitments to fighting climate change. While large swaths of the world's population are becoming aware of climate change and even quite worried about it, most people are not willing to sacrifice a lot of lost income or price hikes to contain this threat. They are also convinced that the fight against climate change cannot but be painful. We clearly underestimate the benefits possibly arising from making our economy more environmentally friendly, energy efficient, and sustainable, starting with the large number of jobs created putting these improvements in place. Worst of all, we do not yet feel the urgency needed to motivate substantial changes in behavior and structures, having a warped sense of time which does not capture the ongoing danger of cumulative greenhouse gas accumulations in the atmosphere. These rapidly building concentrations necessitate radical action now lest they should be allowed to continue growing until it is too late to prevent long-lasting catastrophe. Even if we can adapt to some degree to the damage caused by a large hike in average temperatures, say in excess of 3.5 °C, the effort needed will be painful, incomplete, and very expensive. We are better off trying to prevent such damage to the extent still possible, aiming for the ≤ 2 °C objective of the Paris Agreement.

Urgent action is needed. This is obviously a matter of political leadership. California governor Jerry Brown and French president Emmanuel Macron are living proof that it makes a decisive difference who runs a country or state. This, obviously, works also in reverse as exemplified by Donald Trump. Leaders set policy priorities, and effective leaders can frame a long-term as well as comprehensive climate-mitigation strategy of the kind recently proposed by the Macron Administration with its Plan Climat under the prodding of its iconic environmental minister Nicolas Hulot. Ultimately, it boils down to having enough heads of state make two commitments of decisive importance to our global struggle against climate change. One is to the strong version of the Precautionary Principle whereby we move from putting the burden of proof on those advocating climate actions to taking preventive action irrespective of cost and risk calculations. And the other is the strong version of Sustainability, which prioritizes the preservation of nature rather than accepting that "natural capital" may be degraded as long as we thereby boost human or physical capital in return. Right now, the weak versions of both prevail, justifying halting and slow action on the climate front. If leaders commit to doing everything possible to protect our environment now, as the strong versions of the Precautionary Principle and of sustainability imply, they will find it easier to make the difficult choices they will inevitably face, such as weaning their economy from its dependence on fossil fuels and phasing out subsidies to that effect. But political leadership also involves building popular support for a vision, even using the bully pulpit to educate the electorate as to what needs to be done and how that can best get done. And it is obvious that large portions of the public still lack a sense of urgency when it comes to fighting climate change. It stands to reason that the public, worried about having to face large costs now in order to create benefits which only future generations will enjoy, will be more willing to support climate-mitigation action if such a policy gets embedded in a larger vision of changing capitalism for the better. And the kind of finance-dominated, dirty-energy-driven capitalism pushing us to the brink of environmental catastrophe will indeed need to be profoundly changed if we are to tackle that problem and then keep it contained. What follows are some reflections concerning this necessary transformation of the system.

ACCUMULATION REGIMES IN TRANSITION

As I have already indicated earlier, notably toward the end of Chapter 2, I share a long-term view of how capitalism evolves with the French Régulationists who see the history of that system in terms of consecutive accumulation regimes, each with its own specific "mode of regulation." In that argument, the transition from one regime to another is fueled by systemic financial crises strong enough to expose or cause major underlying structural imbalances in the so-called real economy of production and exchange which can only be resolved with fairly radical and far-reaching reforms.[1] We are arguably in such a transition period. We have had after all a major systemic financial crisis. It started as the "subprime crisis" in the USA in 2007/2008 and then moved on as an asymmetric shock hitting the euro-zone to the point of that region's sovereign-debt-cum-banking crisis of 2009–2012 before collapsing the commodity super-cycle to the detriment of key commodities producers among emerging markets in 2014. The world economy has recovered since hitting its stride in 2017 in the wake of the Obama Administration's deft policy-making and China's rapidly expanding global outreach. Ironically, the renewed worldwide acceleration of economic growth has also reversed progress made in 2015 and 2016 in terms of lowering GHG emissions. Those started rising significantly again in 2017 after two years of hopeful decline. And therein lie the seeds of the next crisis, compounded by the worldwide phasing out of "quantitative easing" programs whereby the world's leading central banks had provided a decade-long artificial stimulus to all kinds of asset markets.

It may well be that rising interest rates will destabilize juiced-up asset markets and so trigger renewed financial turbulence. It is also true that many manufacturing sectors, and even a good many service sectors (not least financial services or business services), suffer from large excess capacities on a global scale which require substantial restructuring and ought to push out weaker companies. This shakeout contributes to the formation of global oligopolies in dozens of key sectors—from steel to food distribution—with much higher degrees of concentration, greater entry barriers, and larger profits margins for the few remaining survivors who at the same time continue to push their global value chains as sources of pressure on smaller suppliers and vulnerable groups of workers. The intensifying battles for market shares across many sectors will at the same time also accelerate the push for automation, made possible on an unprecedented scale by the maturing of artificial intelligence and other labor-saving technologies ("internet of things," blockchains, and nanotechnology), which in turns threatens jobs reaching increasingly into the white-collar domains. The high-tech firms transforming the Internet into a dynamic sphere of value creation (Google, Amazon, Facebook, etc.) use their enormous data-generation power and their control of crucial niches in cyberspace with potentially large network effects for extraction of monopoly rents (i.e., excess profits based on market power) which the financial markets validate by giving these superstars huge valuations. Shareholder value maximization still rules, broadening the impetus for financial income gains linked to monopoly rents. In this finance-dominated economy, profits end up increasingly disassociated from productive investment while the chase for gains today crowds out planning for growth tomorrow. We cannot possibly hope to cope with a long-term challenge necessitating a complete overhaul of our production apparatus and radical shift in our incentive structure, like climate change, with such a biased, concentrated, and uncompetitive economy increasingly controlled by a relatively small number of "superstar" companies focused primarily on extraction of monopoly rents and their validation by very high stock-market valuations.[2]

While the decade-long quantitative easing of leading central banks has been a boon to stock markets, it has squeezed the margins for commercial banks. These have also had their hands full complying with the onerous and complex set of new regulations put into place on a global scale following the systemic financial crisis of 2007/2008. Nonbank financial institutions—hedge funds, finance companies, exchange-traded funds, crowdfunding venues, and others—are filling the void that opened up in the wake of the banks' retrenchment. Those alternative intermediaries are smaller, less capitalized, less diversified, more susceptible to excess, and hence quite vulnerable to be hit by crisis. All of the finance will be subjected soon to new climate-related information gathering, which will also affect the financial statements of listed corporations. These new reporting requirements will intensify already-growing interactions between corporate managers and institutional investors about long-term investment goals, social responsibility, hopefully orderly devaluation of climate-exposed assets, and needed improvements in corporate governance. All this turbulence will unfold against the background of growing concern about rising GHG emissions and temperatures as the world community tries to put the Paris Agreement into effect from 2020 onward. It is this confluence of financial-instability and environmental-degradation forces that may in the end generate a crisis-induced push for much more radical action toward an ecologically oriented accumulation regime (which is what the world arguably needs if it wants to bring the threat of climate change under control).

The Paris Agreement provides the initial reform framework to guide this necessary transition from finance-led capitalism to eco-capitalism in the wake of structural crises pushing in the direction of fundamental change. This accord's intrinsic contradiction between its goal, a temperature hike limited to less than 2 °C which implies zero net GHG emissions by 2060, and its means, a singular focus on market-based solutions of the cap-and-trade type, will need to be resolved one way or another. And if it were to be resolved in favor of preserving the ≤2 °C goal, it might take a climate-related crisis of sufficient depth and width, such as massive flooding in Manhattan or London in the wake of a super-storm, to propel the world community into a much more concerted and ambitious effort. Was should either one of these global financial centers, with Wall Street at the tip of Manhattan already surrounded by water on three sides and both the City and Canary Wharf districts straddling along the Thames River, ever become paralyzed by flood to the point of massively disrupting our financial system, that might very well prompt us to rethink the challenge of climate change. But there are many other catastrophic-weather scenarios imaginable which could trigger a serious global reaction in favor of stronger action to counter the problem. I do think, however, that a crisis combining financial and environmental panic is both possible and at the same time likely to motivate the kind of collective effort needed for our transition to a low-carbon economy.[3]

It would not be the first time that the transition from one accumulation regime to another required a second major crisis to get done. As a matter of fact, each of the accumulation regimes in the past only emerged in the context of a second structural crisis arising in short order to motivate the kind of policy reforms serving as the foundation for the new regime:

- For instance, the first great international financial crisis in 1873 gave rise to a six-year depression in the USA and Europe which later was followed by a second round of crisis-induced debt deflation in either the 1880s (e.g., France) or the 1890s (e.g., USA) before the emerging monopolistic accumulation regime took off in the mid-1890s. At that point, a Britain-led gold-exchange standard had become operational globally to give other rapidly industrializing nations in the Americas and Europe sufficient access to capital, setting the stage for a leap in industrial concentration amidst railroads expanding market size and stock markets encouraging trusts or other large-firm formations.

- The Great Depression of the early 1930s, while itself the engine for a remarkable "New Deal" reform plan amidst the Keynesian Revolution in economic thinking, also destroyed the international economic order and so set the stage for a second systemic crisis of a military nature, World War II, in the wake of which American policy reforms could be extended toward the rest of the industrialized world for the Fordist mass production regime to take root among the other industrial nations, such as Germany, France, or Japan.

- The transition from Fordism to finance-led capitalism was propelled by a crisis-induced and progressively deeper deregulation of money and banking following the collapse of Bretton Woods in August 1971 which freed exchange rates, interest rates, quasi-monies, and bank structures from government regulation. The global stagflation dynamics unleashed by the disintegration of the postwar order in the early 1970s created a dollar-driven credit crunch a decade later which formed the background for the conservative counterrevolution in fiscal and regulatory policies known as "Reaganomics."

We are thus right now in a transition from the regime of finance-led capitalism, which gave us a quarter century of rapid economic growth on an increasingly global scale (1982–2007) before exploding in a deep systemic financial crisis in 2007/2008, to a new accumulation regime that might conceivably organize itself around meeting the challenge of climate change. The Paris Agreement of December 2015 can in this context be seen as the decisively regime-forming policy reform guiding this transition, except that the measures needed to carry out this agreement successfully have yet to be put into place. There are plans in the works how to do so, but these remain largely on the drawing boards. The political will to impose a sufficiently high-carbon price, change our energy mix radically, boost energy efficiency, opt for alternative means of transportation, revamp industrial processes and agriculture, redesign urban landscapes, and climate-proof buildings must await a greater sense of urgency, perhaps even panic, about the changing climate threatening our habitats. Sooner or later, we will get to that point. But will it then be too late to have kept the average temperature hike below the 2 °C threshold?

A POLITICAL PROJECT

These measures required for our necessary transition to a low-carbon economy are feasible, but only if enough of us see them as a priority. Such a consensus, without which there is simply not enough political will to make those difficult choices, demands committed leaders as much as a supportive electorate comprising a majority of voters who make such a transformation of our economy a matter of high urgency. And neither comes about unless carried forward by a political movement mobilized around issues related to climate change. Such political shifts tend to emerge as a necessary complement to the crisis-induced policy reforms with which we manage the transition from one accumulation regime to the next. We could see this happen in the 1890s with the Progressive movement in the USA as much as the Socialist International in Europe both of which endorsed social activism in the interest of improving the social standing, economic protection, and political representation of working-class families. And we saw something similar happen in the 1930s with the Democrats in the USA regrouping around a hugely ambitious reform program known as the New Deal as did the Popular Front governments in France and Spain, while the fascists in Italy, Austria, and Germany practiced different variants of military Keynesianism. Getting to the late 1970s and early 1980s, we could witness conservative counterrevolutions in the UK (propelling the "Iron Lady" Margaret Thatcher to the premiership) and the USA (allowing Ronald Reagan to win), which essentially combined attacks on the "Welfare State" social safety net with large tax cuts aimed disproportionately at the well-to-do and "tight money" policies targeting inflation. That neoliberal program came to be pushed globally, especially onto the less-developed countries finding themselves in a deep debt crisis throughout much of the 1980s, as the so-called Washington Consensus favoring privatization of hitherto state-run enterprises, deregulation of key sectors and the capital account, fiscal austerity, market-determination of the prices of money (exchange rates, interest rates), and flatter income taxes. The most far-reaching of these neoliberal reform suggestions concerned the deregulation of money and banking, out of which emerged finance-led capitalism driving the world economy forward for a quarter of century until the subprime crisis of 2007/2008.

Transitions in accumulation regimes have thus each time been shaped by political movements arising in the context of the second systemic crisis and crystallizing the policy reforms needed to get that transformation done. This time will be no different. Right now, in early 2018, there is a lot of political ferment sweeping over many countries. But the major thrust of that volatility comes from a right-wing nationalist backlash against globalization, mistrusting the elites (media, bankers, even scientists) and focusing its anger above all against immigrants (especially of a different skin color or religion). An ironic by-product of this shift to the right is its impact on our global commitment to fight climate change. Those right-wing populists, whether Trump, the "Leave" camp of Britain's Tories overseeing Brexit, Germany's AfD, Austria's FPÖ, or France's Front National, all are skeptical of climate change and resistant to mitigation action against it, as they profess romantic attachment to a twentieth-century vision of a heavy-industry economy centered on blue-collar workers at the core of their electorate. Center-left parties, notably the Democrats in the USA or Europe's Social Democrats, have either imploded with the rightward shift of their traditional voter base or given way to a more radical anti-capitalist Left as exemplified by Bernie Sanders in the USA, Jeremy Corbyn taking over Britain's Labor Party, or Jean-Luc Mélenchon of the La France Insoumise ("France Unbowed") movement. But that alternative Left's policy program, raising taxes for income redistribution, nationalizing strategic sectors of the economy, and shoring up various income-maintenance programs, seems at times more like a recipe for a past era long gone and at other times finds itself competing with the nationalist Right over who is more protectionist, more anti-immigrant, more hostile to multilateral institutions eroding the much-cherished national sovereignty. This Left too fetishizes a long-lost working-class culture in opposition to ruthless bosses moving manufacturing platforms to lower-wage, non-unionized parts of the world economy. Climate change and/or other issues of global concern requiring internationally coordinated solutions tend to fall by the wayside in this mind-set. Environmentalists, for whom climate change and other ecological concerns are obviously a high priority, have surged here and there locally across the range of advanced capitalist societies. But their popularity has rarely lasted longer than a few years, because they often found it difficult to translate their narrowly focused priorities into a coherent program driving a broader progressive vision of an alternative society and/or they ended up deeply divided fighting each other into political marginalization.

We are thus facing still a political vacuum when it comes to climate change. This is why the issue can be on people's minds, and yet nothing much gets done about it. Politicians of all stripes worry that any up-front costs of climate policy will bury them in a public backlash before the fruits of sacrifice yield any long-term benefits. And the changes needed to get climate change under control will surely end up being profound. They will not only have to transform our industrial apparatus and energy supply, but will also have to alter the priorities of corporations as well as financial institutions. Right now, we still have a capitalist system, which continues to be dominated by the power structure and motivations of finance-led capitalism (notwithstanding its systemic crisis a decade ago), and such a system cannot undertake the needed changes on its own. Of course, we make progress here and there—the spectacular cost declines of solar- and wind-powered energy, steady progress with electric cars, governments all over the world starting to set long-term goals targeting greenhouse gas emissions, the gradually strengthening campaign for ESG criteria in corporate decision-making. But such sprouts of structural change are not enough, occur too slowly, and often lack coordination to form a coherent alternative. Integrating them into a broader vision of a different, more sustainable type of capitalism and pushing them forward as a popular program of policy reforms is an urgent project if we are to overcome the existing rot in our body politic.

A Progressive Vision of Globalization

If we want to address the problem of climate change effectively, we need to embrace the very notion of globalization and then seek ways to push it to a higher, better balanced, more socially and environmentally oriented level. We can only fight this problem and defeat it, if we work together. We thus need a new, progressive vision of globalization, not least to counter the "America First" or "Britain First" types of nationalism so in vogue right now by showing that a differently structured mobilization of activities beyond borders ought to produce properly balanced and distributed benefits which vastly outweigh the costs that may accrue to some. While most people can understand that trade benefits them as consumers by giving them more choice of better-quality products at lower prices or that access to foreign markets offers them better risk-return trade-offs and diversification as investors, many worry about their job security or wage prospects in a much-dreaded "race to the bottom." Globalization also evokes fears of losing cultural identity, of eroding national sovereignty, of greater exposure to problems arising elsewhere, and of others grabbing a piece of your pie. Some of those fears are rooted in fallacies, such as trade being a zero-sum game of winners and losers when in effect it typically benefits both sides by facilitating specialization that boosts their collective output created together. Others are legitimate but overblown, such as losing one's national sovereignty or identity. Be that as it may, these widespread and still-growing concerns about globalization need to be taken seriously and addressed effectively in measured public debate. Much of the resentment is stoked by the fact that globalization bestows huge benefits upon a fairly small group of actors, comprising managers and key personnel of firms thriving in the world economy as well as the institutional investors and bankers backing them, while rendering a much larger number of actors insecure. The worst excesses of this privileged elite must be stopped, be it money laundering, tax evasion, speculation, exploitation, or rent-seeking behavior, and governments must make it a priority to do so. We cannot continue to live in a world where powerful private actors, especially the wealthy and the large corporations, end up paying proportionately far-less taxes than the average citizen while at the same time being able to extort governments for additional gains.[4]

While looking for ways to roll back these excesses to take the sting out of globalization being rightfully seen as a stacked game, we need to convince large numbers of people that we are for the most part better off living in an integrated world economy provided we can find appropriate mechanisms of governance for it. The key issues of today and tomorrow are global and must be addressed as such. We cannot cope with climate change unless it is done together, as we have realized with the Paris Agreement. There are many other issues that we will have to face as a global collective of nation-states:

- The world has a shared interest in nuclear non-proliferation. This is why the 2015 nuclear deal between Iran and six world powers (USA, Russia, China, Britain, France, and Germany), providing a thorough inspection and verification regime under the auspices of the International Atomic Energy Agency, is so important.

- Europe's recent difficulties absorbing a couple of millions of refugees from Syria or sub-Saharan Africa moving through Libya offer valuable lessons to prepare us for a time a few decades from now when hundreds of millions of migrants will be on the move fleeing environmental catastrophe. We know already today that the planet will have to address a demographic imbalance between shrinking populations in the Northern hemisphere and exploding population growth in the Southern hemisphere far outpacing the job creation or food production capacity of its mostly poorer regions. We might as well start planning for it now by mobilizing adequate resources collectively for programmed population transfers and effective integration of new arrivals.
- Food security is already a pressing issue in many regions and likely to become even more so as the warming planet lowers agricultural yields. So we must work together on improving food production and distribution.
- We also have to put into place a more effective global infrastructure to fight pandemics, of which we have already recently seen worrisome signs of spread and intensification (bird flu, Ebola virus, Zika virus)—a pressing issue which is bound to intensify amidst global warming.
- Last but not least is the potentially devastating threat we face from cybersecurity and cyberwarfare, both of which need a collective regime of deterrence capable of dissuading potential miscreants through assured punishment—something we are woefully far from having currently a chance to achieve.

Notwithstanding America's ill-timed withdrawal from global leadership on any and all of these challenges under Trump, we need, if anything, much stronger international institutions and governance mechanisms than those we currently have in place. Most of those problems are handled now through specialized (and often rather small as well as underfunded) agencies of the United Nations, such as the International Atomic Energy Agency, the United Nations High Commissioner for Refugees, the Food and Agriculture Organization, or the World Health Organization. In each of these problem areas, the world community needs to find stronger multilateral agencies with better policy options and larger budgets. Some of these issues may not need to involve all nations at once, as has been attempted with climate policy under the

Paris Agreement. It may be sufficient to have a core of key players set behavioral norms, regulatory standards, or policy priorities, as has happened with the reregulation of banks under the four Basel Agreements sponsored by the Bank for International Settlements.[5] The so-called Group of Twenty (G-20), a forum for twice-a-year meetings of heads of state capable of making commitments and setting policy, may be a good model here, provided these get-togethers produce meaningful and morally binding results. A key question is how to represent the rest of the world, currently aggregated in the so-called Group of 77. Some selected members of the G-20, for instance, Brazil, India, or South Africa, may seek policy consensus among that larger G-77 group and then represent its interests. In the end, however, we will need far more direct representation of smaller "frontier" economies at the forefront of many of the aforementioned problems, starting with climate change, food security, or mass migration. It will be crucially important to give these vulnerable countries enough of a voice and pick effective representatives of that group (e.g., Costa Rica, Botswana, Mongolia) in international deliberations.

Either way, we need to push for a higher level of global policy coordination than the essentially ad hoc and highly uneven framework we have currently in place. Such a push for stronger global governance, especially important in the kind of multipolar world toward which we are drifting gradually, must be rooted in a progressive vision of globalization. Key to this vision is the notion of shared sovereignty. What is "shared" here obviously concerns common policy objectives on challenges which most, if not all, countries agree must be tackled together. But there also has to be widespread acceptance that, for certain key questions affecting the world at large, the national interest can only be realized in a give-and-take process of compromise on the basis of which all countries engaged in such negotiations arrive at a consensus that they then agree to put into effect. We will need a mind-set where needed resource transfers from richer countries to poorer countries are not resented by either side, but fundamentally understood to be beneficial to both. History is replete with examples where such transfers allowing others to catch up or at least to secure a modicum of improvement have had positive results not only for the assisted parties, but also those that did the assisting in terms of finding new allies and/or gaining bigger markets—from America's Marshall Plan in the late 1940s via Bretton Woods during the 1950s and 1960s to American consumers acting as the "buyers of the last resort" in the 1990s and 2000s in support of rapid export-led growth boosting many emerging-market economies. Non-compliant countries, which either refuse to participate in the building of consensus or seek to enjoy free-rider advantages, will have to be made to suffer because of their lack of cooperation, either by marginalization or effective sanctions. The challenge of climate change will be, without a doubt, a serious test case whether the world community can muster sufficient cooperation on a difficult and complex policy question. To the extent that we succeed, it is by all of us thinking globally, organizing supranationally, and acting locally.

A New Social Contract

While we most assuredly will have to reinforce the supranational policy dimension to cope with global challenges like climate change, this will not be enough. We must go further than that, reforming the capitalist system itself. This is not just a technological question, replacing fossil fuels and other sources of greenhouse gases with cleaner alternatives. Nor is it enough to push beyond shareholder value maximization toward a corporate governance framework putting greater stress on social and environmental considerations (the so-called ESG criteria) or find new instruments through which to channel long-term funds for climate-mitigation projects, such as green bonds. The system that gave us global warming cannot just be adjusted on the margins. It will have to undergo a profound socioeconomic and political (in the broadest sense of the term) transformation for it to become a low-carbon economy no longer emitting greenhouse gases at will into the atmosphere. In short, what we need is a new social contract. At the center of that new social contract is a partnership between the public sector, private enterprise, and other social actors to reform the currently configured state-market dichotomy which in many ways has exhausted its (initially more) productive capacity. The task is to affect a transition to a low-carbon economy on a global scale while using this far-reaching transformation as an opportunity to create a more attractive and better balanced version of capitalism.

Even though the private sector will carry out much of this transition on the ground, it lacks the tools and abilities of central coordination which this effort surely requires. That role will fall to the state, in conjunction with other nation-states. To the extent that climate-change mitigation necessitates a makeover of pretty much the entire industrial base, indicative planning should make a comeback as governments define sector-specific and/or technology-altering goals to be reached over certain time periods in pursuit of a downward GHG-emissions path leading to carbon neutrality by, say, 2060.[6] We have already discussed this kind of industrial-policy framework in Chapter 3 when detailing the low-carbon transition plans recently put forth by California's governor Jerry Brown, France's Hollande and Macron Administrations, or China's Xi-Li government. These transition plans will have to be steered with (dis)incentives which governments may deploy as they formulate their energy policies, their agricultural policies, their urban-development policies, their transportation policies, etc. It might make sense to set up a government agency for that purpose, such as France has done (with its ADEME). When it comes to getting the most important transition projects off the ground and bringing them to fruition, this agency might be empowered to offer loan guarantees or engage in joint public-private initiatives. Also crucial in this context should be the funding of a sufficiently large research and development effort which might usefully be designed as a supranational initiative involving many, if not all, countries. Much of the knowledge thereby engendered should be "open access" in recognition of the public-good nature of climate-change mitigation. So we need to rethink intellectual property rights (which may have gone too far in giving innovators monopoly rights) as well as antitrust powers (which need a globally coordinated push against excessive barriers to entry and initiative-stifling concentration). Since so much of the climate-mitigating investment involves energy, transportation, architecture, urban planning, or agriculture, governments should take a very broad view of "infrastructure," make sure that it gets remade in "smart" ways, and then find novel ways to mobilize private capital for that effort in conjunction with public resources. Crucial in that regard will be new kinds of funding instruments. We have already seen the emergence of so-called "green" bonds. But we can imagine going much further than that toward government-backed industrial-development bonds, revenue bonds, or participation shares through which investors benefit from helping to finance the worldwide climate-mitigation effort over the long haul. It is conceivable to let these instruments offer contingent returns so that they guarantee sufficiently attractive minimum returns and then offer better yields the greater their emission-reduction impact.

The private sector too will inevitably have to undergo tremendous change as it settles into a new ecologically oriented and socially embedded accumulation regime, that which we refer to here in this book as

Eco-Capitalism. For one, shareholder value maximization must be replaced by stakeholder-implicating governance which gets enforced by representatives from affected parties (including employees and buyers) and active owners grouped together in mixed-member supervisory boards. Chief executive officers and others in charge of running companies need to be socially responsible leaders keeping in mind to what extent their firms contribute to the public good and how they need to act in the public interest. That aspect should motivate firms as much as the pursuit of profit. There has to be a radical change in the way our industrial leaders think, and this new paradigm needs to be firmly embedded in a normative context, a new culture of what the "bottom line" means.[7] Of course, this change will have to include the remuneration packages of the top corporate managers and their performance incentives. Just as much as we shall have to apply very low discount rates to readily identified climate-related investment projects (as part of our commitment to a strong version of the Precautionary Principle and also to "Strong" Sustainability), so will we have to measure corporate performance beyond narrowly defined profit in terms of its contribution to societal well-being. That is why the emerging "environmental, social, governance" (ESG) criteria we discussed in Chapter 6 are going to be so important at the center of this new corporate evaluation paradigm. We will also have to introduce new performance indicators with which to identify and measure corporate contributions to societal well-being, an effort that has already started as the ESG movement is beginning to take root. And those indicators need to be contextualized in a social-value accounting system tracing the social costs and benefits of corporate activity in all of its dimensions.[8]

The reference in the preceding paragraph to so-called public benefit companies (see note 7) needs to be contextualized further. They could conceivably be just one component, albeit an important one, in a much broader socioeconomic sphere of new types of producers. Earlier, during the late nineteenth century and for much of the twentieth century when socialist thinking still had strong roots, there were lots of cooperatives and mutual societies above all in Europe. Today, we have not-for-profit producers, but they either exist at the margins of the economy (e.g., agriculture), are run by government, or have that status for accounting and tax purposes while in reality pursuing profit (e.g., private American universities). We also have lots of non-governmental organizations (NGOs) offering useful services and making up a vibrant part of civil society. No, what I am talking about here is an entirely different caliber of producers, comprising what gets often referred to as the "social and solidarity economy" (SSE) by which we mean a variety of typically small-scale producers of socially beneficial goods and services made available affordably, if not altogether for free or very little.[9] Such SSE outfits may even comprise NGOs mobilizing around certain issues of public interest. But in addition they can also include partnerships and cooperatives meeting a need that can either be channeled through the marketplace or be made otherwise accessible locally. Increasingly, a key and vital component of this alternative SSE sphere are so-called commons which manage shared resources in which each stakeholder has an equal interest. Such commons engage citizens in a kind of participatory democracy, whose rules and responsibilities are defined in a charter, to tackle issues not properly addressed by state or market. For example, we may envisage an urban commons turning a run-down site currently occupied by squatters into a community center providing hard-to-find services and a meeting place for the neighborhood. Or cleaning up a block. Or helping otherwise marginalized people (e.g., the mentally ill, former prisoners, addicts) to find work and/or needed services. There are obviously many themes around which to organize commons. Some of them will operate in cooperation or even under contract with local governments. They can, and often will, be political in the sense of advancing an agenda targeting the existing power structure. I can imagine that commons can eventually play a huge role in the monitoring, reporting, and verification (MRV) part of the climate-mitigation battle, if not themselves become an active vector for innovative solutions and so end up making an important contribution to render our economy carbon neutral.

Our proposal for a new social contract must address the issue of personal income security. Many people feel nowadays a great deal of insecurity as regards their future and that of their children. There is a lot of economic and financial volatility. Global trade and foreign investment can move large portions of a country's manufacturing capacity abroad. Wages and productivity are stagnant. Technology threatens to eliminate or transform a lot of jobs. Social programs providing needed services are under budgetary pressure. Political turmoil brings forth demagogues able to channel anger and anxiety into populist revolts against the elites representing a failed order and/or minorities targeted as scapegoats. These are not favorable circumstances for launching a globally transformative project centered around climate change. If we are to get the political support and active participation of the citizenry needed for what will have to be a sustained effort lasting decades, we must also create conditions where large numbers of people in otherwise rich countries do not feel existentially threatened and thus ill inclined to act or think beyond tomorrow. For that reason alone, it may behoove us to think of a menu of new income stabilization options, which would also help to rebalance the growing issue of widening income inequality:

- In the face of an imminent wave of automation making potentially many jobs redundant, there has been much talk recently about introducing universal basic income—a kind of welfare check sent to everyone to assure a minimum income level.[10]
- On top of that, we can create additional layers of meaningful protections adapted to the exigencies of a highly dynamic twenty-first-century economy. We can, for instance, envisage a joint private–public scheme to conserve jobs in return for a government subsidy that recipient firms could use to improve job-training facilities or upgrade their productive capacity while workers could also be loaned out for government projects helping the economy at large (as in moving us to a low-carbon economy).
- It would also be helpful to have a federal wage-insurance scheme funded by contributions from employers, workers, and government itself from which workers may draw benefits when forced by trade or technology to accept lower-pay jobs. And workers could also qualify for retraining benefits as an alternative, possibly via vouchers.
- Finally, we could also offer the working-age populations a fourth layer of income protection, consisting of a menu of employment-related options they can draw from such as added parental leave, compensation for voluntary work (as when helping organize commons engaged in MRV work), or seed funding for launching socially useful self-employment or partnership projects.

Eco-capitalism will thus have firms looking beyond their narrowly profit-defined objectives to promote at the same time also public-benefit goals. It could also rely on a vibrant set of social-economy actors pursuing socially useful objectives. And it might offer better income-protection programs for workers who therefore end up endorsing globalization and adapting to rapid technological change more readily. Perhaps the most important feature of eco-capitalism, however, is a much more dynamic public sector where strategic agencies of the state accentuate entrepreneurial initiative, promote competition, and underwrite needed initiatives advancing the transition to a low-carbon economy. Laws and regulations will have to be put into place for needed changes, be they restrictions on the burning of fossil-fuel reserves, launch of climate finance, diffusion of knowledge pertaining to the fight against climate change, sharing of strategic technical know-how, and cooperation among nation-states counterbalancing the innate asymmetry of power relations between them (so that the poorer countries around the equator or small-island nations, all those most vulnerable to climate change, would get proportionately greater resources). It might also be helpful to rethink our tax system, both in terms of assuring its greater de facto progressivity (so that the rich and powerful pay their fair share) and putting the burden of taxes on socially harmful activities (e.g., speculation) or products (e.g., fossil fuels). Some taxes, such as the broadly based and administratively efficient value-added tax, are better than others. A carbon tax might come to occupy a strategically crucial position in such an overhaul of our tax system. But we can also envisage other new taxes, such as a financial-transaction tax on a global scale targeting short-term financial transactions (i.e., speculation) or a robot tax as recently proposed by Microsoft founder Bill Gates and Tesla CEO Elon Musk, to help along the transition from finance-led capitalism to eco-capitalism. It seems also fair to promote efficiency of our tax system by keeping our most strategic taxes, say income taxes or value-added taxes, broadly based so that their rates can be set at low levels. This principle would preclude using tax exemptions, exclusions, credits, and deductions as a way for the state to subsidize or direct certain activities and actors, often in response to successful lobbying efforts by privileged groups. So we need to rethink the state as entrepreneur and fair tax collector in effective pursuit of the public good, allowing eco-capitalism to thrive in a more productive partnership between private industry and the public sector while at the same time being pushed forward by the helpful contributions of a dynamic social-economy sector of dedicated actors.[11]

Rethinking Economics and Finance

I have already discussed on several occasions, most recently at the onset of this concluding chapter, the need to have our political leaders commit to both the strong version of the Precautionary Principle as well as to the strong version of Sustainability. Either one of those commitments, replacing their respective weak versions currently in place, flies in the face of established wisdom as crystallized in standard economic theory or mainstream finance. The "strong" Precautionary Principle obliges us to act, because there is such uncertainty about how bad things might get as climate change intensifies. This obligation to act irrespective of the uncertainty involved suspends the usual risk-return calculations brought to bear when making investment decisions. And the commitment to "strong" sustainability prioritizes preservation of the environment beyond any cost–benefit calculus, whereas standard theory justifies a modicum of environmental degradation provided that negative consequence is counterbalanced by a concomitant increase in physical or human capital in return. So we have already here an inkling, even two, that climate change has implications beyond what mainstream economists think or typical financial analysts frame for their decision-making. And this is indeed the case!

Much of the book has actually been dedicated to illustrating how economics will have to be rethought in the face of climate change. Macroeconomic growth models, for one, will need to incorporate estimates of long-term damage caused by climate change and at the same time give a more accurate picture of how beneficial climate-mitigation activity might be for improving the performance capacity of national economies. Without those extensions, standard models will systematically underestimate both the climate problem and the impact of solutions to that problem. It also seems likely that heterodox macro-models, such as stock-flow consistent models, can give us a lot of rich insights, especially as pertains to how the transition to a low-carbon economy may play out in likely scenarios. These alternative approaches may therefore claim stronger recognition, perhaps even by mainstream macroeconomists who as a group typically resist any challenges to their own way of thinking. We can also surmise that, as we map what a carbon-neutral economy might look like, we shall return to the use of techniques which indicative or central planners used when mobilizing a war economy (as in World War II) or pushing for accelerated industrialization. I am thinking here in particular of input–output models depicting the interdependencies between the different sectors of a national economy or framing the complex dynamic of low-carbon transition scenarios as linear programming problems for a smaller number of solutions to be checked.[12]

Up to now, most economists have tended to define the problem of climate change as if essentially a problem of pollution, the result of industrial activity releasing planet-heating greenhouse gases. As such the problem gets defined as a negative externality, a social cost ignored by those involved in causing this problem, which will need to get internalized. Standard economic theory offers two alternative approaches here, either a market-based solution promoting trade in the externality for there to be a price established (i.e., Coasian bargaining) or by means of an emissions tax equaling the social cost of that negative externality (i.e., Pigovian tax). Consequently, we have ended up basing our entire climate policy so far on either so-called cap-and-trade schemes or carbon taxes, reflecting the solutions pushed by Ronald Coase versus Arthur Pigou. Yet neither has proven so far sufficient to address the problem of climate change adequately, and none will likely succeed to do so in the future. The societal impact of climate change is simply too large and too far-reaching to be reduced simply to an externality the very notion of which implies social costs that are but a fraction, albeit possibly a tangibly important one, of the total costs involved. In reality, climate change goes far beyond that. If we assume "business as usual" going forward, we have to assume that we would likely face catastrophic conditions in many areas of the world within a generation or two from now. Climate change thus represents a systemic risk, far beyond being simply an externality, the avoidance of which must be regarded as a public good. Not making our planet in large parts uninhabitable for our children and grandchildren is a moral duty, especially since those future generations will not have had any say in the matter.

But avoidance of a systemic crisis implies new policy regimes whose breadth and depth must match the nature of the problem they seek to address. We have faced this kind of challenge not long ago. In the aftermath of the global financial crisis of 2007/2008, central banks all over the world felt obliged, after spending literally hundreds of billions of taxpayer funds to bail out ill-behaved bankers, to assure their irate citizens that such a lamentable situation would never recur. Hence, they—the Federal Reserve, the European Central Bank, the Bank of England, the Bank of Japan, etc.—all declared financial stability to be a new primary policy objective and then followed up with a series of new initiatives to make this commitment operational. Central banks took on supervisory responsibilities vis-à-vis their largest banks, introduced a new regulatory approach known as macro-prudential regulation, imposed new safety requirements on banks under Basel III (in terms of capitalization levels, liquidity cushions, and leverage caps), and set up a financial-crisis management regime known as "resolution authority" to avoid taxpayer bailouts.[13] With the Paris Agreement, we have already taken the first step toward such a new policy regime adapted to dealing with a systemic risk on a global scale, requiring countries to commit to increasingly ambitious emission-reduction goals and introduce "flow-consistent" finance.

As concerns such climate finance, it will have to move beyond the key tenets of mainstream financial theory. The currently dominant paradigm pervading corporate governance, shareholder value maximization, implies an emphasis on current earnings and a short planning horizon, neither of which can cope with the challenges of climate change. Those need a very long investment horizon, a willingness to take more costs now for benefits later (and hence accept a temporary cut in profits), and concertation with many stakeholders. All these characteristics will necessarily transform how firms operate and what goals they prioritize, moving them from shareholder value maximization to ESG criteria as has already begun to occur thanks to agitation by some institutional investors realizing the need for different corporate governance priorities in the face of such complex challenges like climate change. But progress in that regard may still be too slow. If stock markets were truly as informationally efficient as mainstream financial economists claim they are, then shareholders would have already begun to devalue the "stranded assets" of high-carbon emitters like the oil companies and by and large done a better job assessing the many implications of climate change than most politicians have been able to. This has not yet happened! Financial markets will end up functioning very differently anyway, if they are to mobilize gigantic sums for long-term projects with considerably uncertain outcomes. At a minimum, investors would have to look beyond just private profits accruing to individual producers and include in their calculation of returns also the generation of social benefits (i.e., here expressed as the avoidance of a social cost, namely GHG emissions).[14] Nor will investors be able to follow the currently widely accepted logic of the capital asset pricing model (CAPM) when calculating the expected return of an asset given its risk. Climate finance will simply not get very far, if it applies such a traditional risk-return trade-off to its investment decisions. Climate-mitigation projects carry major new (physical damage, legal) risks for which we do not yet even have appropriate risk management models.

And those projects also stretch over very long time horizons. If we therefore applied the usual risk-adjusted discount rates to such climate-mitigation projects, we would never undertake them. We need instead to use far-lower discount rates, notwithstanding major climate-related risks. In other words, climate-mitigation investments cannot be subjected to the same profit-driven risk-return logic as the rest of capitalist finance.

Ultimately, it will all come down to how we as investors, or more generally we as a society, conceive of the long term. In its current configuration as finance-led capitalism, there are serious disincentives to think long term or plan for it. Shareholders focus on the bottom line right now, and so do managers motivated by their stock options and performance bonuses. Corporate executives are more inclined to worry about current earnings and short-term returns even though nearly all the activities important to a company's competitive success—building a dedicated and skillful workforce, fostering relationships with reliable suppliers, research and development, investing in plant capacity—are of a long-term nature. Even institutional investors, such as pension funds, are more focused on short-term trading strategies even though their relatively predictable outflows would allow them to pursue a long-term portfolio strategy which might in the end prove more profitable. And finally, we also have policy-making subjected to a relatively short electoral cycle of four to five years, as most politicians resist making painful choices whose eventual benefits arise only later while their costs kick in up-front. This has been precisely the problem with climate policy as well. We are only now beginning to tackle this issue by mapping out the different steps and phases of a multidecade transition to a low-carbon economy. To the extent that much of that transition involves new infrastructure (e.g., power plants, public transportation) or durables in relatively early stages of development (e.g., electric cars), we have to be societally able to plan for the long haul and act accordingly. This is a matter of concertation by responsible state bodies—from public development banks via properly empowered agencies in charge of promoting the low-carbon transition to the energy department. Finance needs to play a constructive role here as well, putting forth new revenue-sharing long-term funding instruments as we have already seen emerge with the "green" bonds. Once ready to prioritize ESG criteria rather than shareholder value maximization, corporations may possibly restructure their incentives, objectives, and planning horizons toward a longer-run view.[15]

Another aspect to rethink is how to mobilize the research and development effort needed for our effort to contain climate change. It will require a lot of innovation across many sectors and applications to put our national economies across the planet onto multidecade paths of steady reductions in GHG emissions. This is not just a matter of assuring a sufficiently large effort and funding it, undoubtedly involving a lot of government support. Even more important may be how best to encourage faster and wider diffusion of knowledge generated by this global R & D effort. To the extent that such knowledge is crucial to the success of our climate-mitigation and adaptation initiatives, it cannot justifiably be kept private. Our currently prevailing regime of intellectual property rights, especially its tough US version, treats innovation as private property endowed with monopoly status and at best being slowly diffused through licensing agreements. Climate-related innovation has to diffuse much more rapidly and cheaply than that. It thus should be accessible on an open-access basis or organized as knowledge commons whose informational contents are collectively owned and managed by a community of users. Here too we have to promote the public-good dimension so that it can be given the space it needs to have as a proper counterweight to our system's otherwise excessive domination by the private profit motive.

All these challenges to the standard notions at the center of mainstream economic theory or finance make a good case for a more fundamental change in how we conceptualize economics. Ever since the "marginalist" revolution in economic thinking in the 1870s, putting an end to a century of Classical Political Economy which had moved from the pro-market musings of Adam Smith to the more troubling interpretations of Karl Marx, we have stripped economics of its social-relational context, replaced historic time with logical time in economic modeling, and reduced any consideration of nature to a question of "land" as a factor of production to be exploited for gainful purposes.[16] If we were to reintegrate climate change properly into economic thinking, a crucially important task in light of what amounts to a systemic threat, we need to re-embed our economy into its societal context and then reconnect both with nature.

Such a dual-integration exercise has many implications of which I will only point out a few. One is abandoning the traditional neoclassical separation of the economy into its "real" sphere of production and exchange and its "nominal" sphere of purely monetary phenomena (interest rates, price level, etc.). This separation, which reduces money to an exogenous stock variable set apart from economic activity, fosters the formulation of (essentially non-monetary) equilibrium conditions. If we bring money back into the picture, we recognize that it functions actually as an endogenous flow variable. It does so by organizing all economic activities as spatially and temporally interwoven cash flows which structure our social space (as a web of social relations of interdependence and conflict between buyers and sellers, workers and their employees, creditors and debtors, and so forth) as well as historic time (where we continuously face uncertainty amidst irreversible path dependencies once we act, having to spend money now in order to make more money later). Money would here also get connected to finance in a positive feedback loop, capturing money creation in connection with debt monetization. Such connection moves our notions of finance beyond mere intermediation between savings and investment toward its money-emitting self-expansion, a crucially helpful step when wishing to address the challenges posited by climate finance and the engine fueling it, carbon money. Our "societally embedded economy" view would surely also prove useful when giving the social-economy layer (of cooperatives, mutuals, associations, and commons) its proper space as a dynamic public–private synthesis beyond the stale dichotomy of market and state.

None of these improvements in how we think economics would suffice in the face of climate change, if we fail to bring in nature as the ecosystem within which our societies live and our economies operate. Thus, we can no longer treat nature as if it were just "natural capital" which we use for gainful purposes or whose degradation we accept as a necessary price to pay for the generation of other types of capital we value more highly because of their immediate contributions to our living standards (see our discussion of "weak" sustainability and Hartwick's Rule in Chapter 4). Nor can we reduce nature to a range of "environmental services," such as getting honey from bees, or letting tourists marvel at coral reefs. Nature is ultimately far more than either, denoting the material and physical world we live in whose sustainability we need to assure if we want to survive as a thriving species. The threat of climate-change forces us to revisit this issue and commit ourselves precisely to that deeper notion of nature for the sake of saving the planet and humankind. Here, we must point to an important distinction dealing with nature. Environmental economics analyzes damage to our environment as an externality and then assesses the costs and benefits of different environmental policies aimed at global warming, air pollution, water quality, solid waste, or toxic substances. By contrast, ecological economics treats the economy as part of our ecosystem in which it is embedded. That notion implies the preservation of nature and, thereby adopting a "strong" notion of sustainability, rejects the mainstream idea that natural capital can be traded off for other types of (human or physical) capital. Our notion of eco-capitalism is centrally rooted in that notion. The key question here is whether we can imagine an ecological-economic vision of (eco-)capitalism that does not require slow or no growth to assure the sustainability of nature, the famous "steady-state" vision of the anti-growth ecological economists discussed in Chapter 4. My view is that we can have "green" growth, but only if we put the essential features of capitalism into a context of sustainable development as the central driving force of such a properly reformed and restructured economy.

SUSTAINABLE DEVELOPMENT GOALS

The world community has grappled with the notion of sustainable development for a quarter of century as a way of setting globally attainable standards to make the planet a better, safer, more just, and cleaner place to live. In 1983, the United Nations set up the World Commission on Environment and Development, also known as the Brundtland Commission, which in seeking to define an agenda for change introduced the notion of sustainable development.[17] This effort led in 1992 to the first United Nations Conference on Environment and Development (UNCED), also referred to as the Earth Summit, in Rio de Janeiro which passed the so-called Agenda 21. This entirely voluntary action plan, for which 178 governments signed up, offered suggestions for sustainable solutions local, state, or national governments could pursue to preserve natural resources, combat poverty, and reduce pollution. Notwithstanding decade-long right-wing agitation against this action plan as imposing a "one-world order" out to destroy America's national sovereignty and its citizens' freedoms, Agenda 21 engendered a worldwide consensus for a set of attainable goals while also explicitly encouraging a large number of stakeholders to get involved in their implementation. At the Millennium Summit of the United Nations in 2000, all 189 members and twenty-two international development organizations signed up to meet eight measurable international development goals by 2015 which included providing universal primary education, halting the spread of AIDS/HIV, reducing child mortality, promoting gender equality, and halving extreme poverty. As we approached that deadline, and in the context of a follow-up conference on sustainable development in 2012 generally referred to as Rio+20, a wide-ranging debate ensued on defining new goals for sustainable development for which the UN set up a 30-member Open Working Group. Two years later, in December 2014, the UN adopted the OWG's proposed seventeen Sustainable Development Goals and also specified 169 targets in pursuit of those goals for its post-2015 agenda. These SDGs and their targets were to be put into effect by 2030.[18]

In contrast to the earlier eight Millennium Development Goals (2000–2015), which defined a single catch-all goal to "ensure environmental sustainability" that in the end focused mostly on giving a lot more poor people in the developing world access to safe drinking water and sanitation services, the seventeen Sustainable Development Goals (2015–2030) laid out a number of more concrete environmental-protection objectives as separate goals. For one, Goal #13 specifies to "[t]ake urgent action to combat climate change and its impacts" which the UN has appropriately tied to full implementation of the Paris Agreement of December 2015. Pertaining to "Life Below Water," Goal #14 asks us to "[c]onserve and sustainably use the oceans, seas, and marine resources" and so address the many threats to this vital resource covering nearly three quarters of our planet's surface. Several of the targets set for this goal are directly tied to fighting the effects of climate change, with an international Ocean Conference in June 2017 initiating a variety of policies such as a global treaty combating overfishing, expansion of conservation areas, restoration of coral reefs, and encouraging marine technology transfer. Goal #15 for "Life On Land" urges us to "[s]ustainably manage forests, combat desertification, halt and reverse land degradation, halt biodiversity loss." These targets are all part and parcel of any comprehensive climate-mitigation policy while extending to concrete steps against poaching or introducing alien species or the unregulated use of genetic resources. Another profoundly climate-related SDG, namely Goal #7 for "Affordable and Clean Energy," pushes greater use of renewable energy and energy-efficiency gains, both already at the core of our global climate-mitigation strategy. We can also make a clear connection between climate policy and Goal #11 for "Sustainable Cities and Communities," especially when it comes to that goal's targets pertaining to buildings and public transportation. Goal #9 covering "Industry, Innovation, and

Infrastructure" pushes for large investments in sustainable infrastructure and environmentally sound industrialization as crucial economic development objectives which can also help fight global warming. Finally, we also have to contextualize Goal #12 for "Responsible Consumption and Production" which asks us to "[e]nsure sustainable consumption and production." That goal, which seeks to "do more and better with less" in terms of greater energy efficiency and less wasteful use of key resources (such as water or food), combines a search for ecologically sound alternative modes of production and consumption patterns with a striving for a better quality of life, as crystallized around the question of "green" jobs or improved consumer education on sustainable lifestyles.

This combination of ecological concerns and quality-of-life matters is ultimately at the heart of sustainability. You cannot have sustainable development if you separate the two. In the end, they both depend on each other, with one depending on the other. You cannot have, for instance, progress in terms of rendering cities more sustainable unless you tackle the issue of slums and squatters at the heart of excessive urbanization in developing countries. You cannot hope to make consumers less wasteful, unless women have more power to make budgetary decisions, run their homes, own property, and get a better education. The emancipation of women is quintessential not only for reducing inequality and combating poverty, but also to slow down population growth which in turn will have a decisive impact on how effectively we cope with climate change over the next three decades or so. The question of women's social standing, still a major issue in many parts of the world in terms of outright discrimination and widely tolerated violence, has been framed as Goal #6 "Gender Equality." While deserving to be a goal in and of itself, it arguably pervades also other SDGs focusing on health, education, food security, poverty, as well as political representation.[19] More basically, ending poverty (Goal #1) and climate action (Goal #13) go hand in hand. Most of the very poor, those making less than $2 per day, live in either Southern Asia or sub-Saharan Africa, both regions that are disproportionately exposed to the negative effects of global warming. Needed resource transfers into those regions should address both of these issues at the same time so that people can also adopt more environmentally sound behavioral norms and practices as prevention is more cost-effective in the end than adapting to environmental damage already done. Countries with many poor people are less stable and more prone to conflict, which makes it more difficult to carry out good climate policies. Poverty eradication and climate action, while interdependent SDGs, also relate each directly to making economic growth more sustainable and providing for large numbers of quality jobs (Goal #8) through adequate levels of productive investments, including in a modern twenty-first-century "smart" infrastructure and in the skill formation of the workforce.

In sum, whether or not we succeed in time to build a low-carbon economy will arguably depend on setting this needed transition within a broader societal context of sustainable development. The 2030 Agenda's seven primarily ecological SDGs—climate action, life below water, life on land, clean water and sanitation, affordable and clean energy, industry, innovation and infrastructure, as well as sustainable cities and communities—cannot be separated from its eight primarily socioeconomic SDGs—no poverty, zero hunger, good health and well-being for people, quality education, gender equality, decent work and economic growth, reduced inequalities, as well as responsible consumption and production. When looking in more detail at the targets specified, we can see that for some of these goals, such as the one pertaining to rendering consumption and production more responsible, the connection between ecological emphasis on sustainability and socioeconomic framing of social welfare gains is direct and explicit. Other goals, such as "ending poverty" or "gender equality," make this same connection more indirectly. Be that as it may, our vision of eco-capitalism, an ecologically oriented type of capitalism built from the bottom-up in the wake of a globally mobilized transition to a low-carbon economy, can take the Sustainable Development Goals of the UN's 2030 Agenda as its central pillars. While most of these SDGs are especially geared toward improving the lot of the developing world, they apply also with still-undiminished relevance to the developed countries.

One interesting aspect of the 2030 Agenda is its emphasis on the political dimension of sustainability, as encapsulated in the last two SDGs calling for "Peace, Justice, and Strong Institutions" (Goal #16) as well as "Partnerships for the Goals" on a global scale between governments at all levels, the private sector and civil society (Goal #17). There is thus recognition, as there was also in the Paris Climate Agreement laying out the goals for climate mitigation and adaptation, that these societally transformative goals require mass mobilization and cooperation among all the major stakeholders affected. They need to work together so as to define the goals in detail, set the targets for meeting these goals, define the empirical indicators measuring progress in pursuit of these targets, build consensus around appropriate policy options likely to help us make the most progress, and undertake needed initiatives to put these policies into effect. This requires a very different type of politics than the currently dominant one rooted in naked exploitation of power by a few and growing polarization of the many. The principles and values needed for a politics of sustainable development should be grounded in the deeper notions of justice, individual capabilities, and societal well-being.[20]

MANAGING A MULTIPOLAR WORLD

We are at this point quite far from such a politics of sustainable development, having matured into a dominant force. While it must count undeniably as progress to have put into place the global Paris Climate Agreement and have the United Nations actively pursue those seventeen Sustainable Development Goals with concrete and ambitious follow-up measures for each, national political debate may well be going in a different direction as country after country finds itself in the grip of right-wing nationalism feeding on widespread anger with stagnant wages, widening income inequality, job insecurity, self-serving elites in power looking out solely for their own advantages, globalization, and immigrants. It is not at all clear whether and, if so, when this trend gets reversed. This will necessitate above all the emergence of an alternative political movement channeling that widespread desire for change into the direction of the SDG pillars of eco-capitalism, based on a regrouping of reformist forces capable of fusing a captivating vision with concrete policy proposals which can muster a majority. This kind of political realignment may come about to the extent that the Left manages to remake itself from a force wedded to a past long gone into a progressive and dynamic movement for sustainable development which at the same time has enough pragmatic sense to propose sensible reforms that have a chance to pass. In my opinion, this movement has to stay away from the demagoguery of nationalism and instead endorse globalization provided it is re-embedded in a progressive policy framework of the kind put forth in the 2030 Agenda and its SDGs.[21] The key here is to balance the national interest with shared sovereignty in the context of global governance through supranational cooperation among many countries, as a growing number of national-interest issues can no longer be isolated or insulated from the actions of other countries. It took a long time, but eventually climate change emerged precisely as such an issue of global governance because it was so obviously a planetary problem. Trump's go-it-alone reaction will in the end only make the imperative of supranational cooperation clearer and stronger. Not only I am pretty sure of that, but I think we will go through similar experiences on many other issues we must face together as a global community of interdependent nation-states—nuclear proliferation, arms race, food security, migration, pandemics, cybersecurity, labor-saving automation, harmonization of regulations pertaining to key sectors, competition policy, education, etc.

We have already begun in recent years to build new institutions and mechanisms of global governance addressing many of these issues— the upgrading of the G-20 to regular meetings of heads of state, the use of the Bank for International Settlements and Financial Stability Board for a global financial reregulation effort, the Comprehensive Plan of Action on Iran's nuclear program put into the place by the world's major powers as a new monitoring and verification model under the auspices of the International Atomic Energy Agency (in this case aimed at Iran), to name a few. A brief look at the United Nation's Web site for its Sustainable Development Goals (www.un.org/sustainabledevelopment/) will impress readers with the range of new institution-building and norm-forming initiatives in pursuit of each of the seventeen SDGs. A key initiative in that regard, to which we have dedicated much discussion in this book, is the Paris Climate Agreement of December 2015 to provide the world with a road map to a low-carbon economy.

As we recognize more and more policy issues in need of globally coordinated approaches, we face at the same time the troubling prospect of a much more complex international constellation of multiple power centers. We are most likely near the end of the century-long Pax Americana. Even without America's self-destructive and ill-conceived retreat from so many multilateral frameworks and institutions under Trump, that trend of America's erosion of superpower dominance was already well underway from the early 2000s onward when the USA got bogged down in two endless wars (Afghanistan and Iraq) while other powers, notably Germany and China, managed to fortify their relative position of strength. We are thus clearly moving toward a sort of tri-polar configuration, centered on the USA, European Union, and China, each of which with its own zone of influence. Moreover, there are also regional power centers emerging. Brazil's role in Latin America, South Africa's position in sub-Saharan Africa, India's domination of South

Asia, Russia's reclaiming of the "near abroad" in an attempt to restore its central role among the neighbors with which it once shared the boundaries of the Soviet Union, or the Middle East as a battleground between Saudi Arabia and Iran are all expressions of that trend.

A multipolar world is a lot more complicated to manage than a unipolar world where the dominant superpower sets the standards and agenda, as was the case in the 1990s after the demise of the Soviet Empire with the USA. Today, the different power centers need to cooperate beyond their narrowly defined interests in order to build global-governance institutions of shared sovereignty, as Obama did with Xi Jinping in the run-up to the Paris COP21 climate negotiations to avoid a repeat of the 2009 debacle in Copenhagen. But in a tri-polar system, there is always the danger that two gang up on one, and we can expect the EU and China to find more common ground especially if America's voluntary abdication of its global leadership role under Trump continues unabated. Marginalized within the US-controlled Bretton Woods institutions (i.e., IMF, World Bank, World Trade Organization), China has in recent years started to build its own network of multilateral organizations such as the Bank of BRICS, the Asia Infrastructure Investment Bank, and the One Belt and One Road Initiative. It stands to reason that these efforts will draw in the European Union, itself a crucially important experiment in supranational governance, as a result of which the USA will need to come to the table with a reformist agenda for remaking global-governance institutions lest it wants to end up even more isolated than it has become amidst Trump's first-year flurry of disengagements.[22]

The area where the evolving multipolarity around three power centers will shape how we deal with the challenges of global governance most decisively is the international monetary system. The combination of Trump's systematic dismantlement of America's "soft power" channels (such as diplomacy, foreign aid and assistance, cultural exchange programs, and strong relationships with allies) and its dysfunctional domestic-policy paralysis may well accelerate the gradual erosion of the US dollar's world-money status. The euro-zone, after having its own systemic crisis (2009–2012) reveals major institutional and policy shortcomings, is now actively seeking to strengthen its cohesion in the wake of many policy reforms, including a fiscal compact, a banking union, a capital-market union, the European Stability Mechanism as a system-wide lender of the last resort, four new EU-wide agencies (e.g., European Banking Authority) grouped together in the European System of Financial Supervision, and important operational changes in the modus operandi of the European Central Bank. At the same time, China has undertaken many important steps over the last decade to internationalize its currency, the Renminbi (or Yuan), a complex process that has included making its exchange-rate regime more flexible and slowly opening up its capital account. We may therefore move toward a distinctly tri-polar configuration over the coming decade, with the USD, the EUR, and the CNY vying for global world-money status as each power center shores up its regional zone of influence. Such a system runs the risk of enormous instability, given the huge volume of short-term cross-border capital flows ("hot money"). It thus needs a modicum of coordination. There are different ways to do that. One would be to keep exchange rates between these three key currencies stabilized within ±10% bands from set levels as we tried for a couple of years among the G-7 countries (USA, Japan, Germany, Britain, France, Italy, and Canada) in the Louvre Agreement of 1987. But unlike that experiment three decades ago, such exchange-rate stabilization would have to go beyond coordinated interest-rate changes and central bank interventions in the currency markets toward a deeper level of cooperation in terms of counterbalancing policy-mix adjustments.[23]

The kind of cooperation and coordination needed as a prerequisite for stabilizing a tri-polar system may very well come about more easily by centering a reform of the international monetary system on the IMF-issued Special Drawing Rights. These SDRs, which are allocated to each IMF member proportionately according to a size-based formula for settlement of official payment obligations or to count as official reserves, are in effect a basket of the key currencies. A SDR comprises US dollars (41.73%), euros (30.93%), renminbi (10.92%), yen (8.33%), and pound (8.09%), and both its exchange rate and interest rate are weighted averages of these five currencies making up its basket. This in itself provides for greater stability, as the fluctuations of each basket currency against that average are smaller than the fluctuations against each other. So, once we have a tri-polar system, we might as well start denominating the entire international monetary system in SDR for stabilization purposes and then build our proposed target-zone system for exchange rates (see note 23) around that SDR anchor. But we can go further than that. Right now, the SDR is not full-fledged money, not least because of severe restrictions on their issue subject to the (typically assured) veto of the US Congress which has up to now severely limited the issue of new SDRs lest they are thus allowed to become an alternative to the US dollar. What if we make the issue of new SDRs no longer dependent on such parliamentary approval by the legislatures of all five basket-component issuers and instead allow them to be created in elastic fashion by the IMF in conjunction with the five underlying central banks involved? Then, the SDRs could gradually start replacing the international circulation of those five key currencies and so emerge as a new form of supranational currency of the kind envisaged by John Maynard Keynes with his Bancor Plan.[24]

It may very well be possible at one point in the unfolding of this international monetary reform to include the aforementioned carbon certificates issued by the International Mitigation Projects Initiator (see Chapter 7) in the SDR basket and so anchor such global carbon money in fully convertible fashion and as part of the target-zone regime for exchange rates. The global carbon price standard would thus become part of the exchange-rate regime through which we hook supranational or domestic climate-mitigation and adaption investment efforts financed by issue of CCs to the rest of the world economy. We need to be bold if we want to achieve zero net carbon neutrality worldwide by 2060, and that means fairly radical reforms on a sustained basis all the way to fully fledged carbon money. We might as well make our capitalist system better, cleaner, more equitable, and more efficient along the way. If there is a will, there will be a way!

NOTES

1. I have highlighted the crucial role of such structural crises in the transition from one accumulation regime to another in R. Guttmann (2015). Those crises start as systemic financial crises and then turn into deeper structural crises before getting resolved with appropriate policy reforms and adjustment mechanisms.

2. The explosive growth in industrial concentration across many sectors has been recently identified as a problematic trend with many unintended consequences by W. Galston and C. Hendrickson (2018) as well as by G. Eggertsson et al. (2018). Both R. Foroohar (2018a) and E. Porter (2018) conclude that this jump in concentration has enabled certain strategic players in the economy to extract massive monopoly rents to a point where this asymmetric power structure implied here has sapped the US economy of its dynamism.

3. There is growing awareness among central bankers, regulators, and academics that we are indeed approaching such a decisive moment of climate "panic" and that this may be the only way to intensify our climate mitigation efforts. See in this context S. Matthews (2017), the report by the European Systemic Risk Board (2016) mentioned earlier, or the work by the Dutch central bank as summarized in the De Nederlandsche Bank (2017).

4. Means of extortion abound. Corruption is rampant in many countries, allowing big favors to be bought for fairly small amounts of compensation. Companies also routinely ask for, and obtain, tax breaks or subsidies or regulatory relief from governments eager for their investments. Investor-state dispute settlement provisions found in many trade agreements make it possible for corporations to sue governments for lots of money in front of special administrative courts that tend to be biased in favor of the private plaintiffs.

5. The Basel-based Bank for International Settlements is in effect a central bank of the world's leading central banks which, among other functions, acts a global rule-maker for commercial banks. Its Basel Committee of Banking Supervision has launched agreements among the BIS' sixty members in 1988, 1996, 2004, and 2010 (with another one envisaged for 2019) for global standards of bank capitalization (as a function of risk-weighted assets), liquidity cushions, leverage caps, risk management, disclosure requirements, and supervisory practices. For more on this body of agreements see R. Guttmann (2016, Chapter 7) or go to the excellent bis.org Web site.

6. Indicative planning, not to be confused with central planning practiced in communist economies of yesteryear, guided the postwar acceleration of industrialization in such fast-growing economies, like France or Japan. Both France's Commissariat du Plan and Japan's Ministry for International Trade and Industry (MITI) successfully promoted domestic capacity building in targeted sectors during the postwar boom to move their domestic economies up the value-added ladder and gain strategic competitive footholds in key industries (e.g., nuclear power, cars, semiconductors).

7. We are beginning to see an important change in this direction. Besides nonprofit corporations pursuing a public good, you have also so-called public-benefit corporations which combine a public-good objective defined in their charter with making a profit. This is a new type of company whose legal status has been introduced in a significant number of countries over the last couple of years (e.g., Australia, France, Israel, Italy, a growing number of states in the USA). For more details on this type of company see T. Howes (2015) as well as M. Geffner (2016).

8. We are now beginning to see suggestions for an alternative system of social-value accounting capturing a wider notion of value which incorporates benefits accruing to affected stakeholders such as greater equality, improvement in human well-being, and strengthened sustainability. For more details see the interesting Web site socialvalueuk.org or L. Mook (2013). This effort needs to be embedded in a broader effort going beyond traditional indicators of economic progress (e.g., GDP) to put forth alternative measures of societal well-being, (in)equality, and quality of life while also taking greater account of the environment and of sustainable development, as so meaningfully suggested by J. Stiglitz et al. (2009).

9. For more on this emerging sphere generally referred to as "social and solidarity economy," a phenomenon taking root in many countries, see International Labour Organization (2011) and P. Utting (2013).

10. The debate about the Universal Basic Income idea has heated up recently not least thanks to a major experiment in Finland which consists of paying 2000 people 560€ per month over a period of two years. For different assessments of Finland's UBI experiment see J. Henley (2018) and M. Sandbu (2018a). R. Foroohar (2018b) presents the idea of a "Digital New Deal" in the wake of a new wave of automation driven by artificial intelligence.

11. It stands to reason that such revitalization of the role of the public sector flies in the face of current political trends whose thrust is to distrust what the state can and should do. An interesting discussion of the need to revive politics in the direction of a more productive and enriching private–public balance can be found in E. Glaser (2018).

12. When I was a graduate student during the 1970s, I was quite impressed with the idea of indicative planning (see, for instance, Lewis 1966 or Meade 1970), with which to shape the national growth dynamic in terms of intersectoral interactions, and the tools used to work out these dynamics, notably the Leontief Input–Output Model and linear programming (see Darst 1991). Both of these techniques are eminently useful to study the implications of different transition paths to a low-carbon economy, as illustrated by J. O'Doherty and R. Tol (2007).

13. Macro-prudential regulation, a new regulatory strategy globally implemented by a G-20 agreement in 2009, involves classifying strategic financial institutions as "systemically important," subjecting those SIFIs to greater oversight and added safety standards (such as larger capital requirements), imposing regular stress tests on those institutions, and deploying a potentially powerful array of selective credit controls to contain asset bubbles or other kinds of excesses in the financial system before they get out of hand. Resolution authority, another aspect agreed to by

the leading nations in 2009, involves a fund financed by special taxes on banks and used to restructure failing or insolvent banks, sharing of losses among creditors ("bail ins"), and special debt instruments that turn into equity under predetermined conditions for recapitalization of banks threatened with insolvency (so-called contingent convertibles or CoCos).

14. In recent years, the emerging movement for greater corporate social responsibility has also pushed for better accounting of social benefits generated by firms which has given rise to first efforts at formulating so-called Social and Environmental Accounting (SEA) standards. For more on this see, for instance, C. Cooper et al. (2005) or R. Gray et al. (2014).

15. I have much appreciated the argument by Q. Jackson (2017) in favor of long-term corporate planning in the interest of sustainability equaling business survivability.

16. Classical Political Economy was primarily concerned with the Labor Theory of Value and the link between growth and distribution. The so-called Marginalist Revolution, brought forth simultaneously by Carl Menger, William Stanley Jevons, and Leon Walras in the early 1870s, established individual optimization amidst resource scarcity and the price mechanism of the market as central concerns.

17. The Brundtland Commission (1987) famously defined sustainable development as "meeting the needs of the present without compromising the ability of future generations to meet their own needs."

18. See United Nations (2017) for a more detailed discussion of these so-called Sustainable Development Goals (SDGs) which are also referred to as the "2030 Agenda for Sustainable Development."

19. Nobel Prize winner Amartya Sen (1990) has systematically stressed the role of gender inequality as a determinant factor in explaining a country's poverty and, by extension, the empowerment of women as the best way to improve living standards. For more on this question along Sen's line of argumentation, see the important article by E. Duflo (2012).

20. See in this context our references in note 25 of Chapter 4 to the notions of justice evoked by John Rawls (1971) and Michael Walzer (1983) in connection with the concept of sustainability. In a pathbreaking work transforming traditional welfare theory and making an important contribution to our understanding of human development, A. Sen (1985) put forth the so-called Capability Approach which focuses on individuals' freedoms to achieve well-being.

21. D. Rodrik (2017, 2018) has addressed precisely this issue of populism arising in response to globalization and how best to overcome this dichotomy with a progressive policy agenda. See also B. DeLong (2017) and M. Sandbu (2018b) on the same issue.

22. R. Haass (2017), who heads the Council of Foreign Relations, insists that Trump's various steps of disengagement under the banner of "America First!" do not so much amount to isolationism as representing a voluntary abdication of a superpower's global responsibilities, bound to create a dangerous leadership vacuum in a period of intensifying turmoil.

23. Such counterbalancing policy-mix adjustments must involve simultaneous fiscal and monetary policy initiatives among both surplus and deficit countries to reduce their respective sectoral imbalances [in the macroeconomic equation $X_n = (S - I) + (T - G)$] for symmetric adjustments designed to keep their respective current-account (i.e., X_n) disequilibria in check. This is the deeper implication of the kind of target-zone regime for exchange rates first proposed by J. Williamson (1986) which the absence thereof explains why the Louvre Agreement of 1987 only lasted a couple of years.

24. The three versions of Keynes' 1941 plan for a supranational currency ("Bancor") issued by an International Currency Union under rules of symmetric adjustments, automatic recycling of surpluses to deficit countries, and fixed, but adjustable exchange rates, which he also proposed at the Bretton Woods Conference of 1944, are reprinted in J. M. Keynes (1980).

REFERENCES

Brundtland Commission. (1987). *Report of the World Commission on Environment and Development: Our Common Future*. New York: United Nations.

Cooper, C., et al. (2005). A Discussion of the Political Potential of Social Accounting. *Critical Perspectives on Accounting, 16*(4), 951–974. https://doi.org/10.1016/j.cpa.2003.09.003.

Darst, R. (1991). *Introduction to Linear Programming: Applications and Extensions*. New York: Marcel Dekker.

DeLong, B. (2017). When Globalization Is Public Enemy Number One. *Milken Institute Review*. http://www.milkenreview.org/articles/when-globalization-is-public-enemy-number-one. Accessed 28 Feb 2018.

De Nederlandsche Bank. (2017, October 5). Increasing Climate-Related Risks Demand More Action from the Financial Sector. *DNBulletin*. https://www.dnb.nl/en/news/news-and-archive/dnbulletin-2017/dnb363837.jsp. Accessed 4 Feb 2018.

Duflo, E. (2012). Women Empowerment and Economic Development. *Journal of Economic Literature, 50*(4), 1051–1079. http://dx.doi.org/10.1257/jel.50.4.1051.

Eggertsson, G., Robbins, J., & Genz Wold, E. (2018). *Kaldor and Piketty's Facts: The Rise of Monopoly Power in the United States* (Working Paper, 2018-02). Washington Center for Equitable Growth. http://equitablegrowth.org/working-papers/kaldor-piketty-monopoly-power/. Accessed 16 Feb 2018.

European Systemic Risk Board. (2016, February). *Too Late, Too Sudden: Transition to a Low-Carbon Economy and Systemic Risk* (Reports of the Advisory Scientific Committee, No. 6). https://www.esrb.europa.eu/pub/pdf/asc/Reports_ASC_6_1602.pdf. Accessed 17 Dec 2017.

Foroohar, R. (2018a, January 14). The Rise of the Superstar Company. *Financial Times.* https://www.ft.com/content/95d16c88-f795-11e7-88f7-5465a6ce1a00. Accessed 16 Feb 2018.

Foroohar, R. (2018b, February 18). Why Workers Need a "Digital New Deal" to Protect Against AI. *Financial Times.* https://www.ft.com/content/0d-263edc-1323-11e8-8cb6-b9ccc4c4dbbb. Accessed 19 Feb 2018.

Galston, W., & Hendrickson, C. (2018, January 5). A Policy at Peace with Itself: Antitrust Remedies for Our Concentrated, Uncompetitive Economy. *Brookings Blog.* https://www.brookings.edu/blog/fixgov/2018/01/05/the-consequences-of-increasing-concentration-and-decreasing-competition-and-how-to-remedy-them/. Accessed 16 Feb 2018.

Geffner, M. (2016, June 10). Could the Public Benefit Company Structure Be Right for Your Business? *dun&bradstreetB2B.* https://b2b.dnb.com/2016/06/10/public-benefit-company-structure-right-business/. Accessed 19 Sept 2017.

Glaser, E. (2018). *Anti-politics: On the Demonization of Ideology, Authority, and the State.* Marquette, MI: Repeater Press.

Gray, R., Brennan, A., & Malpass, J. (2014). New Accounts: Towards a Reframing of Social Accounting. *Accounting Forum, 38*(4), 258–273. https://doi.org/10.1016/j.accfor.2013.10.005.

Guttmann, R. (2015). The Heterodox Notion of Structural Crisis. *Review of Keynesian Economics, 3*(2), 194–212.

Guttmann, R. (2016). *Finance-Led Capitalism: Shadow Banking, Re-regulation, and the Future of Global Markets.* New York: Palgrave Macmillan.

Haass, R. (2017, December 28). America and the Great Abdication. *The Atlantic.* https://www.theatlantic.com/international/archive/2017/12/america-abidcation-trump-foreign-policy/549296/.

Henley, J. (2018, January 12). Money for Nothing: Is Finland's Universal Basic Income Trial Too Good to Be True? *The Guardian.* https://www.theguardian.com/inequality/2018/jan/12/money-for-nothing-is-finlands-universal-basic-income-trial-too-good-to-be-true. Accessed 20 Feb 2018.

Howes, T. (2015, December 11). With a Public Benefit Corporation, Profit and Good Karma Can Coexist. *Entrepreneur.* https://www.entrepreneur.com/article/253059. Accessed 19 Sept 2017.

International Labour Organization. (2011). *Social and Solidarity Economy: Our Common Road Toward Decent Work*. http://ilo.org/empent/units/cooperatives/WCMS_166301/lang–en/index.htm.

Jackson, Q. (2017, August 8). Short-Term Thinking Is Killing Us in the Long Run. *Eco-Business*. http://www.eco-business.com/opinion/short-term-thinking-is-killing-us-in-the-long-run/. Accessed 24 Nov 2017.

Keynes, J. M. (1980). *The Collected Writings of John Maynard Keynes* (Vol. 25, pp. 42–139), ed. D. Moggridge. Cambridge: Cambridge University Press.

Lewis, A. (1966). *Development Planning: The Essentials of Economic Policy*. London: George Allen & Unwin.

Matthews, S. (2017, July). Alarmism Is the Argument We Need to Fight Climate Change. *Slate*. http://www.slate.com/articles/health_and_science/science/2017/07/we_are_not_alarmed_enough_about_climate_change.html. Accessed 4 Feb 2018.

Meade, J. (1970). *The Theory of Indicative Planning*. Manchester, UK: Manchester University Press.

Mook, L. (2013). *Accounting for Social Value*. Toronto: University of Toronto Press.

O'Doherty, J., & Tol, R. (2007). *An Environmental Input-Output Model for Ireland* (ESRI Working Paper, No. 178). http://dx.doi.org/10.2139/ssrn.964473.

Porter, E. (2018, February 13). Big Profits Drove a Stock Boom. Did the Economy Pay a Price? *New York Times*. https://www.nytimes.com/2018/02/13/business/economy/profits-economy.html. Accessed 13 Feb 2018.

Rawls, H. (1971). *A Theory of Justice*. Cambridge, MA: Harvard University Press.

Rodrik, D. (2017). *Populism and the Economics of Globalization* (NBER Working Paper, No. 23559). http://www.nber.org/papers/w23559. Accessed 28 Feb 2018.

Rodrik, D. (2018, February 21). What Does a True Populism Look Like? It Looks Like the New Deal. *New York Times*. https://www.nytimes.com/2018/02/21/opinion/populism-new-deal.html?partner=rss&emc=rss. Accessed 28 Feb 2018.

Sandbu, M. (2018a, February 28). Welfare Lessons from Finland. *Financial Times*.

Sandbu, M. (2018b, February 28). Are There Good Types of Populism? *Financial Times*.

Sen, A. (1985). *Commodities and Capabilities*. Amsterdam: North-Holland.

Sen, A. (1990). More Than 100 Million Women Are Missing. *New York Review of Books, 37*(20). http://www.nybooks.com/articles/1990/12/20/more-than-100-million-women-are-missing/?printpage=true.

Stiglitz, J., Sen, A., & Fitoussi, J. (2009). *Report by the Commission on the Measurement of Economic Performance and Social Progress*. http://ec.europa.eu/eurostat/documents/118025/118123/Fitoussi+Commission+report.

United Nations. (2017). *Sustainable Development Goals: 17 Goals to Transform Our World*. New York: United Nations Department of Public Information. http://www.un.org/sustainabledevelopment/sustainable-development-goals/.

Utting, P. (2013, April 30). What Is Social and Solidarity Economy and Why Does It Matter? *People, Spaces, Deliberation Blog*. World Bank. http://blogs.worldbank.org/publicsphere/what-social-and-solidarity-economy-and-why-does-it-matter.

Walzer, M. (1983). *Spheres of Justice*. New York: Basic Books.

Williamson, J. (1986). Target Zones and the Management of the Dollar. *Brookings Papers on Economic Activity*, *1986*(1), 165–174. https://www.brookings.edu/wp-content/uploads/1986/01/1986a_bpea_williamson.pdf.

BIBLIOGRAPHY

Aglietta, M. (1979). *A Theory of Capitalist Regulation: The US Experience.* London: Verso. First Published as *Régulation et crises du capitalisme* (Paris: Calmann-Levy), 1976.

Aglietta, M. (1998). La globalisation financière. In CEPII, *L'économie mondiale 2000* (pp. 52–67). Paris: La Decouvérte.

Aglietta, M. (2000). Shareholder Value and Corporate Governance: Some Tricky Questions. *Economy and Society, 29*(1), 146–159. http://dx.doi.org/10.1080/030851400360596.

Aglietta, M. (2015a, January). *The Quality of Growth: Accounting for Sustainability* (AFD Research Papers, Agence Française de Developpement, No. 2015-01). https://www.afd.fr/sites/afd/files/2017-09/01-papiers-recherche-Quality-of-Growth_Aglietta.pdf.

Aglietta, M. (2015b). What Is the Quality of Growth? Sustainability and Inclusiveness. In L. Haddad, H. Hiroshi, & N. Meisel (Eds.), *Growth Is Dead, Long Live Growth: The Quality of Growth and Why It Matters* (pp. 19–56). Tokyo: JICA Research Institute. https://www.jica.go.jp/jica-ri/publication/booksandreports/jrft3q0000002a2s-att/Growth_is_Dead_Long_Live_Growth.pdf.

Aglietta, M., & Espagne, E. (2016, April). *Climate and Finance Systemic Risks, More Than an Analogy? The Climate Fragility Hypothesis* (CEPII Working Paper, No. 2016–10). http://cerdi.org/uploads/sfCmsNews/html/2958/201705_Espagne_Climate.pdf.

Aglietta, M., Espagne, E., & Perrissin Fabert, B. (2015, February). *A Proposal to Finance Low Carbon Investment in Europe.* La Note d'Analyse, No. 24, France Strategie. http://www.strategie.gouv.fr/sites/strategie.gouv.fr/files/atoms/files/bat_notes_danalyse_n24_-_anglais_le_12_mars_17_h_45.pdf.

© The Editor(s) (if applicable) and the Author(s) 2018
R. Guttmann, *Eco-Capitalism*,
https://doi.org/10.1007/978-3-319-92357-4

Allen, K. (2017, May 25). Sellers of Green Bond Face a Buyer's Test of Their Credentials. *Financial Times*.

Arrhenius, S. (1896). On the Influence of Carbonic Acid in the Air Upon the Temperature on the Ground. *Philosophical Magazine and Journal of Science, 41*(5), 237–276.

Aton, A. (2017, October 3). Most Americans Want Climate Change Policies. *Scientific American*. https://www.scientificamerican.com/article/most-americans-want-climate-change-policies/.

Authers, J. (2015a, October 23). Vote of No Confidence in Shareholder Capitalism. *Financial Times*.

Authers, J. (2015b, December 15). Climate Talks Mark Turning Point for Investors. *Financial Times*.

Baker, K., & Ricciardi, V. (2014, February–March). How Biases Affect Investor Behavior. *European Financial Review*. www.europeanfinancialreview.com/?p=512.

Baker, K., Filbeck, G., & Ricciardi, V. (2016, December–January). How Behavioral Biases Affect Finance Professionals. *European Financial Review*. http://www.europeanfinancialreview.com/?p=12492.

Bhandary, R. (2017, August 25). Markets and Non-market Approaches for International Cooperation in the Paris Agreement: Open Questions in the International Negotiations. *Climate Policy Lab*. https://www.climatepolicy-lab.org/news/2017/8/25/markets-and-non-market-approaches-for-interna-tional-cooperation-in-the-paris-agreement-open-questions-in-the-internation-al-negotiations. Accessed 25 Dec 2017.

Battiston, S., et al. (2017). A Climate Stress-Test of the Financial System. *Nature Climate Change, 7*, 283–288. https://www.nature.com/articles/nclimate3255?WT.feed_name=subjects_business.

Bauerle, N. (2017). What Is Blockchain Technology? *CoinDesk Blockchain 101*. https://www.coindesk.com/information/what-is-blockchain-technology/. Accessed 31 Dec 2017.

Benford, J., et al. (2009). Quantitative Easing. Bank of England *Quarterly Bulletin*, 90–100. http://www.bankofengland.co.uk/publications/Documents/quarterlybulletin/qb090201.pdf.

Black, F., Jensen, M., & Sholes, M. (1972). The Capital Asset Pricing Model: Some Empirical Tests. In M. Jensen (Ed.), *Studies in the Theory of Capital Markets* (pp. 79–121). New York: Praeger.

Bloomberg, L. P. (2017, May 17). *2016 Impact Report*. https://data.bloomb-erglp.com/company/sites/28/2017/05/17_0516_Impact-Book_Final.pdf. Accessed 8 July 2017.

Bodansky, D. (2010). *The International Climate Change Regime: The Road from Copenhagen*. Policy Brief. Harvard Project on International Climate Agreements, Belfer Center for Science and International Affairs, Harvard Kennedy School.

Boulding, K. (1966). The Economics of the Coming Spaceship Earth. In H. Jarrett (Ed.), *Environmental Quality in a Growing Economy* (pp. 3–14). Baltimore, MD: Johns Hopkins University Press. http://dieoff.org/page160.htm. Accessed 8 Dec 2016.

Boyer, R. (2000). Is a Finance-Led Growth Regime a Viable Alternative to Fordism? A Preliminary Analysis. *Economy and Society, 29*(1), 111–145. http://dx.doi.org/10.1080/030851400360587.

Boyer, R., & Saillard, Y. (2001). *Regulation Theory: The State of Art*. London: Routledge. First Published as *Théorie de le régulation: l'état de saviors* (Paris: La Decouverte), 1995.

Bozikovic, A. (2017, October 17). Google's Sidewalk Labs Signs Deal for 'Smart City' Makeover of Toronto's Waterfront. *The Globe and Mail*. https://www.wired.com/story/google-sidewalk-labs-toronto-quayside/. Accessed 15 Nov 2017.

Brown, J., & Jacobs, M. (2011). Leveraging Private Investment: The Role of Public Sector Climate Finance. *ODI Background Note*. Overseas Development Institute. https://www.odi.org/sites/odi.org.uk/files/odi-assets/publications-opinion-files/7082.pdf.

Brundtland Commission. (1987). *Report of the World Commission on Environment and Development: Our Common Future*. New York: United Nations.

Cai, Y., Judd, K., & Lontzek, T. (2013). *The Social Cost of Stochastic and Irreversible Climate Change* (NBER Working Papers, No. 18704). National Bureau of Economic Research. http://www.nber.org/papers/w18704.

Caiani, A., Godin, A., Caverzasi, E., Gallegati, M., Kinsella, S., & Stiglitz, J. (2016). Agent Based Stock-Flow Consistent Macroeconomics: Towards a New Benchmark. *Journal of Economic Dynamics and Control, 69*(1), 375–408.

Callaghan, I. (2017, October 28). Blockchain to Unblock Climate Finance? *NDCI Global*. http://ndci.global/blockchain-unblock-climate-finance/. Accessed 31 Dec 2017.

Campiglio, E., Godin, A., & Kemp-Benedikt, E. (2017). *Networks of Stranded Assets: A Case for a Balance Sheet Approach* (AFD Research Papers, No. 2017-54).

Campiglio, E., Godin, A., Kemp-Benedikt, E., & Matikainen, S. (2017). The Tightening Links Between Financial Systems and the Low-Carbon Transition. In M. Sawyer & P. Arestis (Eds.), *Economic Policies Since the Global Financial Crisis* (pp. 313–356). London: Palgrave Macmillan. https://doi.org/10.1007/978-3-319-60459-6_8.

296

Camuzeaux, J., & Medford, E. (2016, December 12). How Companies Set Internal Prices on Carbon. *Climate 411* Blog. Environmental Defense Fund. http://blogs.edf.org/climate411/2016/12/12/how-companies-set-internal-prices-on-carbon/. Accessed 4 Oct 2017.

Canfin-Grandjean Commission. (2015). *Mobilizing Climate Finance: A Roadmap to Finance a Low-Carbon Economy.* http://www.cdcclimat.com/IMG/pdf/exsum-report_canfin-grandjean_eng.pdf.

Cao, J. (2010). *Beyond Copenhagen: Reconciling International Fairness, Economic Development, and Climate Protection.* Discussion Paper 2010-44, Harvard Project on International Climate Agreements, Belfer Center for Science and International Affairs, Harvard Kennedy School.

Carbon Disclosure Project. (2016). *Embedding a Carbon Price into Business Strategy.* https://b8f65cb373b1b7b15feb-c70d8ead6ced550b4d987d7c03f-cdd1d.ssl.cf3.rackcdn.com/cms/reports/documents/000/001/132/original/CDP_Carbon_Price_report_2016.pdf?1474899276. Accessed 25 Nov 2017.

Carbon Disclosure Project. (2017). *Commit to Putting a Price on Carbon.* https://www.cdp.net/en/campaigns/commit-to-action/price-on-carbon. Accessed 4 Oct 2017.

Carbon Market Watch. (2013, March 4). Joint Implementation: CDM's Little Brother Grew Up to Be Big and Nasty. *Newsletter #2.* https://carbonmarketwatch.org/2013/03/04/joint-implementation-cdms-little-brother-grew-up-to-be-big-and-nasty/. Accessed 19 Dec 2017.

Carbon Market Watch. (2015). *Avoiding Hot Air in the 2015 Paris Agreement* (Working Paper, No. 8). https://carbonmarketwatch.org/wp-content/uploads/2015/08/Hot-air-in-the-2015-Paris-agreement_clean_FINAL.pdf. Accessed 13 Dec 2017.

Carbon Tracker. (2011). *Unburnable Carbon—Are the World's Financial Markets Carrying a Carbon Bubble?* http://www.carbontracker.org/wp-content/uploads/2014/09/Unburnable-Carbon-Full-rev2-1.pdf. Accessed 23 Apr 2015.

Carbon Tracker. (2013). *Things to Look Out for When Using Carbon Budgets!* (Working Papers, No. 08-2014). http://www.carbontracker.org/wp-content/uploads/2014/08/Carbon-budget-checklist-FINAL-1.pdf. Accessed 12 Oct 2016.

Carbon Trade Watch. (2009, December). *Fact Sheet 2: Carbon Offsets.* http://www.carbontradewatch.org/downloads/publications/factsheet02-offsets.pdf. Accessed 22 Sept 2017.

Carlson, A. (2016, August 24). Does AB 197 Mean the End of Cap and Trade in California? *LegalPlanet.* http://legal-planet.org/2016/08/24/does-ab-197-mean-the-end-of-cap-and-trade-in-california/. Accessed 17 Aug 2017.

Carney, M. (2015, September 29). *Breaking the Tragedy of the Horizon—Climate Change and Financial Stability.* Speech at Lloyd's of London. http://www.bankofengland.co.uk/publications/Pages/speeches/2015/844.aspx. Accessed 2 Nov 2015.

Caverzasi, E., & Godin, A. (2014). Post-Keynesian Stock-Flow-Consistent Modelling: A Survey. *Cambridge Journal of Economics, 39*(1), 157–187. https://doi.org/10.1093/cje/beu021.

Center for Clean Air Policy. (2016). *CCAP Submission on Internationally Transferred Mitigation Outcomes.* New York: UNFCCC. https://unfccc.int/files/parties_observers/submissions_from_observers/application/pdf/696.pdf. Accessed 23 Sept 2017.

Chen, H. (2015, December 12). Paris Climate Agreement Explained: Climate Finance. *National Resources Defence Council Blog.* https://www.nrdc.org/experts/han-chen/paris-climate-agreement-explained-climate-finance. Accessed 8 Oct 2016.

Cheng, L., Trenberth, K., Palmer, M., et al. (2016). Observed and Simulated Full-Depth Ocean Heat-Content Changes for 1970–2005. *Ocean Science, 12,* 925–935. http://www.ocean-sci.net/12/925/2016/. Accessed 13 Oct 2016.

Clark, D. (2012, November 26). Has the Kyoto Protocol Made Any Difference to Carbon Emissions? *The Guardian.* www.theguardian.com/environment/blog/2012/nov/26/kyoto-protocol-carbon-emissions.

Clark, G. L., Feiner, A., & Viehs, M. (2014). *From the Stockholder to the Stakeholder: How Sustainability Can Drive Financial Outperformance.* Oxford: Smith School of Enterprise and Environment, Oxford University. https://doi.org/10.2139/ssrn.2508281.

Clark, P. (2015a, December 15). Climate Deal: Carbon Dated? *Financial Times.*

Clark, P. (2015b, March 15). Global Carbon Emissions Stall in 2015. *Financial Times.*

Clark, P. (2017, May 18). The Big Green Bang: How Renewable Energy Became Unstoppable. *Financial Times.*

Climate Action Tracker. (2016). *Tracking (I)NDCs: Assessment of Mitigation Contributions to the Paris Agreement.* http://climateactiontracker.org/indcs.html. Accessed 4 Aug 2017.

Clouse, C. (2016, October 27). The Carbon Bubble: Why Investors Can No Longer Ignore Climate Risks. *The Guardian.* https://www.theguardian.com/sustainable-business/2016/oct/27/investment-advice-retirement-portfolio-tips-climate-change-financial-risk.

Coady, D., Parry, I., Sears, L., et al. (2015). *How Large Are Global Energy Subsidies?* (IMF Working Papers, WP/15/105). Washington, DC: International Monetary Fund. http://www.imf.org/external/pubs/ft/wp/2015/wp15105.pdf. Accessed 14 Oct 2016.

Coase, R. (1960). The Problem of Social Cost. *Journal of Law and Economics,* *3*(1), 1–44.

Cooper, C., et al. (2005). A Discussion of the Political Potential of Social Accounting. *Critical Perspectives on Accounting, 16*(4), 951–974. https://doi.org/10.1016/j.cpa.2003.09.003.

COMEST (World Commission on the Ethics of Scientific Knowledge and Technology). (2005). *The Precautionary Principle.* Paris: UNESCO. http://unesdoc.unesco.org/images/0013/001395/139578e.pdf.

Corea, J., & Jaraite, J. (2015). *Carbon Pricing: Transaction Costs of Emissions Trading vs. Carbon Taxes* (CERE Working Papers, 2015: 2). https://gupea.ub.gu.se/handle/2077/38073. Accessed 15 Sept 2017.

Cuff, M. (2017). New Blockchain-Based 'Carbon Currency' Aims to Make Carbon Pricing Mainstream. *BusinessGreen.* https://www.businessgreen.com/bg/news/3017564/new-blockchain-based-carbon-currency-aims-to-make-carbon-pricing-mainstream.

Curtis, M. (2017, October 3). Guest Blog: Are European Taxpayers Funding Land Grabs and Forest Destruction? *FERN Blog.* http://www.fern.org/node/6380. Accessed 4 Oct 2017.

Dafermos, Y., Nikolaidi, M., & Galanis, G. (2017). A Stock-Flow-Fund Ecological Macroeconomic Model. *Ecological Economics, 131*(1), 191–207.

Daly, H. (1980). *Economics, Ecology, Ethics. Essays Towards a Steady-State Economy.* San Francisco, CA: W. H. Freeman.

Daly, H. (2015). *Economics for a Full World.* Great Transition Initiative. http://www.greattransition.org/images/Daly-Economics-for-a-Full-World.pdf. Accessed 9 Dec 2016.

Darst, R. (1991). *Introduction to Linear Programming: Applications and Extensions.* New York: Marcel Dekker.

De Brunhoff, S. (1978). *The State, Capital and Economic Policy.* London: Pluto Press.

DeConto, R., & Pollard, D. (2016, March 31). Contribution of Antarctica to Past and Future Sea-Level Rise. *Nature, 531,* 591–611.

DeLong, B. (2017). When Globalization Is Public Enemy Number One. *Milken Institute Review.* http://www.milkenreview.org/articles/when-globalization-is-public-enemy-number-one. Accessed 28 Feb 2018.

Deutsche Asset Management. (2017). *Sustainable Finance Report, Issue #2.* https://www.db.com/newsroom_news/2017/cr/deutsche-asset-management-report-the-gathering-forces-of-esg-investing-en-11553.htm. Accessed 14 July 2017.

Dietz, S., & Stern, N. (2015). Endogenous Growth, Convexity of Damage, and Climate Risk: How Nordhaus' Framework Supports Deep Cuts in Carbon Emissions. *Economic Journal, 125*(583), 574–620. https://doi.org/10.1111/ecoj.12188.

De Nederlandsche Bank. (2017, October 5). Increasing Climate-Related Risks Demand More Action from the Financial Sector. *DNBulletin*. https://www.dnb.nl/en/news/news-and-archive/dnbulletin-2017/dnb363837.jsp. Accessed 4 Feb 2018.

Döll, S. (2009). *Climate Change Impacts in Computable General Equilibrium Models: An Overview* (HWWI Research Paper, No. 1-26). Hamburg Institute of International Economics. https://www.econstor.eu/bitstream/10419/48201/1/64016790X.pdf.

Douthwaite, R. (1999). *The Ecology of Money*. Cambridge, UK: Green Books. http://www.feasta.org/documents/moneyecology/contents.htm.

Drupp, M., et al. (2015). *Discounting Distangled: An Expert Survey on the Determinants of the Long-Term Social Discount Rate* (Working Paper, No. 195). Leeds University's Centre for Climate Change Economics and Policy. http://piketty.pse.ens.fr/files/DruppFreeman2015.pdf.

Duflo, E. (2012). Women Empowerment and Economic Development. *Journal of Economic Literature, 50*(4), 1051–1079. http://dx.doi.org/10.1257/jel.50.4.1051.

Edmans, A., Fang, V. W., & Lewellen, K. (2017). Equity Vesting and Investment. *Review of Financial Studies, 30*(7), 2229–2271. https://doi.org/10.1093/rfs/hhx018.

Eggertsson, G., Robbins, J., & Genz Wold, E. (2018). *Kaldor and Piketty's Facts: The Rise of Monopoly Power in the United States* (Working Paper, 2018-02). Washington Center for Equitable Growth. http://equitablegrowth.org/working-papers/kaldor-piketty-monopoly-power/. Accessed 16 Feb 2018.

Ehrlich, P., & Harte, J. (2015). Biophysical Limits, Women's Rights and the Climate Encyclical. *Nature Climate Change, 5,* 904–905. https://doi.org/10.1038/nclimate2795.

Eichengreen, B. (2011). *Exorbitant Privilege: The Rise and Fall of the Dollar and the Future of the International Monetary System*. New York: Oxford University Press.

Ellingboe, K., & Koronowski, R. (2016, March 8). Most Americans Disagree with Their Congressional Representative On Climate Change. *ThinkProgress*. https://thinkprogress.org/most-americans-disagree-with-their-congressional-representative-on-climate-change-95dc0eee7b8f#.7swbc9wd3.

Energy Transitions Commission. (2017). *Better Energy, Greater Prosperity*. http://energytransitions.org/sites/default/files/BetterEnergy_Executive%20Summary_DIGITAL.PDF.

Engelman, R. (2009, June 1). Population and Sustainability: Can We Avoid Limiting the Number of People? *Scientific American*. https://www.scientificamerican.com/article/population-and-sustainability/. Accessed 9 Dec 2016.

Enkvist, P.-A., Nauclér, T., & Rosander, J. (2007, February). A Cost Curve for Greenhouse Gas Reduction. *McKinsey Quarterly*. (https://www.mckinsey.com/business-functions/sustainability-and-resource-productivity/our-insights/a-cost-curve-for-greenhouse-gas-reduction. Accessed 11 Apr 2017.

Environmental Protection Agency. (2016). *A Student's Guide to Global Warming.* https://www3.epa.gov/climatechange/kids/index.html. Accessed 24 Sept 2016.

Epstein, G. (2005). *Financialization and the World Economy.* Cheltenham, UK: Edward Elgar.

European Systemic Risk Board. (2016, February). *Too Late, Too Sudden: Transition to a Low-Carbon Economy and Systemic Risk* (Reports of the Advisory Scientific Committee, No. 6). https://www.esrb.europa.eu/pub/pdf/asc/Reports_ASC_6_1602.pdf. Accessed 17 Dec 2017.

EUROSIF. (2016). *European SRI Study 2016* (7th ed.). https://www.eurosif.org/sri-study-2016/. #SRIStudy2016 #ESG. Accessed 14 July 2017.

Evans, S., Pidcock, R., & Yeo, S. (2017, February 14). Q & A: The Social Cost of Carbon. *Explainers* Blog, Carbon Brief. https://www.carbonbrief.org/qa-social-cost-carbon. Accessed 24 Nov 2017.

Evershed, N. (2017, January 19). Carbon Countdown Clock: How Much of the World's Carbon Budget Have We Spent? *The Guardian.* https://www.theguardian.com/environment/datablog/2017/jan/19/carbon-countdown-clock-how-much-of-the-worlds-carbon-budget-have-we-spent. Accessed 31 May 2017.

Fama, E. (1965). The Behavior of Stock Market Prices. *Journal of Business, 38*(1), 35–104. https://doi.org/10.1086/294743.

Fama, E. (1970). Efficient Capital Markets: A Review of Theory and Empirical Work. *Journal of Finance, 25*(2), 383–417. https://doi.org/10.2307/2325486.

Fama, E., & Jensen, M. (1983). Separation of Ownership and Control. *Journal of Law and Economics, 26*(2), 301–325.

Farmer, J. D., Hepburn, C., Mealy, P., & Teytelboym, A. (2015). A Third Wave in the Economics of Climate Change. *Environmental and Resource Economics, 62*(2), 329–357.

Ferron, C., & Morel, R. (2014). *Smart Unconventional Monetary Policies (SUMO): Giving Impetus to Green Investment* (Climate Report, No. 46). http://www.cdcclimat.com/IMG/pdf/14-07_climate_report_no46__smart_unconventional_monetary_policies-2.pdf. Accessed 24 Mar 2016.

Fischer, E. M., & Knutti, R. (2015). Anthropogenic Contribution to Global Occurrence of Heavy-Precipitation or High-Temperature Extremes. *Nature Climate Change, 5*(6), 560–565. http://iacweb.ethz.ch/staff/fischer/download/etc/fischer_knutti_15.pdf.

Floods, C. (2017, May 8). Green Bonds Need Global Standards. *Financial Times.*

Food and Agriculture Organization. (2008). *Climate Change and Food Security: A Framework Document.* Rome: United Nations. http://www.fao.org/forestry/15538-079b31d45081fe9c3dbc6ff34de4807e4.pdf.

Foroohar, R. (2018a, January 14). The Rise of the Superstar Company. *Financial Times.* https://www.ft.com/content/95d16c88-f795-11e7-88f7-5465a6ce1a00. Accessed 16 Feb 2018.

Foroohar, R. (2018b, February 18). Why Workers Need a "Digital New Deal" to Protect Against AI. *Financial Times.* https://www.ft.com/content/0d-263edc-1323-11e8-8cb6-b9ccc4c4dbbb. Accessed 19 Feb 2018.

Fortune. (2017, August 29). *This Hong Kong Tech Startup Wants to Build a Natural Capital Marketplace.* http://fortune.com/2017/08/29/hong-kong-tech-environmental-credit/0.

Forum pour l'Investissement Responsable. (2016). *Article 173-VI: Understanding the French Regulation on Investor Climate Reporting* (FIR Handbook, No. 1). http://www.frenchsif.org/isr-esg/wp-content/uploads/Understanding_article173-French_SIF_Handbook.pdf. Accessed 21 June 2017.

Fouché, G. (2008, April 29). Sweden's Carbon-Tax Solution to Climate Change Puts It Top of the Green List. *The Guardian.* https://www.theguardian.com/environment/2008/apr/29/climatechange.carbonemissions. Accessed 5 Oct 2017.

Fountain, H. (2016, October 6). Over 190 Countries Adopt Plan to Offset Air Travel Emissions. *New York Times.* http://www.nytimes.com/2016/10/07/science/190-countries-adopt-plan-to-offset-jet-emissions.html?_r=0. Accessed 7 Oct 2016.

Friedman, M. (1956). *Studies in the Quantity Theory of Money.* Chicago: University of Chicago Press.

Friedman, M. (1962). *Capitalism and Freedom.* Chicago: University of Chicago Press.

Friedman, M. (1970, September 13). The Social Responsibility of Business Is to Increase Its Profits. *New York Times Magazine.* http://www.colorado.edu/studentgroups/libertarians/issues/friedman-soc-resp-business.html. Accessed 30 June 2017.

Frontier Economics. (2008). *Modeling Climate Change Impacts Using CGE Models: A Literature Review.* http://www.garnautreview.org.au/CA25734E0016A131/WebObj/ModellingClimateChangeImpacts/$File/Modelling%20Climate%20Change%20Impacts%20-%20Frontier%20Economics.pdf.

Frostenson, S. (2017, April 4). We Knew Trump Wanted to Gut the EPA. A Leaked Plan Shows How It Would Be Done. *Vox.* https://www.vox.com/energy-and-environment/2017/4/4/15161156/new-budget-documents-trump-gut-epa. Accessed 11 Nov 2017.

Gallagher, K., et al. (2016), *Fueling Growth and Financing Risk: The Benefits and Risks of China's Development Finance in the Global Energy Sector* (Working Paper 5-16). Boston University Global Economic Governance Initiative.

Galston, W., & Hendrickson, C. (2018, January 5). A Policy at Peace with Itself: Antritrust Remedies for Our Concentrated, Uncompetitive Economy. *Brookings Blog.* https://www.brookings.edu/blog/fixgov/2018/01/05/the-consequences-of-increasing-concentration-and-decreasing-competition-and-how-to-remedy-them/. Accessed 16 Feb 2018.

Gati, A. (2017, August 31). Environmental Fintech Veridium Labs Announces Crowdsale. *Bankless Times.* https://www.banklesstimes.com/2017/08/31/environmental-fintech-veridium-labs-announces-crowdsale/. Accessed 2 Jan 2018.

Geffner, M. (2016, June 10). Could the Public Benefit Company Structure Be Right for Your Business? *dun&bradstreetB2B.* https://b2b.dnb.com/2016/06/10/public-benefit-company-structure-right-business/. Accessed 19 Sept 2017.

Georgescu-Roegen, N. (1971). *The Entropy Law and the Economic Process.* Cambridge, MA: Harvard University Press.

Glaser, E. (2018). *Anti-politics: On the Demonization of Ideology, Authority, and the State.* Marquette, MI: Repeater Press.

Gleckler, P., Durack, P., Stoufferii, R., et al. (2016). Industrial-Era Global Ocean Heat Uptake Doubles in Industrial Era. *Nature Climate Change, 6,* 394–398.

Global Carbon Project. (2017). *Global Carbon Budget 2016—Presentation.* http://www.globalcarbonproject.org/carbonbudget/16/presentation.htm. Accessed 31 May 2017.

Global Commission on Economy and Climate. (2014). *Better Growth, Better Climate* (The New Climate Economy Project—2014 Report). http://newclimateeconomy.report.

Global Commission on the Economy and Climate. (2016). *The Sustainable Infrastructure Imperative.* https://newclimateeconomy.report/2016/.

Godley, W., & Lavoie, M. (2007). *Monetary Economics: An Integrated Approach to Credit, Money, Income, Production and Wealth.* Basingstoke, UK: Palgrave Macmillan.

Gordon, R. (2015). Secular Stagnation: A Supply-Side View. *American Economic Review, 105*(5), 54–59. https://www.aeaweb.org/articles?id=10.1257/aer.p20151102. Accessed 25 May 2017.

Gordon, S. (2017, March 29). Juncker's European Investment Plan: The Rhetoric Versus the Reality. *Financial Times.*

Gray, R., Brennan, A., & Malpass, J. (2014). New Accounts: Towards a Reframing of Social Accounting. *Accounting Forum, 38*(4), 258–273. https://doi.org/10.1016/j.accfor.2013.10.005.

Green, F. (2015). *Nationally Self-Interested Climate Change Mitigation: A Unified Conceptual Framework* (Working Paper, No. 224). Centre for Climate Change Economics and Policy.

Grene, S. (2012, September 7). Early Days But Cautious Optimism on ESG. *Financial Times.*

Grene, S. (2016, January 24). Quants Are New Ethical Investors. *Financial Times.*

Guardian Environment Network. (2011, July 26). *What is the Clean Development Mechanism?* https://www.theguardian.com/environment/2011/jul/26/cleandevelopment-mechanism.

Guttmann, R. (1994). *How Credit-Money Shapes the Economy: The United States in a Global System.* Armonk, NY: M. E. Sharpe.

Guttmann, R. (1997). *Reforming Money and Finance: Towards a New Monetary Regime.* Armonk, NY: M. E. Sharpe.

Guttmann, R. (2003). *Cybercash: The Coming Reign of Electronic Money.* New York: Palgrave Macmillan.

Guttmann, R. (2008, December). *A Primer on Finance-Led Capitalism and Its Crisis* (Revue de la Régulation, No. 3/4). regulation.revues.org/document5843.html.

Guttmann, R. (2009). Asset Bubbles, Debt Deflation, and Global Imbalances. *International Journal of Political Economy, 38*(2), 46–69.

Guttmann, R. (2015). The Heterodox Notion of Structural Crisis. *Review of Keynesian Economics, 3*(2), 194–212.

Guttmann, R. (2016). *Finance-Led Capitalism: Shadow Banking, Re-regulation, and the Future of Global Markets.* New York: Palgrave Macmillan.

Haass, R. (2017, December 28). America and the Great Abdication. *The Atlantic.* https://www.theatlantic.com/international/archive/2017/12/america-abidcation-trump-foreign-policy/549296/.

Hambel, C., Kraft, H., & Schwartz, E. (2015). *Optimal Carbon Abatement in a Stochastic Equilibrium Model with Climate Change* (NBER Working Papers, No. 21044). National Bureau of Economic Research. http://www.nber.org/papers/w21044.

Hamrick, K., & Gallant, M. (2017). *Unlocking Potential: State of the Voluntary Carbon Markets 2017.* Ecosystem Marketplace. https://www.cbd.int/financial/2017docs/carbonmarket2017.pdf.

Hardt, L., & O' Neill, D. (2017). Ecological Macroeconomic Models: Assessing Current Developments. *Ecological Economics, 134,* 198–211. https://doi.org/10.1016/j.ecolecon.2016.12.027

Harris, J. (2009). *Twenty-First Century Macroeconomics: Responding to the Climate Challenge.* Cheltenham, UK: Edward Elgar.

Hartwick, J. (1977). Intergenerational Equity and the Investing of Rents from Exhaustible Resources. *American Economic Review, 67*(5), 972–974.

Harvey, F. (2015, December 14). Paris Climate Change Agreement: The World's Greatest Diplomatic Success. *The Guardian.* https://www.theguardian.com/environment/2015/dec/13/paris-climate-deal-cop-diplomacy-developing-united-nations. Accessed 16 Dec 2015.

Hein, E. (2012). *The Macroeconomics of Finance-Dominated Capitalism—And Its Crisis.* Cheltenham, UK: Edward Elgar.

Heindl, P. (2012). *Transaction Costs and Tradable Permits: Empirical Evidence from the EU Emissions Trading Scheme* (ZEW Discussion Paper, No. 12-021). http://ftp.zew.de/pub/zew-docs/dp/dp12021.pdf. Accessed 15 Sept 2017.

Henley, J. (2018, January 12). Money for Nothing: Is Finland's Universal Basic Income Trial Too Good to Be True? *The Guardian*. https://www.theguardian.com/inequality/2018/jan/12/money-for-nothing-is-finlands-universal-basic-income-trial-too-good-to-be-true. Accessed 20 Feb 2018.

Hinkel, J., Brown, S., Exner, L., et al. (2012). Sea-Level Rise Impacts on Africa and the Effects of Mitigation and Adaptation: An Application of DIVA. *Regional Environmental Change, 12*(1), 207–224.

Hodgson, P. (2015, June 22). Top CEOs Make More Than 300 Times the Average Worker. *Fortune*.

Hood, C., & Soo, C. (2017). *Accounting for Mitigation Targets in Nationally Determined Contributions Under the Paris Agreement*. OECD/IEA Climate Change Expert Group Papers. http://dx.doi.org/10.1787/63937a2b-en. Accessed 26 Dec 2017.

Hope, C. (2006). The Marginal Impact of CO_2 from PAGE2002: An Integrated Assessment Model Incorporating the IPCC's Five Reasons for Concern. *Integrated Assessment, 6*(1), 19–56.

Hope, C. (2013). Critical Issues for the Calculation of the Social Cost of CO_2: Why the Estimates from PAGE09 Are Higher Than Those from PAGE2002. *Climatic Change, 117*(3), 531–543.

Hornby, L. (2017, May 4). China Leads World on Green Bonds But Benefits Are Hazy. *Financial Times*.

Hourcade. J.-C. (2015). Harnessing the Animal Spirits of Finance for a Low-Carbon Transition. In S. Barrett. C. Carraro, & J. De Melo (Eds.), *Towards a Workable and Effective Climate Regime* (pp. 497–514). Washington, DC: VOXeBook, Center for Economic and Policy Research. http://voxeu.org/sites/default/files/file/hourcade.pdf.

Howes, T. (2015, December 11). With a Public Benefit Corporation, Profit and Good Karma Can Coexist. *Entrepreneur*. https://www.entrepreneur.com/article/253059. Accessed 19 Sept 2017.

Huber, D., & Gulledge, J. (2011). *Extreme Weather and Climate Change: Understanding the Link and Managing the Risk*. Arlington, VA: Center of Climate and Energy Solution. http://www.c2es.org/publications/extreme-weather-and-climate-change.

Iansiti, M., & Lakhani, K. (2017, January–February). The Truth About Blockchain. *Harvard Business Review*. https://hbr.org/2017/01/the-truth-about-blockchain. Accessed 21 July 2017.

IETA. (2016). *A Vision for the Market Provisions of the Paris Agreement*. Geneva: International Emissions Trading Association. http://www.ieta.org/resources/

Resources/Position_Papers/2016/IETA_Article_6_Implementation_
Paper_May2016.pdf. Accessed 23 Sept 2017.

Ingraham, C. (2015, March 1). How to Steal an Election: A Visual Guide.
Washington Post. https://www.washingtonpost.com/news/wonk/wp/2015/
03/01/this-is-the-best-explanation-of-gerrymandering-you-will-ever-
see/?utm_term=.3263f0af3fcd. Accessed 8 Nov 2017.

Institute for Policy Integrity. (2017, February). *Social Costs of Greenhouse Gases—
Fact Sheet*. https://www.edf.org/sites/default/files/social_cost_of_green-
house_gases_factsheet.pdf.

International Capital Markets Association. (2017). *The Green Bond Principles 2017*.
Paris: ICMA. https://www.icmagroup.org/assets/documents/Regulatory/
Green-Bonds/GreenBondsBrochure-JUNE2017.pdf. Accessed 25 July 2017.

International Energy Agency. (2014). *450 Scenario: Method and Policy Framework*.
Paris: IEA. http://www.worldenergyoutlook.org/media/weowebsite/2014/
Methodologyfor450Scenario.pdf.

International Energy Agency. (2015). *World Energy Outlook 2015 Factsheet*.
Paris: IEA. http://www.worldenergyoutlook.org/media/weowebsite/2015/
WEO2015_Factsheets.pdf.

International Labour Organization. (2011). *Social and Solidarity Economy: Our
Common Road Toward Decent Work*. http://ilo.org/empent/units/coopera-
tives/WCMS_166301/lang—en/index.htm.

International Monetary Fund. (2016). *IMF and the Environment*. Washington,
DC. http://www.imf.org/external/np/fad/environ/.

International Monetary Fund. (2017). *Climate, Environment, and the IMF*. Factsheet.
https://www.imf.org/en/About/Factsheets/Climate-Environment-and-the-IMF.

IPCC. (2014). *Fifth Assessment Report: Climate Change 2014—Synthesis Report*.
Cambridge, UK: Cambridge University Press. http://www.ipcc.ch/report/
ar5/.

IRENA. (2015). *Renewable Energy and Jobs—Annual Review 2015*. http://www.
irena.org/menu/index.aspx?mnu=Subcat&PriMenuID=36&CatID=
141&SubcatID=585.

Jackson, Q. (2017, August 8). Short-Term Thinking Is Killing Us in the Long
Run. *Eco-Business*. http://www.eco-business.com/opinion/short-term-think-
ing-is-killing-us-in-the-long-run/. Accessed 24 Nov 2017.

Jackson, T. (2009/2017). *Prosperity Without Growth: Economics for a Finite
Planet*. London: Routledge. Published as a Much Revised Second Edition
Prosperity Without Growth: Foundations for the Economy of Tomorrow in 2017.

Jenkins, H. (2016, June 28). Climate Denial Finally Pays Off. *Wall Street Journal*.
http://www.wsj.com/articles/climate-denial-finally-pays-off-1467151625.
Accessed 12 Oct 2016.

Jensen, M., & Meckling, W. (1976). Theory of the Firm: Managerial Behavior, Agency Costs, and Ownership Structure. *Journal of Financial Economics, 3*(4), 305–360.

Jergler, D. (2015, March 12). U.S. Dominates Climate Change Litigation. *Climate Control*. http://www.insurancejournal.com/news/national/2015/03/12/360370.htm. Accessed 17 June 2017.

Jopson, B. (2015, December 30). US Views on Climate Change Pose Test for 2016 Candidates. *Financial Times*.

Kahneman, D., & Tversky, A. (1979). Prospect Theory: An Analysis of Decision Under Risk. *Econometrica, 47*(2), 263–292. https://doi.org/10.2307/1914185.

Kaminska, I. (2017, June 14). Blockchain's Governance Paradox. *Financial Times*.

Keynes, J. M. (1930/1933). Economic Possibilities for Our Grandchildren. Reprinted in Ibid. *Essays in Persuasion*. London: Macmillan.

Keynes, J. M. (1936). *A General Theory of Employment, Interest and Money*. London: Macmillan.

Keynes, J. M. (1980). *The Collected Writings of John Maynard Keynes* (Vol. 25, pp. 42–139), ed. D. Moggridge. Cambridge: Cambridge University Press.

King, D., Browne, J., Layard, R., et al. (2015). *A Global Apollo Programme to Combat Climate Change*. http://cep.lse.ac.uk/pubs/download/special/Global_Apollo_Programme_Report.pdf.

Klein, N. (2014). *This Changes Everything: Capitalism vs. The Climate*. New York: Simon & Schuster.

Kondratiev, N. (1925/1984). *The Major Economic Cycles* (in Russian), Moscow. Translated as *The Long Wave Cycle*. New York: Richardson & Snyder.

Kondratiev, N. (1926/1936). Die langen Wellen der Konjunktur. *Archiv für Sozialwissenschaft und Sozialpolitik, 56*, 573–609. Published in English as "The Long Waves in Economic Life." *Review of Economic Statistics, 17*(6), 105–115.

Kriebel, D., Tickner, J., Epstein, P., et al. (2001). The Precautionary Principle in Environmental Science. *Environmental Health Perspectives, 109*(9), 871–876. https://www.ncbi.nlm.nih.gov/pmc/articles/PMC1240435/.

Krippner, G. (2005). The Financialization of the American Economy. *Socio-Economic Review, 3*(2), 173–208.

Krugman, P. (2013, September 25). Bubbles, Regulation, and Secular Stagnation. *New York Times*. https://krugman.blogs.nytimes.com/2013/09/25/bubbles-regulation-and-secular-stagnation/.

Kynge, J. (2014, April 24). Sustainable Investors Outstrip Emerging Market Benchmarks, *Financial Times*.

Lagarde, C. (2015, December). Ten Myths About Climate Change Policy. *Finance and Investment*, 64–67. http://www.imf.org/external/np/fad/environ/pdf/011215.pdf. Accessed 22 Sept 2017.

Lazonick, W., & O'Sullivan, M. (2000). Maximizing Shareholder Value: A New Paradigm for Corporate Governance. *Economy and Society, 29*(1), 13–35.

League of Conservation Voters. (2016). *In Their Own Words: 2016 Presidential Candidates on Climate Change.* http://www.lcv.org/assets/docs/presidential-candidates-on.pdf. Accessed 15 Sept 2016.

Le Page, M. (2015, December 19). Paris Climate Deal Is Agreed—But Is It Really Good Enough? *New Scientist*, Issue 3052. https://www.newscientist.com/issue/3052%20/. Accessed 9 Oct 2016.

Lewis, A. (1966). *Development Planning: The Essentials of Economic Policy.* London: George Allen & Unwin.

Lewis, M. (2007, August 26). In Nature's Casino. *New York Times Magazine.* http://www.nytimes.com/2007/08/26/magazine/26neworleans-t.html?pagewanted=print. Accessed 18 June 2017.

Lipietz, A. (1985). *The Enchanted World: Inflation, Credit and the World Crisis.* London: Verso. First Published as *Le monde enchanté: De la valeur a l'envol inflationniste* (Paris: F. Maspero), 1983.

Lipietz, A. (1987). *Mirages and Miracles: The Crisis of Global Fordism.* London: Verso. First Published as *Mirages et Miracles* (Paris: La Decouverte), 1985.

Magdoff, F., & Foster, J. B. (2014). Stagnation and Financialization: The Nature of the Contradiction. *Monthly Review, 66*(1), 4–25.

Mandel, E. (1980). *Long Waves of Capitalist Development: A Marxist Interpretation.* Cambridge, UK: Cambridge University Press.

Marshall, A. (2017, October 19). Alphabet Is Trying to Reinvent the City, Starting with Toronto. *Wired.* https://www.wired.com/story/google-sidewalk-labs-toronto-quayside/.

Marshall, Al. (1890). *Principles of Economics.* London: Macmillan.

Marx, K. (1867/1992). *Capital* (Vol. 1). London: Penguin Classics.

Matthews, S. (2017, July). Alarmism Is the Argument We Need to Fight Climate Change. *Slate.* http://www.slate.com/articles/health_and_science/science/2017/07/we_are_not_alarmed_enough_about_climate_change.html. Accessed 4 Feb 2018.

Mazzucato, M. (2013). *The Entrepreneurial State: Debunking Public vs. Private Sector Myths.* London: Anthem Press.

McGee, P. (2017, November 8). Electric Cars' Green Image Blackens Beneath the Bonnet. *Financial Times.*

McKinsey. (2013, September). *Pathways to a Low-Carbon Economy: Version 2 of the Global Greenhouse Gas Abatement Cost Curve.* Report. http://www.

mckinsey.com/business-functions/sustainability-and-resource-productivity/our-insights/pathways-to-a-low-carbon-economy.

McKinsey Global Institute. (2016, June). *Bridging Global Infrastructure Gaps*. Report. https://www.mckinsey.com/industries/capital-projects-and-infrastructure/our-insights/bridging-global-infrastructure-gaps.

McSweeny, R., & Pearce, R. (2016, May 19). Analysis: Only Five Years Left Before 1.5C Carbon Budget Is Blown. *CarbonBrief*. https://www.carbonbrief.org/analysis-only-five-years-left-before-one-point-five-c-budget-is-blown. Accessed 30 May 2017.

Meade, J. (1970). *The Theory of Indicative Planning*. Manchester, UK: Manchester University Press.

Megerian, C., & Dillon, L. (2016, September 8). Gov. Brown Signs Sweeping Legislation to Combat Climate Change. *Los Angeles Times*.

Megerian, C., & Mason, M. (2016, August 22). 'An Exercise in Threading the Needle': Lawmakers Perform Balancing Act to Move Climate Legislation Forward. *Los Angeles Times*.

Mill, J. S. (1848). *Principles of Political Economy*. London: John W. Parker.

Milman, O. (2016, October 6). World Needs $90tn Infrastructure Overhaul to Avoid Climate Disaster, Study Finds. *The Guardian*. https://www.theguardian.com/environment/2016/oct/06/climate-change-infrastructure-coalplants-green-investment. Accessed 17 Jan 2017.

Minsky, H. (1964). Longer Waves in Financial Relations: Financial Factors in the More Severe Depressions. *American Economic Review, 54*(3), 324–335.

Molla, R. (2014, September 15). What Is the Most Efficient Source of Energy? *Wall Street Journal*. http://blogs.wsj.com/numbers/what-is-the-most-efficient-source-of-electricity-1754/.

Mook, L. (2013). *Accounting for Social Value*. Toronto: University of Toronto Press.

Mooney, A. (2016, May 11). Academics Back Exxon and Chevron Climate Openness Vote. *Financial Times*.

Mooney, C. (2017, October 11). New EPA Document Reveals Sharply Lower Estimate of the Cost of Climate Change. *Washington Post*. https://www.washingtonpost.com/news/energy-environment/wp/2017/10/11/new-epa-document-reveals-sharply-lower-estimate-of-the-cost-of-climate-change/?utm_term=.eeb7ab8e5d02. Accessed 25 Nov 2017.

Montgomery, D. W. (1972). Markets in Licenses and Efficient Pollution Control Programs. *Journal of Economic Theory, 5*(3), 395–418.

Murray. S. (2015, June 2). Sustainability Measurement: Index Looks to Connect Investors. *Financial Times*.

National Climate Assessment. (2014). *Climate Change Impacts in the United States*. Washington, DC: U.S. Global Change Research Program. http://nca2014.globalchange.gov.

Neumayer, E. (2003). *Weak Versus Strong Sustainability: Exploring the Limits of Two Opposing Paradigms*. Cheltenham, UK: Edward Elgar.

New Climate Institute. (2016a). *Climate Initiatives, National Contributions, and the Paris Agreement.* Berlin. https://newclimate.org/2016/05/23/climate-initiatives-national-contributions-and-the-paris-agreement/. Accessed 27 May 2017.

New Climate Institute. (2016b). *Conditionality of Intended Nationally Determined Contributions.* https://www.transparency-partnership.net/documents-tools/conditionality-intended-nationally-determined-contributions-indcs. Accessed 23 Dec 2017.

Nikiforos, M., & Zezza, G. (2017). *Stock-Flow Consistent Macroeconomic Models: A Survey* (Working Paper, No. 891). Levy Institute. http://www.levyinstitute.org/publications/stock-flow-consistent-macroeconomic-models-a-survey.

Nikolova, A., & Phung, T. (2017, February 19). Microsoft's Carbon Fee: Going Beyond Carbon Neutral. *Our Stories.* Yale Center for Business and the Environment. http://cbey.yale.edu/our-stories/microsoft's-carbon-fee-going-beyond-carbon-neutral. Accessed 10 May 2017.

Nordhaus, W. (1992). *The DICE Model: Background and Structure* (Cowles Foundation Discussion Papers, No. 1009). https://cowles.yale.edu/sites/default/files/files/pub/d10/d1009.pdf.

Nordhaus, W. (2013). *The Climate Casino: Risk, Uncertainty, and Economics for a Warming World.* New Haven, CT: Yale University Press.

Nordhaus, W. (2016). Revisiting the Social Cost of Carbon. *Proceedings of the National Academy of Sciences of the United States, 114*(7), 1518–1523. https://doi.org/10.1073/pnas.1609244114.

Nordhaus, W. (2017). *Evolution of Modeling of the Economics of Global Warming: Changes in the DICE Model, 1992–2017* (Cowles Foundation Discussion Papers, No. 2084). https://cowles.yale.edu/sites/default/files/files/pub/d20/d2084.pdf.

Nuccitelli, D. (2016, October 6). Pew Survey: Republicans Are Rejecting Reality on Climate Change. *The Guardian.* https://www.theguardian.com/environment/climate-consensus-97-per-cent/2016/oct/06/pew-survey-republicans-are-rejecting-reality-on-climate-change. Accessed 10 Dec 2016.

Odendahl, T. (2016, January 22). The Failures of the Paris Climate Change Agreement and How Philantropy Can Fix Them. *Stanford Social Innovation Review.* https://ssir.org/articles/entry/the_failures_of_the_paris_climate_change_agreement_and_how_philanthropy_can. Accessed 10 Oct 2016.

O'Doherty, J., & Tol, R. (2007). *An Environmental Input-Output Model for Ireland* (ESRI Working Paper, No. 178). http://dx.doi.org/10.2139/ssrn.964473.

Office of Management and Budget. (2003). *Circular A-4: Regulatory Analysis.* White House. https://www.transportation.gov/sites/dot.gov/files/docs/OMB%20Circular%20No.%20A-4.pdf.

Organization of Economic Development and Cooperation. (2016a). *OECD Environmental Performance Reviews—France 2016.* Paris: OECD. http://dx.doi.org/10.1781/9789264252714-en. Accessed 16 Aug 2017.

Organization of Economic Development and Cooperation. (2016b). *Green Investment Banks: Scaling Up Private Investment in Low-Carbon, Climate-Resilient Infrastructure.* http://www.oecd.org/environment/cc/green-investment-banks-9789264245129-en.htm.

Organization of Economic Cooperation and Development. (2017). *Investing in Climate, Investing in Growth.* http://www.oecd.org/environment/cc/g20-climate/.

Palmer, B. (2016, April 28). Should You Buy Carbon Offsets? National Resources Defense Council. *Our Stories.* https://www.nrdc.org/stories/should-you-buy-carbon-offsets. Accessed 30 Dec 2017.

Pelenc, J., Ballet, J., & Dedeurwaerdere, T. (2015). Weak Sustainability Versus Strong Sustainability. *Brief for GSDR 2015.* Global Sustainable Development Report, United Nations. https://sustainabledevelopment.un.org/content/documents/6569122-Pelenc-Weak%20Sustainability%20versus%20Strong%20Sustainability.pdf. Accessed 24 Nov 2017.

Perez, C. (2003). *Technological Revolutions and Financial Capital: The Dynamics of Bubbles and Golden Ages.* Cheltenham, UK: Edward Elgar.

Perez, C. (2009). *Technological Revolutions and Techno-Economic Paradigms* (Working Papers in Technology Governance and Economic Dynamics, No. 20). The Other Canon Foundation. http://edu.hioa.no/pdf/technological_revolutions.pdf. Accessed 15 Nov 2017.

Perry, I. (2015). *Implementing a US Carbon Tax: Challenges and Debates.* Washington, DC: International Monetary Fund. http://dx.doi.org/10.5089/9781138825369.071.

Pfeiffer, A., Millar, R., Hepburn, C., et al. (2016). The '2°C Capital Stock' for Electricity Generation: Committed Cumulative Carbon Emissions from the Electricity Generation Sector and the Transition to a Green Economy. *Applied Energy, 179,* 1395–1408. http://dx.doi.org/10.1016/j.apenergy.2016.02.093. Accessed 14 Oct 2016.

Pidcock, R. (2017, November 15). Analysis: What Global Emissions in 2016 Mean for Climate Change Goals. *Carbon Brief.* https://www.carbonbrief.org/what-global-co2-emissions-2016-mean-climate-change. Accessed 24 Nov 2017.

Pigou, A. (1920). *The Economics of Welfare.* London: Macmillan.

Pimentel, D. (2017). Blue Foundation Creates New Blockchain Token for Zero-Emission Projects. *BlockTribune.* http://blocktribune.com/blue-foundation-creates-new-blockchain-token-zero-emission-projects/. Accessed 5 Jan 2018.

Pindyck, R. (2017). The Use and Misuse of Models for Climate Policy. *Review of Environmental Economics and Policy, 11*(1), 100–114.

Pisany-Ferry, J. (2015, December 4). Finance Can Save the World from Climate Change. *European CEO.* https://www.europeanceo.com/business-and-management/finance-can-save-the-world-from-climate-change/. Accessed 19 Sept 2017.

Pollin, R. (2015). *Greening the Global Economy.* Cambridge, MA: MIT Press.

Pooley, E. (2017, June 14). What Has Trump Done to the Environment so Far? A Quick Rundown. *EDF Voices.* https://www.edf.org/blog/2017/06/14/what-has-trump-done-environment-so-far-quick-rundown. Accessed 11 Nov 2017.

Popper, N. (2017, October 27). An Explanation of Initial Coin Offerings. *New York Times.* https://www.nytimes.com/2017/10/27/technology/what-is-an-initial-coin-offering.html. Accessed 31 Dec 2017.

Porter, E. (2016, March 1). Does a Carbon Tax Work? Ask British Columbia. *New York Times.* https://www.nytimes.com/2016/03/02/business/does-a-carbon-tax-work-ask-british-columbia.html. Accessed 6 Oct 2017.

Porter, E. (2018, February 13). Big Profits Drove a Stock Boom. Did the Economy Pay a Price? *New York Times.* https://www.nytimes.com/2018/02/13/business/economy/profits-economy.html. Accessed 13 Feb 2018.

Price, R., Thornton, S., & Nelson, S. (2007a). *The Social Cost of Carbon and the Shadow Price of Carbon: What They Are, and How to Use Them in Economic Appraisal in the UK* (MPRA Paper, No. 74976). https://mpra.ub.uni-muenchen.de/74976/1/MPRA_paper_74976.pdf. Accessed 10 May 2017.

Price, R., Thornton, S., & Nelson, S. (2007b). *The Social Cost of Carbon and the Shadow Price of Carbon: What They Are, and How to Use Them in Economic Appraisal in the UK.* DEFRA. https://www.gov.uk/government/uploads/system/uploads/attachment_data/file/243825/background.pdf. Accessed 28 Nov 2017.

Principles for Responsible Investment. (2016). *French Energy Transition Law: Global Investor Briefing.*

Prudential Regulation Authority. (2015). *The Impact of Climate Change on the UK Insurance Sector.* London: Bank of England. http://www.bankofengland.co.uk/pra/Documents/supervision/activities/pradefra0915.pdf. Accessed 4 June 2017.

Qiu, J. (2016, June 2). Investigating Climate Change the Hard Way at Earth's Icy "Third Pole." *Scientific American.* www.scientificamerican.com.

Radic, V., & Hock, R. (2011). Regionally Differentiated Contribution of Mountain Glaciers and Ice Caps to Future Sea-Level Rise. *Nature Geoscience.* www.nature.com/naturegeoscience.

Rai, V., & Henry, A. D. (2016). Agent-Based Modelling of Consumer Energy Choices. *Nature Climate Change, 6*(June), 556–562. https://doi.org/10.1038/nclimate2967.

Randall, T. (2016, April 6). Wind and Solar Are Crushing Fossil Fuels. *Bloomberg News.* http://www.bloomberg.com/news/articles/2016-04-06/wind-and-solar-are-crushing-fossil-fuels. Accessed 15 Oct 2016.

Rattani, V. (2017, November 8). COP 23: Are Countries Ready for the New Market Mechanism? *Down To Earth.* http://www.downtoearth.org.in/news/

cop-23-are-countries-ready-for-new-market-mechanism—59061. Accessed 25 Dec 2017.

Rawls, H. (1971). *A Theory of Justice*. Cambridge, MA: Harvard University Press.

Reynolds, F. (2015, June 7). Hedge Funds Warm to Responsible Investment Principles. *Financial Times*.

Rezai, A. (2011). The Opportunity Cost of Climate Policy: A Question of Reference. *Scandinavian Journal of Economics, 113*(4), 885–903.

Ricardo, R. (1817). *On the Principles of Political Economy and Taxation*. London: John Murray.

Rodrik, D. (2017). *Populism and the Economics of Globalization* (NBER Working Paper, No. 23559). http://www.nber.org/papers/w23559. Accessed 28 Feb 2018.

Rodrik, D. (2018, February 21). What Does a True Populism Look Like? It Looks Like the New Deal. *New York Times*. https://www.nytimes.com/2018/02/21/opinion/populism-new-deal.html?partner=rss&emc=rss. Accessed 28 Feb 2018.

Rogelj, J., den Elzen, M., et al. (2016). Paris Agreement Climate Proposals Need a Boost to Keep Warming Well Below 2 °C. *Nature, 534*, 631–639. https://doi.org/10.1038/nature18307.

Roos, M. (2017). Endogenous Economic Growth, Climate Change and Societal Values: A Conceptual Model. *Computational Economics, 50*(1), 1–34.

Ross, S. (1973). The Economic Theory of Agency: The Principal's Problem. *American Economic Review, 63*(1), 134–139.

Rust, S. (2016, February 1). France Aims High with First-Ever Investor Climate-Reporting Law. *Investment and Pensions Europe*. https://www.ipe.com/countries/france/france-aims-high-with-first-ever-investor-climate-reporting-law/10011722.fullarticle. Accessed 21 June 2017.

Sandbag. (2011, July 26). What Is the Clean Development Mechanism (CDM)? *The Guardian*. https://www.theguardian.com/environment/2011/jul/26/clean-development-mechanism. Accessed 11 Nov 2017.

Sandbu, M. (2018a, February 28). Welfare Lessons from Finland. *Financial Times*.

Sandbu, M. (2018b, February 28). Are There Good Types of Populism? *Financial Times*.

Sala, L. (2017, December 5). The Swedish Carbon Tax: How to Tackle Climate Change in an Efficient Way. *Traileone*. http://www.traileoni.it/?p=4915.

Sato, M., Ciszewska, M., & Laing, T. (2016). *Demand for Offsetting and Insetting in the EU Emissions Trading Scheme* (Working Paper, No. 237). Grantham Research Institute on Climate Change and the Environment. http://www.lse.ac.uk/GranthamInstitute/wp-content/uploads/2016/06/Working-Paper-237-Sato-et-al.pdf. Accessed 28 Dec 2017.

Say, J.-B. (1803/1821). *A Treatise on Political Economy*. London: Longman. First Published in French as *Traité d'économie politique*.

Schumacher, E. F. (1973). *Small Is Beautiful: Economics as If People Mattered*. New York: Harper & Row.

Schumpeter, J. (1911/1934). *Theory of Economic Development: An Inquiry into Profits, Capital, Credit, Interest, and the Business Cycle*. Cambridge, MA: Harvard University Press. First Published in German as *Theorie der wirtschaftlichen Entwicklung*.

Schumpeter, J. (1939). *Business Cycles*. New York and London: McGraw-Hill.

Schwartz, J. (2017, February 7). 'A Conservative Climate Solution': Republican Group Calls for Carbon Tax. *New York Times*. https://www.nytimes.com/2017/02/07/science/a-conservative-climate-solution-republican-group-calls-for-carbon-tax.html. Accessed 6 Oct 2017.

Scott, M. (2013, March 17). Market for Green Buildings Warms Up. *Financial Times*.

Scott, M. (2014, November 16). Low Carbon Becomes a High Priority. *Financial Times*.

Scott, M. (2015, December 15). Investors Seek Ethical Benchmarks. *Financial Times*.

Scott, M., & Hansen, K. (2016, September 16). Sea Ice Overview. *NASA Earth Observatory*. http://earthobservatory.nasa.gov/Features/SeaIce/?src=features-hp&eocn=home&eoci=feature.

Sen, A. (1985). *Commodities and Capabilities*. Amsterdam: North-Holland.

Sen, A. (1990, December 20). More Than 100 Million Women Are Missing. *New York Review of Books, 37*(20). http://www.nybooks.com/articles/1990/12/20/more-than-100-million-women-are-missing/?printpage=true.

Seppecher, P., Salle, I., & Lavoie, M. (2017). *What Drives Markups? Evolutionary Pricing in an Agent-Based Stock-Flow Consistent Macroeconomic Model* (Document de travail du CEPN, No. 2017-3). Centre d'Economie de l'Université Paris Nord. https://hal.archives-ouvertes.fr/hal-01486597/document.

Sharpe, W. (1964). Capital Asset Prices: A Theory of Market Equilibrium Under Conditions of Risk. *Journal of Finance, 19*(3), 425–442.

Simmel, G. (1900/1978). *The Philosophy of Money*. Abingdon, UK: Routledge.

Sirkis, A. (2016). Preface: The Challenge of Moving the Trillions. In A. Sirkis et al. (Eds.), *Moving the Trillions: A Debate on Positive Pricing of Mitigation Actions* (pp. 10–20). Paris: CIRED. http://www2.centre-cired.fr/IMG/pdf/moving_in_the_trillions.pdf.

Skeptical Science. (2016). *Quotes by James Inhofe vs. What the Science Says*. https://www.skepticalscience.com/skepticquotes.php?s=30. Accessed 15 Sept 2016.

Smith, A. (1776). *The Wealth of Nations*. London: W. Strahan and T. Cadell.

Smith, S., & Braathen, N. (2015). *Monetary Carbon Values in Policy Appraisal: An Overview of Current Practice and Key Issues* (OECD Environment Working Papers, No. 92). http://dx.doi.org/10.1787/5jrs8st3ngvh-en.

Solow, R. (1986). On the Intergenerational Allocation of Natural Resources. *Scandinavian Journal of Economics, 88*(1), 141–149. https://doi.org/10.2307/3440280.

Soulami, A. B. H. (2017, May 17). Using Scenario Analysis to Mitigate Climate Risks. *Environmental Finance.* https://cib.bnpparibas.com/sustain/a-four-step-process-for-modelling-climate-risk_a-3-946.html. Accessed 21 June 2017.

Spors, K. (2017, November 1). The Lowdown on Buying Carbon Offsets. *Green Business.* https://smallbiztrends.com/2010/09/lowdown-on-carbon-offsets.html. Accessed 30 Dec 2017.

Stavins, R. (1995). Transaction Costs and Tradable Permits. *Journal of Environmental Economics and Management, 29*(1), 133–148. http://scholar.harvard.edu/files/stavins/files/transaction_costs_jeem.pdf. Accessed 15 Sept 2017.

Steckel, J. C., Edenhofer, O., & Jacob, M. (2015). Drivers for the Renaissance of Coal. *PNAS (Proceedings of the National Academy of Sciences of the United States of America), 112*(29), E3775–E3781. http://www.pnas.org/content/112/29/E3775.abstract. Accessed 14 Oct 2016.

Stein, J. (2012, October 11). Evaluating Large-Scale Asset Purchases. *Speech.* https://www.federalreserve.gov/newsevents/speech/stein20121011a.htm.

Stern, N. (2006). *Stern Review: The Economics of Climate Change.* London: H.M. Treasury. http://unionsforenergydemocracy.org/wp-content/uploads/2015/08/sternreview_report_complete.pdf. Last accessed 18 Nov 2017.

Stern, N. (2015). *Why Are We Waiting? The Logic, Urgency, and Promise of Tackling Climate Change.* Cambridge, MA: MIT Press.

Stern, N. (2016). Current Climate Models Are Grossly Misleading: Nicholas Stern Calls on Scientists, Engineers and Economists to Help Policy-Makers by Better Modelling the Immense Risks to Future Generations, and the Potential for Action. *Nature, 530*(7591), 407–409.

Stiglitz, J., Sen, A., & Fitoussi, J. (2009). *Report by the Commission on the Measurement of Economic Performance and Social Progress.* http://ec.europa.eu/eurostat/documents/118025/118123/Fitoussi+Commission+report.

Stiglitz, J., & Stern, N. (2017). *Report of the High-Level Commission on Carbon Prices.* Washinton, DC: World Bank Publishing. https://www.carbonpricing-leadership.org/report-of-the-highlevel-commission-on-carbon-prices/.

Stockhammer, E. (2004). Financialisation and the Slowdown of Accumulation. *Cambridge Journal of Economics, 28*(5), 719–741.

Strand, J. (2017). *Unconditional and Conditional NDCs Under the Paris Agreement: Interpretations and Their Relations to Policy Instruments* (CREE Working Paper, No. 09/2017). https://www.cree.uio.no/publications/CREE_working_papers/pdf_2017/strand_carbon_pricing_cree_wp09_2017.pdf. Accessed 23 Dec 2017.

Struzik, E. (2014, July 10). Loss of Snowpack and Glaciers in Rockies Poses Water Threat. *Yale Environment 360.* http://e360.yale.edu/feature/loss_of_snowpack_and_glaciers_in_rockies_poses_water_threat/2785/. Accessed 29 Sept 2016.

Summers, L. (2016, March/April). Secular Stagnation: What It Is and What to Do About It. *Foreign Affairs, 95*(2). http://larrysummers.com/2016/02/17/the-age-of-secular-stagnation/. Accessed 25 May 2017.

Swedish Institute. (2016). *Sweden Tackles Climate Change.* https://sweden.se/nature/sweden-tackles-climate-change/. Accessed 5 Oct 2017.

Tabuchi, H. (2017, March 24). California Upholds Auto Emissions Standards, Setting Up Face-Off with Trump. *New York Times.*

Tett, G. (2016, September 22). Energy Companies Must Act to Avoid Banks' Mistakes. *Financial Times.*

Teulings, C. & Baldwin, R. (Eds.). (2014). *Secular Stagnation: Facts, Causes, and Cures.* London: CEPR Books. https://scholar.harvard.edu/files/farhi/files/book_chapter_secular_stagnation_nov_2014_0.pdf. Accessed 25 May 2017.

TFCD. (2016, December 14). *Recommendations of the Task Force on Climate-Related Financial Disclosures.* Basel, CH: Financial Stability Board. https://www.fsb-tcfd.org/wp-content/uploads/2016/12/16_1221_TCFD_Report_Letter.pdf. Accessed 18 Feb 2017.

The Economist. (2014, July 31). *British Columbia's Carbon Tax: The Evidence Mounts.* https://www.economist.com/blogs/americasview/2014/07/british-columbias-carbon-tax. Accessed 6 Oct 2017.

The Economist. (2015a, March 9). *What Is Quantitative Easing?* https://www.economist.com/blogs/economist-explains/2015/03/economist-explains-5. Accessed 23 Mar 2017.

The Economist. (2015b, October 31). *The Trust Machine: The Promise of the Blockchain.* https://www.economist.com/news/leaders/21677198-technology-behind-bitcoin-could-transform-how-economy-works-trust-machine. Accessed 11 Nov 2015.

The Economist. (2017, November 9). *The Meaning in the Madness of Initial Coin Offerings.* https://www.economist.com/news/leaders/21731161-there-ico-bubble-it-holds-out-promise-something-important-meaning. Accessed 15 Nov 2017.

Thwaites, J., Amerasinghe, N. M., & Ballesteros, A. (2015, December 18). What Does The Paris Agreement Do for Finance? *World Resources Institute Blog.*

http://www.wri.org/blog/2015/12/what-does-paris-agreement-do-finance. Accessed 8 Oct 2016.

Tol, R. (2009). The Economic Impact of Climate Change. *Journal of Economic Perspectives, 23*(2), 29–51.

Trancik, J. (2014). Renewable Energy: Back the Renewables Boom. *Nature, 507*(7492). http://www.nature.com/news/renewable-energy-back-the-renewables-boom-1.14873. Accessed 2 Oct 2016.

Trenberth, K., Fasullo, J. T., Von Schuckmann, K., et al. (2016). Insights into Earth's Energy Imbalance from Multiple Sources. *Journal of Climate, 29*(20), 7495–7505. http://dx.doi.org/10.1175/JCLI-D-16-0339.1. Accessed 13 Oct 2016.

U.K. Treasury. (2003). *The Green Book: Appraisal and Evaluation in Central Government.* https://www.gov.uk/government/uploads/system/uploads/attachment_data/file/220541/green_book_complete.pdf.

UN-HABITAT. (2008). *Latin American and Caribbean Cities at Risk Due to Sea-Level Rise.* http://www.preventionweb.net/english/professional/maps/v.php?id=5649.

Union of Concerned Scientists. (2013). *Causes of Sea Level Rise: What the Science Tells Us.* http://www.ucsusa.org/global_warming/science_and_impacts/impacts/causes-of-sea-level-rise.html#.V-rp9TKPDBI. Accessed 27 Sept 2016.

United Nations. (2017). *Sustainable Development Goals: 17 Goals to Transform Our World.* New York: United Nations Department of Public Information. http://www.un.org/sustainabledevelopment/sustainable-development-goals/.

United Nations Development Programme. (2011). *Ethiopia's Climate-Resilient Green Economy.* http://www.undp.org/content/dam/ethiopia/docs/Ethiopia%20CRGE.pdf.

United Nations Environment Programme. (2016). *Climate Change.* http://www.unep.org/climatechange/. Accessed 15 Oct 2016.

United Nations Environment Programme. (2017). *The Status of Climate Change Litigation—A Global Review.* http://columbiaclimatelaw.com/files/2017/05/Burger-Gundlach-2017-05-UN-Envt-CC-Litigation.pdf. Accessed 17 June 2017.

United Nations Framework Convention on Climate Change. (2014a). *Kyoto Protocol.* http://unfccc.int/kyoto_protocol/items/2830.php.

United Nations Framework Convention on Climate Change. (2014b). *Green Climate Fund.* http://unfccc.int/bodies/green_climate_fund_board/body/6974.php.

United Nations Framework Convention on Climate Change. (2015). *Adoption of the Paris Agreement.* http://unfccc.int/resource/docs/2015/cop21/eng/l09r01.pdf.

United Nations Framework Convention on Climate Change. (2016a). *The Interim NDC Registry*. http://unfccc.int/focus/ndc_registry/items/9433. php.

United Nations Framework Convention on Climate Change. (2016b). *Aggregate Effect of the Intended Nationally Determined Contributions: An Update.* http://unfccc.int/resource/docs/2016/cop22/eng/02.pdf.

United Nations Framework Convention on Climate Change. (2017, January 23). *Microsoft's Climate Fee—A Shining Example of Corporate Climate Action.* http://newsroom.unfccc.int/climate-action/how-microsofts-carbon-fee-is-driving-climate-action-forward/. Accessed 4 Oct 2017.

U.S. Energy Information Administration. (2015). *International Energy Outlook 2016.* http://www.eia.gov/forecasts/ieo/. Accessed 15 Oct 2016.

U.S. Energy Information Administration. (2016). *Annual Energy Outlook 2016 with Projections to 2040.* https://www.eia.gov/outlooks/aeo/pdf/0383(2016).pdf.

USSIF. (2016). *US Sustainable, Responsible, and Impact Investing Trends.* Washington, DC: US SIF Foundation.

Utting, P. (2013, April 30). What Is Social and Solidarity Economy and Why Does It Matter? *People, Spaces, Deliberation Blog.* World Bank. http://blogs.worldbank.org/publicsphere/what-social-and-solidarity-economy-and-why-does-it-matter.

Van Der Sluijs, J., & Turkenburg, W. (2007). Climate Change and the Precautionary Principle. In E. Fisher, J. Jones, & R. von Schomberg (Eds.), *Implementing the Precautionary Principle: Perspectives and Prospects* (pp. 245–269). Cheltenham, UK: Edward Elgar.

Van Ommen Kloeke, E. (2014). How Will Climate Change Affect Food Security? *Elsevier Connect.*

Van Vuuren, D., et al. (2009). Comparison of Top-Down and Bottom-Up Estimates of Sectoral and Regional Greenhouse Gas Emission Reduction Potentials. *Energy Policy, 37,* 5125–5139.

Victor, P. (2012). Growth, Degrowth and Climate Change: A Scenario Analysis. *Ecological Economics, 84,* 206–212. https://doi.org/10.1016/j.ecolecon.2011.04.013.

Vivid Economics. (2013, May). *The Macroeconomics of Climate Change.* A Report Commissioned by the UK Department for Environment, Food and Rural Affairs.

Wagner, G., & Weitzman, M. (2015). *Climate Shock: The Economic Consequences of a Hotter Planet.* Princeton, NJ: Princeton University Press.

Walzer, M. (1983). *Spheres of Justice.* New York: Basic Books.

Weart, S. (2008). *The Discovery of Global Warming.* Cambridge, MA: Harvard University Press.

Wicksteed, P. (1910). *The Common Sense of Political Economy*. London: Macmillan.

Williamson, J. (1986). Target Zones and the Management of the Dollar. *Brookings Papers on Economic Activity, 1986*(1), 165–174. https://www.brookings.edu/wp-content/uploads/1986/01/1986a_bpea_williamson.pdf.

Woerdman, E. (2001). Emission Trading and Transaction Costs: Analyzing the Flaws in the Discussion. *Ecological Economics, 38*(2), 293–304. https://doi.org/10.1016/S0921-8009(01)00169-0. Accessed 16 Sept 2017.

Wolf, M. (2014a, November 11). An Unethical Bet in the Climate Casino. *Financial Times*.

Wolf, M. (2014b, June 14). A Climate Fix Would Ruin Investors. *Financial Times*.

Wolf, M. (2014c, September 23). Clean Growth Is a Safe Bet in the Climate Casino. *Financial Times*.

Wolf, M. (2015a, December 15). The Paris Climate Change Summit Is But One Small Step for Humankind. *Financial Times*.

Wolf, M. (2015b, June 9). Why Climate Uncertainty Justifies Actions. *Financial Times*.

Wolf. M. (2015c, March 3). The Riches and Perils of the Fossil-Fuel Age. *Financial Times*.

World Bank. (2016a). *State and Trends of Carbon Pricing, 2016*. Washington, DC. documents.worldbank.org.

World Bank. (2016b). *The NDC Platform: A Comprehensive Resource on National Climate Targets and Action*. http://www.worldbank.org/en/topic/climatechange/brief/the-ndc-platform-a-comprehensive-resource-on-national-climate-targets-and-action.

World Food Programme. (2016). *Climate Impacts on Food Security*. Rome: United Nations. https://www.wfp.org/climate-change/climate-impacts.

Zwick, S. (2016a, February 1). *The Road from Paris: Green Lights, Speed Bumps, and the Future of Carbon Markets*. Ecosystem Marketplace. http://www.ecosystemmarketplace.com/articles/green-lights-and-speed-bumps-on-road-to-markets-under-paris-agreement/. Accessed 11 Nov 2016.

Zwick, S. (2016b, January 29). *Building on Paris, Countries Assemble the Carbon Markets of Tomorrow*. Ecosystem Marketplace. http://www.ecosystemmarketplace.com/articles/building-on-paris-countries-assemble-the-carbon-markets-of-tomorrow/. Accessed 11 Nov 2016.

INDEX

CPSIA information can be obtained
at www.ICGtesting.com
Printed in the USA
LVHW081819200619

621870LV00010B/209/P

9 783319 923567